ADAPTIVE AND FUNCTIONAL POLYMERS, TEXTILES AND THEIR APPLICATIONS

ADAPTIVE AND FUNCTIONAL POLYMERS, TEXTILES AND THEIR APPLICATIONS

Jinlian Hu

The Hong Kong Polytechnic University, Hong Kong

ICP

Imperial College Press

Published by

Imperial College Press
57 Shelton Street
Covent Garden
London WC2H 9HE

Distributed by

World Scientific Publishing Co. Pte. Ltd.
5 Toh Tuck Link, Singapore 596224
USA office: 27 Warren Street, Suite 401-402, Hackensack, NJ 07601
UK office: 57 Shelton Street, Covent Garden, London WC2H 9HE

British Library Cataloguing-in-Publication Data
A catalogue record for this book is available from the British Library.

ADAPTIVE AND FUNCTIONAL POLYMERS, TEXTILES AND THEIR APPLICATIONS

ISBN-13 978-1-84816-475-8
ISBN-10 1-84816-475-0

Printed in Singapore by Mainland Press Pte Ltd.

Preface

As smart materials, adaptive polymers can show sharp, controllable and predictable changes as responses to the external stimuli and have attracted more and more interests in their properties, structures, innovative design and applications. Many technological breakthroughs have been achieved during the past several decades. As the demands in new functions and enhanced performance of industrial and consumer products, some functional polymers have also been extensively explored and shown tremendous application potential.

In order to provide a resource for people in different professions in searching for knowledge and applications in adaptive and versatile functional polymers, this book has covered their latest developments as well as their applications in textiles, medical and cosmetics fields. The concepts and adaptive mechanisms are introduced. Their characteristics through different application fields have been presented. The applications of adaptive polymers to textiles have been highlighted. One of the unique features of this book is its interdisciplinary in nature.

In Chapter 1, basic concepts are suggested and definitions of adaptive polymers and textiles are reviewed. Following that, the characteristics, structures and potential applications of adaptive polymers are summarised. In Chapter 2, shape memory polymers according to their molecular structures, trigger patterns and applications are introduced based on literature review. One of the novel functional shape memory polymers with substrate bonded antibacterial activity is detailed and a new design of supramolecular shape memory polymers is discussed. Finally, shape memory fibers are compared with other existing man-made fibers to clarify the uniqueness of such smart textile materials.

Adaptive polymeric gels are a class of smart materials and such hydrogels can respond to changes in different environment such as temperature and pH. It is an exciting feature of intelligent materials. Due to their hydrophilic and biocompatible nature, most of them are used as biomaterials in medical and cosmetics areas. In Chapter 3, we have detailed this type of adaptive polymeric materials.

Adaptive polymeric particles can be made in different forms. One of the major features, which affect their properties and applications, is their particle size. The structure of core-shell and microencapsulation has made the adaptive polymeric particles incomparable position. The new developments of synthesis and characterization techniques of adaptive polymers have provided various ways in textiles and cosmetics. Therefore, Chapter 4 has focused on the discussions of polymeric particles and their applications.

Some groundbreaking contributions to the advancement of adaptive polymers, particularly in relation to textiles, have been made in the Hong Kong Polytechnic University. Towards the goal, fundamental adaptive textiles researches were carried out and some related models have been developed. The novel applications are the issues of Chapter 5 in adaptive textiles.

Many advances of adaptive composites with adaptive polymers are made in the past several decades. As a newborn, adaptive polymeric composite has brought new applications in modern academics and industries. The research into the interrelationship of processing, structure, property and performance is now the essence of adaptive polymeric composites. We have detailed the progress in this area in Chapter 6.

In Chapter 7, fabrication of nanofibres and nanofilms by electrospinning and self-assembly methods are introduced. At the same time, the principles and parameters of electrospinning and self-assembly are discussed. The morphologies and structures of nanofibres and nanofibrous nonwovens are described briefly. A comprehensive understanding of adaptive nanofibres and nanofilms can be obtained in this chapter.

Many original developments of cosmetics are founded on the combination of materials, biomedical sciences and technologies. Adaptive polymeric materials would greatly promote their innovation of cosmetics applications. The research efforts in adaptive polymers can improve the quality of life for beauty in a substantial trend. In our Chapter 8, we have detailed the above contents.

The technological advances of life sciences in the last decades significantly accelerated the medical research and developments. New horizons for the applications of adaptive and functional polymers in medical kingdom have been opened and more and more methods are developed to meet the practical requirements. Modern adaptive polymers and textiles are tailor-made to meet a wide variety of innovative applications including the biomimic and intelligent textiles. The medical applications of adaptive and functional polymers have been discussed and an outlook into the potential future developments is provided in Chapter 9.

The subsequent contents of the book focus on the four types of versatile functional polymers: dendrimers, hyperbranched polymers, chitosan and cyclodextrin. The major advancements in relative fields are detailed. Various application examples are presented. Their characteristics are highlighted and challenges to be addressed in the near future are pointed out.

The following are the major features of this book:

1. Focused on adaptive polymers. There are many smart and functional materials and adaptive systems, but we have mainly focused on adaptive polymers with some functional ones, which we have experience with and which we believe are important for new generations of materials and products.
2. Integrative in nature. The book has covered molecular design, synthesis, structure and properties, characterisation method and applications of adaptive polymers.
3. Cross-disciplinary teams' research work. Since the book needs to cover wide spectrum of contents from a vertical and horizontal points of view cross-disciplinary knowledge, understanding is very important. This is reflected in the book and is from our research work and team in the Shape Memory Textile Center in the Hong Kong Polytechnic University where we have more than 20 people in

the group every year with research expertise in polymer synthesis, physics, engineering and applications, biological and medical and cosmetics, textiles and apparel processing and evaluation.

4. Application oriented. Over the past ten years, we have been looking for materials for textile applications and once we started to research into Shape Memory Polymers, we have been working on different applications of adaptive polymers. Every week we have at least one meeting to review, brainstorm, design, discuss and develop different applications of adaptive materials and products. This application oriented research can benefit the readers with a good understanding of the potentials of these adaptive polymers.
5. Extensive partnership with different organiszations, particularly companies. In the past ten years, we have been working with local and international companies (including market players P&G, TAL, etc.) to develop new materials and products. These cover textile, apparel and intimate products, medical devices, cosmetics, etc. In addition, because of our pioneering work, companies visited us and we visited over 200 including market players such as BASF, Clariant, Smith and Nephew. This experience has given us a very unique opportunity to penetrate into the market and understand the needs of industries. This is a very important driving force for us to explore adaptive polymers day and night for many years and this partially reflected in the book.
6. Long term and pioneering research in adaptive polymers, particularly shape memory polymers.

With the above features and contents, we believe that this book will be a valuable reference to many people in different professions and at different levels/stages of their research careers in companies, universities and research institutes.

Prof J. L. HU
The Hong Kong Polytechnic University
August 8, 2009

Contents

Preface v

Acknowledgements xxi

Chapter 1 Introduction to Adaptive Polymers and Textiles 1

1.1 Concepts of adaptive polymers and textiles 2
1.1.1 Introduction to adaptive polymers 2
1.1.2 Introduction to adaptive textiles 5
1.2 Adaptive polymers 6
1.2.1 Shape memory polymers 8
1.2.2 Adaptive polymeric hydrogels 9
1.2.3 Adaptive polymeric particles 10
1.3 Adaptive textiles 12
1.3.1 Adaptive nanofibres and nanofilms 13
1.3.2 Shape memory fibres and fabrics 15
1.4 Framework of the book 16
1.5 Future aspects and summary 18
References 19

Chapter 2 Shape Memory Polymers 24

2.1 Introduction 25
2.2 Principles of shape memory function in SMPs 25
2.3 Classification of SMPs 27
2.3.1 Molecular structures of SMPs 27
2.3.1.1 Covalent bond between switches and net-points 27
2.3.1.2 Non-covalent bond between switches and net-points 29
2.3.2 Trigger patterns of SMPs 30

2.3.2.1 Direct thermal senstitive shape memory effect 30
2.3.2.2 Indirect thermal sensitive shape memory effect 31
2.3.2.3 Other external stimulus 33
2.3.2.3.1 Light as external stimulus 33
2.3.2.3.2 Multi-external stimulus in hydrogels 34
2.3.3 Applications of SMPs 35
2.3.3.1 SMPs used for bio-medical areas 36
2.3.3.1.1 Fixity and orthodontic application 36
2.3.3.1.2 Smart implant materials 37
2.3.3.1.3 Micro-acturator and others in biomedical areas 38
2.3.3.2 SMPs applied to textiles 38
2.3.3.3 Usage in thermal recording, sensors, actuators and others 39
2.4 SMPs with substrate bonded antibacterial activity 41
2.4.1 Molecular structure of SMP with antibacterial activity 42
2.4.2 Characterisation of substrate bonded antibacterial activity 44
2.4.3 Shape memory function 47
2.5 Supramolecular SMPs 49
2.5.1 Molecular design of supramolecular SMPs 49
2.5.2 Characterisation of supramolecular SMPs with AFM, DMA 51
2.5.3 Shape memory function of supramolecular SMP 53
2.6 Shape memory fibres 57
2.6.1 One way shape memory fibre 57
2.6.2 Two way shape memory fibres 60
References 61

Chapter 3 Adaptive Polymeric Gels and Applications 72

3.1 Introduction 73
3.1.1 Introduction of gels 73
3.1.2 Introduction of polymeric hydrogels 74
3.2 Classification and molecular structure of polymeric gels 77
3.2.1 Classification of polymeric gels 77
3.2.2 Gels formation by intermolecular bonding 79
3.2.3 Gels design criteria for hydrogels 80
3.2.4 Theory and principles of gelation 81
3.3 Synthesis of adaptive polymeric gels 82
3.3.1 Fundamentals of gelation 82

3.3.2 Swelling theory of hydrogels 82
3.3.3 Adaptive sensitive hydrogels 83
3.3.4 Preparation of adaptive hydrogels 85
3.3.4.1 Grafting methods onto cellulose based nonwovens 85
3.3.4.2 Plasma treatment 86
3.3.4.3 γ-ray pre-irradiation 86
3.3.4.4 UV-irradiation 87
3.3.4.5 Free radical polymerisation 87
3.4 Properties and applications of adaptive polymeric gels 90
3.4.1 Volume phase transition of polymeric gels 90
3.4.2 Properties of adaptive hydrogels 92
3.4.3 Applications of hydrogels 96
3.4.4 Applications of adaptive hydrogels 97
3.5 Summary 99
References 100

Chapter 4 Adaptive Polymeric Particles and Applications 107

4.1 Introduction of polymeric particles 108
4.2 Classification of adaptive polymeric particles 109
4.3 Properties of adaptive polymeric particles 111
4.3.1 Surface properties of polymeric particles 112
4.3.2 Size of polymeric particles 113
4.3.3 Shape of polymeric particles 114
4.3.4 Modification of polymeric particles 116
4.4 Manufacturing of polymeric particles 116
4.4.1 Aggregation theory 117
4.4.2 Manufacturing polymeric particles 118
4.4.2.1 Sol-gel process 120
4.4.2.2 Surface modification of particles 120
4.4.2.3 Fabrication of core-shell structure polymeric particles 121
4.4.2.4 Micro-encapsulation technology 121
4.5 Applications of adaptive particles 124
4.5.1 Application to textiles and fibres 125
4.5.1.1 Nanoparticple modification on textiles 125
4.5.1.2 Self cleaning modification on textiles 126
4.5.1.3 Shape memory finishing on textiles 126
4.5.1.4 Micro-encapsulation on textiles 128

4.5.1.5 Phase change material particles 129
4.5.2 Applications in medicine 130
4.5.2.1 Biodegradable polymer micro-hydrogels 130
4.5.2.2 Nanoparticles in gene therapy 131
4.5.2.3 Particles in drug delivery 131
4.5.3 Application to cosmetics 131
4.5.3.1 Particles on makeup cosmetics 132
4.5.3.2 Particles on skin cosmetics 133
4.6 Summary 134
References 135

Chapter 5 Adaptive Textiles Using Adaptive Polymers 142

5.1 Adaptive textiles for thermoregulation 143
5.1.1 Types of phase change materials 143
5.1.2 Incorporation method of PCMs in textiles 144
5.1.2.1 Fibre technology 144
5.1.2.2 Coating method 146
5.1.2.3 Laminating method 146
5.1.3 Applications in adaptive garments 147
5.2 Shape memory polymeric textiles 148
5.2.1 Shape memory polymeric fibres 149
5.2.2 Shape memory function from fibre to fabrics 152
5.2.2.1 Self-adaptability of shape memory fabrics 152
5.2.2.2 Self-repairability of shape memory fabrics 152
5.2.3 Shape memory garments 155
5.2.4 Shape memory finishing fabrics and garments 157
5.3 Adaptive chameleon textiles 158
5.3.1 Types of chameleon materials for adaptive textiles 159
5.3.2 Application methods of chameleon materials in textiles 161
5.4 Luminescent adaptive textiles 163
5.4.1 Introduction of luminescent materials 163
5.4.2 Applications of luminescent in adaptive textiles 164
5.5 Conductive polymer textile 165
5.5.1 Conducting polymer fibres, fabrics and sensors 166
5.5.2 Applications in textiles 168
5.6 Other functional textiles 169
5.6.1 Nano-technology applications in textiles 169
5.6.1.1 Water repellence 169

5.6.1.2 Antibacterial properties 170
5.6.1.3 Self cleaning 170
5.6.1.4 UV protection 171
5.6.1.5 Antistatic properties 172
5.6.2 Intelligent impact protection textiles 172
5.7 Perspective 173
References 173

Chapter 6 Adaptive Polymeric Composites and Applications 180

6.1 Introduction 181
6.2 Thermal adaptive polymeric composites 181
6.2.1 Introduction of thermal adaptive polymeric composites 181
6.2.2 Thermal adaptive SMP reinforcement 182
6.2.3 Polymer/polymer blending with thermal adaptive effect 183
6.2.4 SMP/thermoexpanded graphite (TEG) 184
6.2.5 Thermal expansion of SMP composites 185
6.2.6 Thermal conductivity of SMP composites 185
6.2.7 Micro-mechanics of SMP composites 186
6.2.8 SMP composites with special functions 186
6.2.9 Development trends of thermal adaptive SMP composites 187
6.2.9.1 SMP composites based on non-thermal phase transition 188
6.2.9.2 Two-way SMP composites 189
6.3 Electro adaptive polymeric composites 189
6.3.1 Introduction of electro adaptive polymeric composites 189
6.3.2 Electrostrictive graft elastomers 190
6.3.3 Electrostrictive cellulose paper 191
6.3.4 Electro-viscoelastic elastomers 192
6.3.5 Ferroelectric electro adaptive polymers 192
6.3.6 Dielectric electro adaptive polymers 193
6.3.7 All-organic electro adaptive polymers 193
6.3.8 Electro adaptive liquid crystal elastomers 194
6.3.9 Electro adaptive conductive polymers 194
6.3.10 Electro adaptive composites from conductive thermal adaptive SMPs 195
6.3.11 Electro adaptive ionomeric polymer-metal composites 195
6.3.12 Electro adaptive ionic polymer gels 196
6.3.13 Electro adaptive carbon nanotubes 197

6.4 Light adaptive polymeric composites 197
6.4.1 Introduction of light adaptive polymeric composites 197
6.4.2 Photodeformability induced by photoisomerization 198
6.4.3 Photodeformablity induced by photoinduced ionic dissociation 198
6.4.4 Photodeformability induced by photoreactive molecules 198
6.4.5 Light adaptive composites from thermal adaptive SMPs 199
6.5 Magnetic adaptive polymeric composites 199
6.5.1 Introduction of magnetic adaptive polymeric composites 199
6.5.2 Magnetic adaptive elastomers 199
6.5.3 Magnetic adaptive gels 199
6.5.4 Magnetic adaptive SMPs 201
6.6 Moisture/water adaptive polymeric composites 201
6.6.1 Introduction of moisture/water adaptive polymeric composites 201
6.6.2 Moisture/water adaptive polymer with decreased switch temperature 201
6.6.3 Moisture/water adaptive polymer with hydrophilic segment 202
6.7 Applications to textiles 202
6.7.1 Thermal adaptive polymer composites for textile applications 203
6.7.2 Electro adaptive polymeric composites for textile applications 203
6.7.3 Light adaptive polymeric composites for textile applications 203
6.7.4 Magnetic adaptive composites for textile applications 204
6.7.5 Water/moisture adaptive polymeric composites for textile applications 204
6.8 Summary 205
References 205

Chapter 7 Adaptive Polymeric Nanofibre and Nanofilm 215

7.1 Introduction 216
7.2 Fabrication technologies 216
7.2.1 Electrospinning 218
7.2.1.1 Definition of electrospinning 218
7.2.1.2 Principle of electrospinning 219

7.2.1.3 Parameters of electrospinning 220
7.2.1.3.1 Solution concentrations/viscosity 220
7.2.1.3.2 Electrical voltage/applied voltage 221
7.2.1.3.3 Feeding rate 222
7.2.2 Self-assembly 223
7.2.2.1 Definition of self-assembly 223
7.2.2.2 Principle of self-assembly 223
7.2.2.2.1 Non-covalent interactions 224
7.2.2.2.2 Intermolecular packing 225
7.2.2.2.3 Biological self-assembly 225
7.2.2.2.4 Nanomotor 225
7.2.2.3 Advantages of self-assembly technology 226
7.3 Morphologies and structure 227
7.3.1 Structure of nanofibres 227
7.3.1.1 Huge surface area to volume ratio 227
7.3.1.2 Orientation structure 228
7.3.1.3 Surface structures of nanofibres 228
7.3.1.4 Core-shell nanofibres 229
7.3.2 Structure in nanofibrous nonwoven 229
7.3.2.1 Fibre arrangement 229
7.3.2.2 Variation in diameter 230
7.3.2.3 Pore structure 230
7.4 Development of adaptive nanofibres and nanofilms 231
7.4.1 Shape memory polymer nanofibres/non-woven 232
7.4.2 PH-responsive nanofibre/nanofilms 233
7.4.3 Adaptive phase change nanofibres 234
7.4.4 Nonlinear optical nanofibres 234
7.4.5 Conducting nanofibres/nanofilms 236
7.4.6 Optoelectronic nanofilms 237
7.4.7 Bacterial cellulose nanofibres 238
7.4.8 Adaptive composites 239
7.4.9 Others 240
7.5 Applications of nanofibres, nanofilms of adaptive polymers 241
7.5.1 Protective clothing applications 241
7.5.2. Filtration applications 241
7.5.3 Biomedical application 242
7.5.3.1 Delivery of drug and pharmaceutical composition 242
7.5.3.2 Tissue engineering 243

7.5.3.3 Wound dressing 244
7.5.4 Other functional application 244
References 245

Chapter 8 Cosmetics Applications of Adaptive and Functional Polymers 252

8.1 Introduction 253
8.2 Definition, history and classification of cosmetics 253
8.2.1 Definition of cosmetics 253
8.2.2 Development history of cosmetics 254
8.2.2.1 Original cosmetics 254
8.2.2.2 Emulsion cosmetics 255
8.2.2.3 Functional cosmetics 255
8.2.2.4 Bionic cosmetics 255
8.2.3 Classification of cosmetics 256
8.2.4 Adaptive and functional polymers in cosmetics 257
8.2.4.1 Adaptive and functional polymers with unique advantages 257
8.2.4.2 Increasing maturing technology in polymer industry 257
8.2.4.3 Consumer expectations 258
8.3 Cosmetic applications of adaptive and functional polymers 258
8.3.1 Hydrogels 258
8.3.2 Polysaccharides 261
8.3.3 Cyclodextrin 265
8.3.4 Shape memory polymers 268
8.3.5 Nanopolymer particles 270
8.4 Glamourous prospects in the cosmetic field 274
8.4.1 Multifunctional cosmetics 275
8.4.2 Smart cosmetics 275
8.4.2.1 Adaptive hydrogel cosmetics 275
8.4.2.2 Shape memory cosmetics 276
8.4.3 Nanoparticle cosmetics 276
8.5 Summary 277
References 278

Chapter 9 Medical Applications of Adaptive Polymers 284

9.1 Introduction 285

9.2 Aims and requirements of medical adaptive polymers 285
9.2.1 Biocompatibility 286
9.2.2 Functionality 286
9.2.3 Nontoxicity 287
9.2.4 Mechanical properties 287
9.2.5 Other properties 288
9.3 Adaptive polymeric materials for medical applications 288
9.3.1 Adaptive polymeric materials for tissue engineering 288
9.3.1.1 Electroactive polymer for artificial muscles 288
9.3.1.2 Biodegradable scalffold for tissue engineering 291
9.3.2 Adaptive polymeric materials for diagnosis and medical treatments 294
9.3.2.1 Adaptive polymers in drug delivery 294
9.3.2.2 Smart polymer nanoparticles for diagnostics 298
9.3.2.3 Adaptive polymeric materials in biosensors 301
9.3.2.4 Adaptive polymeric materials for eye related materials 304
9.3.3 Adaptive polymeric materials for medical devices 307
9.4 Summary 310
References 311

Chapter 10 Special Adaptive and Functional Polymers and Their Applications 319

10.1 Structure, synthesis, properties and applications of dendrimer macromolecules 319
10.1.1 Introduction 319
10.1.2 Synthesis of dendrimer macromolecules 320
10.1.2.1 Divergent synthesis 320
10.1.2.2 Convergent synthesis 322
10.1.2.3 Other approaches with improved efficiency 323
10.1.3 Properties and applications 325
10.1.3.1 Medical applications 325
10.1.3.2 Nanoscale catalyst 326
10.1.3.3 Dendritic sensors 326
10.1.3.4 Energy ‘funnels’ 327
10.1.4 Applications in textiles 327
10.1.4.1 Dye extraction 327
10.1.4.2 Dyeing of PP fibres 328

10.1.4.3 Dyeing of cotton 329
10.1.4.4 Multi-functional finishing of textiles 329
10.1.5 Summary 330
10.2 Structure, synthesis, properties and applications of hyperbranched macromolecules 330
10.2.1 Introduction 330
10.2.2 Synthesis of hyperbranched macromolecules 331
10.2.2.1 Step-growth polycondensation of monomer ABx 331
10.2.2.2 Self-condensing vinyl polymerisation 332
10.2.2.3 Ring-opening multibranching polymerisation of latent ABx 333
10.2.2.4 Other strategies for the synthesis of hyperbranched macromolecules 335
10.2.2.4.1 The A_2+B_2 route to hyperbranched macromolecules 335
10.2.2.4.2 Graft 'onto' and graft 'from' approaches 335
10.2.3 Properties and applications of hyperbranched macromolecules 336
10.2.3.1 Coating 336
10.2.3.2 Medicine 337
10.2.3.3 Polymer blending 337
10.2.3.4 Toughening for epoxy-based composites 338
10.2.3.5 Nonlinear optics (NLO) 338
10.2.4 Applications in textiles 339
10.2.4.1 Dyeing 339
10.2.4.2 Spinning 339
10.2.5 Summary 339
10.3 Structure, properties of chitosan and its applications 340
10.3.1 Definition and structure of chitosan 340
10.3.2 Properties of chitosan 343
10.3.2.1 Degree of acetylation 345
10.3.2.2 Solubility 345
10.3.2.3 Crystallinity 345
10.3.2.4 Swelling/deswelling (pH sensitivity) 346
10.3.2.5 Antibacterial activity 346
10.3.3 Applications of Chitosan 347
10.3.3.1 Chitosan based wound-dressing 349

10.3.3.2 Chitosan modified PNIPAAm hydrogels 350
10.3.3.3 Chitosan antibacterial finishing on cotton 351
10.3.3.4 Chitosan durable press finishing on cotton 352
10.3.3.5 Chitosan on dye applications 352
10.3.3.6 Chitosan finishing on silk 353
10.3.3.7 Chitosan fibres 353
10.3.4 Chitosan used in cosmetics 353
10.3.5 Summary 354
10.4 Cyclodextrins 355
10.4.1 Structure and properties of CDs 355
10.4.1.1 Structure of CDs 356
10.4.1.2 Physical and chemical properties of CDs 357
10.4.1.3 Applications of CDs and CD complexes 359
10.4.2 Applications of CDs to textiles 359
10.4.2.1 Applications in textile dyeing 360
10.4.2.2 Applications on textile finishing 361
10.4.2.3 Applications in textile functioning 362
10.4.2.4 Applications in textile monitoring 363
10.4.3 Applications of CDs to skin/healthcare products 364
10.4.3.1 Applications in improving stability and bioavailability 364
10.4.3.2 Applications in change solubility 366
10.4.3.3 Applications in control release 366
10.4.3.4 Applications in cahnge penetrability 367
10.4.3.5 Applications in reduing inflammation and irritation 368
10.4.4 Summary 369
References 369

Index 385

Acknowledgements

The contents of this book are based partially on the original and integrated work produced in my research group (Shape Memory Textile Center in the Hong Kong Polytechnic University) over the past ten years. A number of people have contributed to and assisted with the preparation of the manuscript and the author is especially grateful to the following: Dr Baohua Liu, who helped with format checking and editing of the manuscript throughout; Ms Susanna Li helped with organising and managing the files and correspondence. In general, I am deeply indebted to my research students, postdoctoral fellows and research assistants, who have made contributions to the major part of the contents and made the book possible.

Some original data and part of the contents are taken from the PhD theses written under my supervision by Dr Yong Zhu, Ms Haitao Zhuo, Dr Baohua Liu, Mr Qinghao Meng and Dr Xuemei Ding. In particular, Dr Yong Zhu has prepared Chapter 2 mainly based on his PhD thesis and some new developments. Dr Baohua Liu has provided chapters on Adaptive Polymeric Gels and Adaptive Polymeric Particles and Applications. Dr Jing Lu has integrated the contents of Chapter 5 on Adaptive Textiles made by adaptive polymers. Based on his PhD study, Mr Qinghao Meng edited the chapter on Adaptive Polymeric Composites. Adaptive Polymeric Nano-fiber and Nano-film is from Zhuo Hai Tao's PhD work. Cosmetics Applications of Adaptive and Functional Polymers was edited by Liu Ge, who was working with us and now moved to Australia for further studying. Medical Applications of Adaptive Polymers was co-edited by Zeng Wei and Wu Xinbo. The last chapter on Special Functional Polymers and Their Applications has combined the work of different people. The dendrimer and the hyperbranched polymers by Qinghao Meng, the Chitosan is partially from the work of Dr Baohua Liu and Cyclodextrin is provided by Wu Xingbo.

In addition, I would like to acknowledge the funding support of the Hong Kong Textile and Apparel industry as well as the Hong Kong Government through the Innovation and Technology Commission over the past years. From these industry-guided projects, development of shape memory garments/fabrics and development of shape memory fibers, we can realise the significance of shape memory polymers to the textile and apparel industry as well as to the academic world through the knowledge accumulated during these developments.

Chapter 1

Introduction to Adaptive Polymers and Textiles

This chapter will briefly introduce the molecular design, synthesis, structure and properties, characterisation method and applications of adaptive polymers. There are many smart and functional materials, and adaptive polymers are one such example. This book mainly focuses on adaptive polymers and textiles. First, some concepts will be suggested and definitions of adaptive polymers and textiles are reviewed. Following that, the characteristics, structures and potential applications of adaptive polymers will be summarised. Developing novel smart adaptive polymers and textiles are challenging, and some future trends are suggested at the end.

Keyword: adaptive polymers; adaptive textiles; adaptive polymeric gel; adaptive polymeric particles; nanofibres and nanofilm; medical; cosmetics; drug delivery; dendrimer; hyperbranched polymer; chitosan; cyclodextrin

1.1 Concepts of Adaptive Polymers and Textiles

First of all, the concept of adaptive polymers should be defined. As a general viewpoint, adaptive polymers are a type of polymer which is designed to respond to different external stimuli in a controlled and predictable manner. Adaptive polymers can display distinct and great changes when responding to a small stimuli.[1,2] As a counterpart, adaptive textiles are a type of smart textile which displays a smart response to a narrow range of external stimuli. Accordingly, adaptive textiles display preplanned/predesigned responses when undergoing small environmental changes (e.g. pH, temperature, ionic strength, chemical, light, etc.). Intensive studies in the relative fields of adaptive/functional materials commenced several decades ago in an attempt to understand their structures, properties, preparations and applications. Recently, there are increasingly more studies in the field of the adaptive/functional materials. With time, much of the focus in research has been on the design of materials with 'smart' or 'intelligent' behaviours. A number of approaches have been employed to reach these goals, including: (a) the synthesis of functional polymers with specific compositions and architecture; (b) blending of polymer materials with special molecular additives; (c) designing special structures of required materials and (d) modifications by various chemical/physical treatments. Significant efforts have also been made to prepare, characterise, and understand the structure/property relationships of adaptive/functional materials.[3] This chapter offers a broad overview of the current status in research on adaptive and functional polymer/textiles. The characteristics, structure, design, modification and applications of adaptive/functional polymers and textiles will be introduced briefly in the following sections and details will be addressed in the following chapters. This chapter also emphasises recent results and developments in relevant fields.

1.1.1 *Introduction to adaptive polymers*

Adaptive polymers, also known as 'stimuli-responsive' polymers or so called 'intelligent/smart' polymers, have been presented as stimuli-sensitive, intelligent, smart, or environmentally sensitive polymers[4] that respond with large property changes to small physical or chemical stimuli. The concept of adaptive polymers is shown in Fig. 1.1.

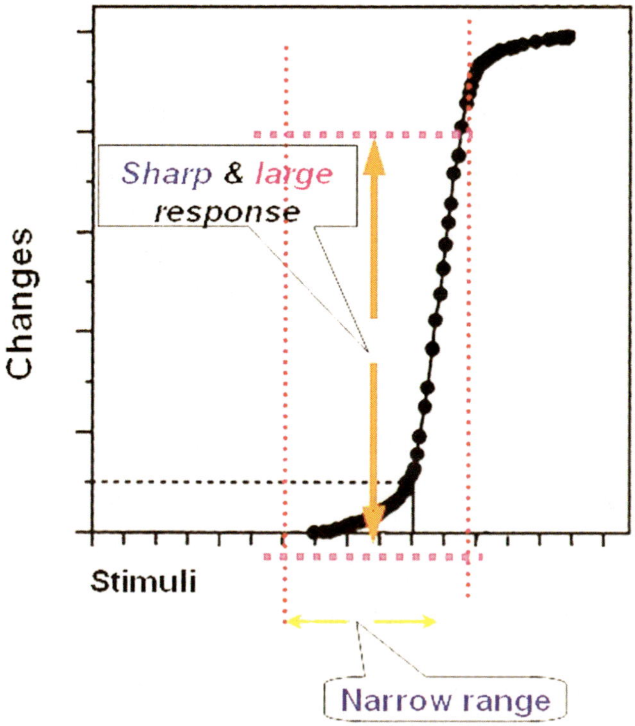

Fig. 1.1 Concept of adaptive polymers. (Reprinted with permission from the presentation entitled '"Smart" Polymers and Smart Hydrogels: Design, Synthesis and Applications', presented by Prof. Allan S. Hoffman in HKPOLYU on 24 Oct 2007).

Many attempts have been made by scientists to design smart materials which can respond to different external stimuli, such as temperature, pH, light, electric field, chemicals and ionic strength. These responses are manifested as dramatic changes in one or more ways of the following: shape, surface characteristics, solubility, colour, forms, and formation of intricate molecular self-assembly or a sol-to-gel transition.[5]

Adaptive polymers are a very interesting class of materials since they can exhibit marked and rapid conformational changes in response to external stimuli. In other words, stimuli sensitive polymers (SSPs) are polymeric materials which can change size, shape or characteristics based on the stimuli. These polymers are able to change their physical shapes or chemical properties when small changes in the surrounding environment occur. The adaptive polymer might recognise a stimulus as a signal, and then judge the degree of this signal, and consequently, change their conformation as a direct response.[6] As adaptive materials, they acquire unique properties and offer the possibility of achieving control over biocompatibility, wettability, permeability, swelling, transparency and so on. Adaptive polymers have also attracted considerable attention in biotechnologies due to their outstanding properties. Successful applications have already been demonstrated in sensing, intelligent textiles, bio-separation and drug delivery systems.[2] Different stimulations have listed in Table 1.1.

Table 1.1 Environmental stimulation.

pH Value	Enzyme substrates	Biochemical
Temperature	Ionic strength	Sonic radiation
Electrical and magnetic fields	Chemical agents and solvents	Affinity ligands
Mechanical stress and strain	Electromagnetic radiation (UV, visible)	Other biochemical agents

There are a number of adaptive materials that exist, for example, adaptive polymeric hydrogels, adaptive polymeric particles, shape memory polymers, smart film/fibres, and smart textiles. Also, some derivatives and copolymers of cyclodextrin and chitosan, and some unique adaptive structures, such as hyperbranched polymers, dendrimer, supramoleculars, can be included the field of adaptive materials. A detailed review will be summarised in the following chapters. A schematic diagram of smart hydrogel response to certain stimuli, such as changes in temperature or pH, is shown in Fig. 1.2. Prominent examples of smart hydrogels are polyacrylic acid (pH-sensitive, Fig. 1.2) and poly-N-isopropylacrylamide(PNIPAAm, temperature-sensitive, Fig. 1.2) with a transition temperature of about 32°C.[7]

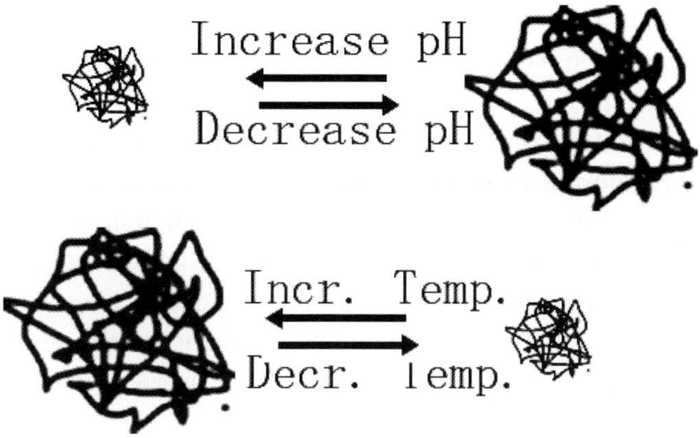

Fig. 1.2 Schematic of 'smart' hydrogel response with temperature/Ph.[7]

1.1.2 *Introduction to adaptive textiles*

The emergence of new and innovative technologies and changes in demographics and lifestyles of consumers have generated an upsurge in the development and production of multifunctional and adaptive textiles and clothing. Adaptive polymer textiles mainly refer to textiles which are made of adaptive polymer alone or a combination of adaptive polymer and other natural, synthesis materials that will sense and react to environmental conditions or stimuli from mechanical, thermal, chemical, electrical or magnetic sources.

The need for multifunctional textiles has resulted in the investigation of fibres and textiles traditionally used in many areas, such as biology, medicine, engineering, geo-textiles and space technology. These diversified requirements have led to new methods and approaches in designing, producing and applying novel textiles. For example, by incorporating SSP into textile structures, it is possible to combine the mechanical properties of textiles with the environmental responsiveness of SSPs.[8] Coating, radiation grafting, high-intensity UV-curing and wet spinning of crosslinked gels into fibres are techniques that generate such structures. Upon exposure to a specific stimulus, for example, change in pH of the surrounding solution, the SSP dramatically swells or collapses. SSP can thus function as a reservoir for a chemical compound to be released on demand. Possible applications are controlled delivery of functional substances (drugs, nutrients, herbicides, etc.), temperature and moisture regulation, separation, communication, robotic muscles, sensors and quality control.[8]

Adaptive textiles may be used for additional functions specific to an adverse or extreme climate, job or professional environment or to enhance adaptability and/or productivity of the user. When textile assumes an additional function over and above the conventional purpose of covering the human body for protection, it may be regarded as an adaptive textile. If this additional functionality can adjust to changes in usage conditions, then the textile may be regarded as active smart or intelligent.

1.2 Adaptive Polymers

Recent developments in the field of polymeric science have produced adaptive polymers or SSPs that possess the unique capability to change structure in response to small environmental changes, such as temperature, pH, salt, light, electrical fields and stress. The changes of adaptive polymers could be driven by non-covalent interactions, including coulombic force, hydrogen-bonding, van der Waals force, and hydrophobic interactions. In accordance to external stimuli, adaptive polymers can be divided into those that have temperature and pH-sensitivity, and those that have electrical, ion, moisture, and stress responsiveness. With respect to material forms, adaptive polymers can be classified into adaptive polymeric gel, adaptive polymeric particles, shape memory polymers, adaptive polymeric film and adaptive polymeric fibre. The adaptive materials can also be classified as adaptive composite, polymer and textiles according to their components.

Temperature sensitive polymers have a phase transition that occurs at a certain temperature. The pH-responsive adaptive polymers are formed of ionic polymers with weak acid groups, such as carboxyls. The pH responsive polymers are highly swollen at an alkaline pH, because the carboxyl groups are ionised and mutual repulsion by the negatively charged carboxyl groups forces the polymer chains apart, bringing large amounts of water into the gel. At a low pH, the carboxyls lose their charge and the reduced repulsion allows the gel to collapse. A list of various adaptive polymers that can be used is presented in Table 1.2.

Table 1.2 Adaptive polymers[9]

Type of Stimulus	Responsive Polymers
pH	Dendrimers
	Poly (L-lysine) ester
	Poly (hydroxyproline)
	Lactose PEG grafted poly (L-lysine) nanoparticle
	Poly (L-lysine)-g-poly (histidine)
	Poly (propyl acrylic acid)
	Poly (ethacrylic acid)
	Polysilamine
	Eudragit S-100
	Eudragit L-100
Ca^{2+}	Alginate
Mg^{2+}	Chitosan
Organic solvent	Eudragit S-100
Temperature	PNIPAAm
Magnetic field	PNIPAAm hydrogels containing ferromagnetic material
	PNIPAAm-co-acrylamide
Ru^{2+A} Ru^{3+} (redox reaction)	PNIPAAm hydrogesl containing Tris (2-2 bipyridil) ruthenium (II)
Temperature (sol-gel transition)	Poloxamers
	Chitosan-glycerol phosphate water
	Prolastin
	Hybrid htdrogels of polymer and protein domains
Electric potential	Polythiophen gel
IR radiation	Poly (N-vinyl carbazole) composite
UV radiation	Polyacrylamide crosslinked with 4-(methacryloylamino) azobenzene
	Polyacrykamide-trphenylmathane leuco derivatives
Ultrasound	Dodecyl isocyanate-modifies PEG-grafted poly (HEMA)
Dual-Stimuli-Sensitive Polymers	
Ca2+ and PEG	Carbocymethyl cellullose
Ca2+ and temperature	Eudragit S-100
Ca2+ and acetonytrile	Eudragit S-100
32°C and 36°C	Hydrogels of oligoNIPAAm and oligo (N-vinylcaprolactum)
pH and temperature	Poly (N-acryloyl-N-propyl piperazine)
Light and Temperature	Poly (vinyl alcohol)-graft-polyacrylamide-triphenylmethane leucocyanide derivatives

1.2.1 *Shape memory polymers*

Materials are said to show a shape memory effect if they can be deformed, fixed into a temporary shape and then recover their original, permanent shape only by exposure to an external stimulus,[10] such as heating, lighting or pH. This reversible ability opens up numerous applications in smart textiles,[11] adaptive facilities, sensors and actuators.

Shape memory polymers may sense thermal, mechanical, magnetic or electric stimulus and exhibit actuation or some predetermined response, making them possible to tune into technical parameters, such as shape, position, strain, stiffness, natural frequency, damping, friction and other static and dynamical characteristics. Shape memory properties provide very attractive insight into materials science, opening unexplored horizons and giving access to unconventional functions in every material class. During past decades, shape memory polymers, as a group of the most applicable smart materials, have developed rapidly not only in academics, but also in the industry. Shape memory polymers have excellent characteristics, such as low cost, good processing ability, high shape recoverability, large stretchability and a wide range of shape recovery temperatures in comparison to shape memory alloy.[12,13] Shape memory polymers are largely developed functional polymers which find wide applications in various fields. The relations between shape memory effects and structures are investigated based on current research.[14] Shape memory polymers have received increasing attention because of their scientific and technological significance. In principle, a thermally induced shape memory effect can be activated by an increase in temperature.

Shape memory function can be realised in many polymer systems. In a molecular structure, there are covalent and non-covalent bonded shape memory polymers. Shape memory ability can be triggered by temperature. In an indirect trigger pattern, some researchers have explored a variety of new types of shape memory polymers which can respond to novel external stimuli, such as water immersion,[15] light,[10] electricity[16] and so on.

Supramolecular shape memory polymers possess novel quadruple hydrogen bonding and were only studied in academics until now.[17]

Shape memory polymers are one of the key smart materials that are developed widely. Shape memory fibres have been engineered systematically at the Hong Kong Shape Memory Textiles Research Centre[18] for applications of shape memory polymers. The uniqueness of such smart fibres and fabrics are highlighted based on experimental results. Due to the reversibility of shape memory polymers, they can be used in multiple areas, such as biomedicine, bone fixity materials, orthodontic applications, intelligent implant materials, micro actuators and sensors. Shape memory polyurethanes feel good to touch, have anti-wrinkle properties and feasible forms of application, which can be applied in various special textile fields.[19]

1.2.2 *Adaptive polymeric hydrogels*

Hydrogels are water-swollen polymeric materials with a distinct three-dimensional structure. A three-dimensional network is formed by crosslinking polymer chains. A polymeric hydrogel is a network of hydrophilic polymers that can swell in water and hold a large amount of water while maintaining structure. Adaptive polymeric hydrogels are a special classification of hydrogels which have water as the solvent and show different changes under specific stimuli. In current developments of adaptive polymeric hydrogels, they are widely applied in many fields, including drug delivery, sensors, cosmetics, medicine, bio-engineering, etc.

Adaptive polymeric hydrogels have the unique capability to change structure in response to small environmental changes, such as temperature, pH, salt, light, electrical fields, and stress.[20] Most adaptive polymeric hydrogels are temperature or pH sensitive, meaning that a phase transition occurs at a certain temperature or pH value. Ionic polymers with weak acid groups, such as carboxyls, have been produced as pH sensitive hydrogels. These polymeric hydrogels are highly swollen at an alkaline pH, because the carboxyl groups are ionised and mutual repulsion by the negatively charged carboxyl groups forces the polymer chains apart, bringing large amounts of water into the hydrogel.[21]

Forms of adaptive polymeric hydrogels cover crosslinked (permanent) hydrogels, reversible hydrogels, micelles, modified interfaces, conjugated solutions and so on. Stimuli-responsive polymers and hydrogels show their response properties within setting conditions. Hence, there are a large variety of different approaches. The responses can also be surface, mechanical, dissolution/precipitation, degradation, drug release, changes in hydration state, swelling/collapsing, hydrophilic/hydrophobic changes in shape, conformational changes and micellisation (Fig. 1.3).

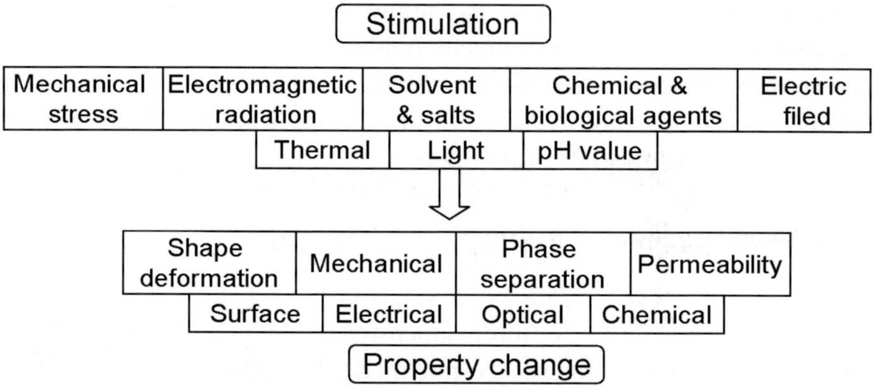

Fig. 1.3 Potential stimuli and responses of synthetic polymers.

The classification of adaptive polymeric gels may be based on the source, including natural or synthetic gels; nature of the crosslinking, such as covalent or physical gels; nature of the network, such as homopolymer, copolymer, interpenetrating or double networks; presence of pores, such as homogeneous (optically transparent), microporous and macroporous hydrogels; and finally, on their fate in an organism, such as degradable and nondegradable hydrogels. Each type of adaptive polymeric hydrogel has a special application in various fields.

1.2.3 *Adaptive polymeric particles*

Adaptive polymeric particles can exhibit a number of unique properties due to their tiny forms and differences from normal particles. People may combine the benefits of smart materials and particle materials to create a new generation product that can be seamlessly integrated into a complex system as an adaptive polymeric particle. Commonly, a polymer particle is a spheriform matter and can be prepared through a polymerisation process. Adaptive polymeric particles are a type of particles which can express distinct or other changes within a narrow range of external stimuli, such as light, heat or temperature, pH or ion, electric fields, chemicals and other stimuli.[22-24] As a special class of adaptive polymeric particles, nanoparticles are microscopic particles which are sized between 1 and 100 nanometers. Nanoparticles are generally classified based on their dimensionality, morphology, composition, uniformity and agglomeration.[25]

In order to serve special applications, different adaptive and/or functional particles could be obtained by controlling the synthesis conditions, selection of composition, optimisation of technology and so on. These properties of particles include size, morphology, surface group attachment, composition, dispersion, etc. Collections of many particles exist as powders, suspension or colloids.

There are many important chemical and physical parameters of adaptive polymeric particles, such as size, shape, morphology, light reflection/diffraction, viscosity, solvent ability, absorbency, groups and so on. Many interesting properties of adaptive polymeric particles are the result of their dimension, which lies between atomic and bulk. The surface area and size are two important aspects of particles. The aggregation of particles must also be taken into consideration when applied. Two important implications of the size range of particles are: (a) particle materials have enormous surface areas and surface energies, and (b) the properties of particles are not always those of the corresponding bulk matter or corresponding atoms or molecules. The surface properties of polymeric particles include hydrophobicity/hydrophilicity, surface activity, van der Vaals force, reflection and diffraction, contact angle, absorbency and so on. The shape of the particle has many influences on its properties. The properties of polymeric particles can be changed as desired through chemical, physical, mechanical modifications or other techniques. The surface structure, surface energy, electrical property, surface absorbency, and reaction ability can be modified by medication to serve the requirements of new materials, techniques and technology. The modification of polymeric particles can broaden its application fields or endorsement with novel characteristics. Modifications include coating, the coupling agent, precipitation reaction, surface chemistry, grafting, plasma or ion treatment, micro-encapsulation and many others.

There are various chemical and physical methods to obtain adaptive polymeric particles, such as sol-gel process,[26] and surface medication. As well, the core-shell structure,[27] micro-sphere,[28] and microencapsulation techniques[26] are widely used to prepare adaptive polymeric particles.

Adaptive polymeric particles have many applications in various fields, including nanoparticle modifications on textiles,[29] surface modifycations,[30] self cleaning on cotton modifications,[31] medication on wool fabrics,[32] nanofinishing,[33] shape memory finishing on cotton fabrics,[34] and micro-encapsulation on textiles.[35] In the medical fields, adaptive polymeric hydrogels also have numerous applications, such as gene therapy,[36] and drug delivery.[37] Furthermore, there are many applied aspects in cosmetics.[38] Precisely shaped polymeric particles and structures could be widely used for applications in photonic materials, micro-electro-mechanical systems (MEMS), biomaterials and self-assembly.

By combining research in the design, preparation, characterisation and application of adaptive polymeric particles, there may be the following research prospects for the future. The first is the relationship between the structure and physical/chemical properties of adaptive polymeric particles. The second is the fundamental theory research of adaptive polymeric particles. Finally, there are the industrial techniques and applications of polymeric particles. If all of these aspects are able to be developed in the near future, there may be a greater and deeper understanding and application of adaptive polymeric particles.

1.3. Adaptive Textiles

The design and development of functional textiles have attracted increasingly more attention in recent years. The number of attempts in this field is extensively parallel to research that focus on electronics, several solar energy-based systems, self cleaning, fire retarding, quick-drying, etc. Chameleon fibres[39] are a type of adaptive fibres which change their colours reversibly according to external environmental conditions. In accordance to the difference in functions, textiles could be classified in the following categories:[40] shape memory, chameleonic, water-resistant and vapour permeable (hydrophilic/non porous), heat storage, thermo regulated, vapour absorbing, heat evolving fabric and electrically heated suits.[11,13] Chameleon textiles and fibres are bionic applications with respect to the colour change theory. Self-adaptability of shape memory fabrics made from shape memory polyurethanes have been developed by the Shape Memory Textile Centre at the Hong Kong Polytechnic University since 2003.[11,41] The details will be introduced in Chapter 6.

The successful applications of adaptive textiles are limited and still under investigation. A remarkable group of them is related to the manufacture of microcapsules so that spherical bi-component particles consist of shells and surrounded core material.[42]

Shape changing fibres, yarns and fabrics have also been developed using a novel class of stimuli sensitive copolymers that can be readily processed and chemically integrated into textile substrates. Several types of shape memory polymers could be employed to prepare shape memory polymer fibres. With the introduction of shape changing materials, it has become possible to develop new composite yarns and fabrics which can change shapes reversibly by capturing moisture from the environment depending on the temperature and pH. Environmentally responsive textile production has become a reality with such materials and fabrics so that alterations in water-vapour transmission or porosity along with changes in environment conditions have been developed for some specific applications. In addition, numerous products may be developed using such materials both for apparel and technical applications.[43]

Adaptive textiles provide potential and enormous opportunities in the textile industry both in the fashion and clothing sectors, as well as in the technical textile sector. Moreover, these developments will be the result of active collaboration between people from a whole variety of backgrounds, such as engineering, science, design, and process development.

1.3.1 *Adaptive nanofibres and nanofilms*

Adaptive polymeric materials have developed greatly in many fields with the advancement of nanotechnology during the past decades. The combining of suitable fabrication techniques, various nanofibres, nanofilms, and nanoparticles of adaptive polymers can, therefore, be achieved. For example, nanofibres with a shape memory effect may be prepared from shape memory polymers by the electro spinning method, and pH responsive nanofilms could be obtained from the self-assembly technique. When adaptive polymers are fabricated with nanostructures, many unique properties could be obtained due to the surface properties and orientation structure of nano materials. Some novel functions which cannot be obtained in traditional fibres can now be achieved with nanofibres or nanofilms. Therefore nanofibres and nanofilms of adaptive polymers can be applied in many fields.

Electrospinning is a versatile method for processing solutions or melts into continuous fibres with diameters ranging from a few micrometers to a few nanometers. Electrospinning can be applicable to many types of soluble or fusible polymers. Also, polymers can be chemically modified and tailored with additives ranging from simple carbon-black particles to complex species, such as enzymes, viruses, and bacteria.[44] Electrospinning depends on a multitude of molecular, process, and technical parameters. The typical fabrication technologies of nanofibres and nanofilms will be discussed in the following chapters.

Many polymer systems, including water-soluble polymers, biopolymers and their derivatives, solvent-based and multiphase polymer systems, such as polymer blends, have been reported to obtain nanofibres or nanofibrous nonwoven membranes through electrospinning.[44] Preliminary investigations have indicated that the electrospun nanofibres present both minimal impedance to moisture vapour diffusion and extreme efficiency in trapping aerosol particles in comparison with conventional textiles.[45,46] The pore size of the nanofilm could be adjusted as designed.[47]

The fabrics made from nanofibres and nanofilms have shown some unique properties due to the speciallty of nano materials. Protective clothing can be prepared through electrospinning by using adaptive nanofibres and nanofilms. Consequently, electrospinning is a good choice to prepare lightweight and breathable fabrics because of their great surface area to volume. Fabrics composed of nanofibres are capable of neutralizing chemical agents without influencing air and water vapour permeability. Nonwoven materials prepared by layer-by-layer deposition during electrospinning have high porosity with tiny pores. The nano structure could provide good resistance to the penetration of chemically harmful agents in aerosol forms. In the biomedical field, adaptive nanofibres and nanofilms can be used in tissue engineering,[48] scaffolding,[49] wound dressing[50] and many other functional applications. Also, some cosmetics materials, such as substrates of masks, can be obtained from adaptive nanofilm.

1.3.2 *Shape memory fibres and fabrics*

Shape memory polymer fibres and fabrics can be prepared from shape memory polymers,[51] which include shape memory polyurethanes,[52,53] polyethylene terephthalate-polyethylene oxide copolymers, polystyrene-poly(1,4-butadiene) copolymers, polyethylene/nylone-6-graft copolymers, triblock copolymers made from poly(tetrahydrofuran) and poly(2-methyl-oxazoline), thermoplastic polynorbornenes, and other polymers that show shape memory effects by cross linking after spinning, such as polyethylene, poly(vinyl chloride), and polyethylene-poly(vinyl chloride) copolymers. The adaptive fibres can be obtained by wet, dry, melt, reaction and electric spinning.[54]

Shape memory yarns include shape memory fibres alone or blended yarn of shape memory polymer fibres and natural, regenerated, or synthetic fibres. The blended yarns can be core spun, friction and fancy yarns. Textile articles also include sewing threads of shape memory polymer fibres and ordinary natural or synthetic fibres.[55] Shape memory fabrics include those that are woven, knitted, braided and nonwoven. The above-mentioned fibres, yarns and fabrics can easily return to their original shape when they are heated above the shape memory switch temperature, even if they are wrinkled or deformed.

Shape memory fabrics can be used for wrinkle-free effects, reform recovery, and many kinds of designs that are fancy in art. The fabrics can be applied to textiles depending on the properties and uses that are desired, such as to the collars, cuffs of shirts and any other apparel which needs shape fixity. As well, there can be applications to the elbows and knees of apparels, and any other cloths which need to recover from bagging, as well as shape fixity of denim, velvet, cord, knitting fabrics or any other fabrics.

Shape memory finishing fabrics can be acquired with coating shape memory emulsion or combining shape memory film. The Hong Kong Polytechnic University has studied shape memory finishing chemicals and technologies for cotton fabrics,[56] wool fabrics and garment finishing[55] during the past several years.

1.4 Framework of the Book

This book will outline the most updated studies/achievements in adaptive and some very useful functional polymers, such as chitosan, cyclodextrin, dendrimer and those that are hyper branched. The principles of adaptiveness and their properties, molecular design and characterisation with emphasis on relevance to applications, particularly in textiles and related areas, will be reviewed. The whole book is comprised of ten chapters.

Chapter 1 presents a general background of the book, as well as concepts of adaptive polymers and textiles, their progress and applications in different fields. A general introduction of different adaptive polymeric polymers and adaptive textiles are reported as well in this chapter.

An original and novel adaptive polymeric, shape memory polymer, is systemically outlined in Chapter 2. The principles, classification and molecular design of shape memory polymers are reviewed and summarised. Different trigger patterns of shape memory polymers are also provided in this chapter. Novel directions, such as antibacterial and supramolecular shape memory polymers, are introduced and reviewed. The preparation and application of one-way and two-way shape memory fibres are provided at the end of the chapter.

A comprehensive review in the development of adaptive polymeric gels and hydrogels, their classification, properties and preparation methods, and application aspects are included systematically in Chapter 3. The application of adaptive polymeric gels in different fields will be discussed and conclusions will be provided.

In Chapter 4, adaptive polymeric particles (including nanoparticles) will be discussed. First, polymeric and adaptive polymeric particles are introduced. Then, classification, properties and surface medication of adaptive polymeric particles are summarised. Their manufacturing methods and application fields are reviewed in the last section.

Adaptive textiles made from adaptive polymers are discussed in Chapter 5. First, fabrication techniques of adaptive textiles are provided. Then, the adaptive mechanisms of textiles are reviewed in different classes. The application of luminescent adaptive membranes and nanotechnology in adaptive textiles will follow. Some applications of adaptive textiles will be provided.

Adaptive polymeric composites make up mainly the content of Chapter 6. This chapter analyses the classification and properties of adaptive polymeric composites comprehensively. Thermal, electro, light, magnetic and moisture adaptive polymeric composites are reviewed respectively. The development direction and future trends of adaptive polymeric composites are concluded at the end of the chapter.

Chapter 7 will discuss adaptive polymeric nanofibres and nanofilms. Electrospinnings as a powerful fabrication technology, will be practically discussed. New developments in manufacturing methods and self-assembling are also reviewed. Following that, the morphology and structure of nanofibres and nanofilms are summarised. Some important applications of adaptive polymeric nanofibres and nanofilms are also discussed in this chapter.

Chapter 8 focuses on applications of adaptive and functional polymers in cosmetics. The definition of cosmetics, the development process and classification are first given. Some advantages of adaptive polymers applied in cosmetics are also discussed. Then, various functional and adaptive polymers, such as adaptive polymeric hydrogels, cyclodextrin, polysaccharide, shape memory polymers and polymeric nanoparticles, are reviewed in the sections that follow. Additionally, smart and multifunctional cosmetics are reviewed. Finally, the development trends of smart cosmetics will be discussed.

Then, the medical applications of adaptive and functional polymers are reviewed. First, the aims and requirements of medically adaptive polymers are pointed out. Some excellent advantages can be exhibited when applying adaptive polymers as medical products due to special properties, such as biocompatibility, functionality, non-toxicity, mechanical properties and others. The applications of adaptive and functional polymers in the medical field are then reviewed. Finally, the development directions are provided as concluding remarks.

Some special and widely researched adaptive polymers, hydra-branched polymers, dendrimer, chitosan and cyclodextrin, are discussed in the conclusion of the book.

1.5 Future Aspects and Summary

There is an increasing requirement in society for materials that are better suited for contemporary needs, such as better response to applications, multiple uses and less stress to the environment. In recent years, scientists all over the world have focused on the synthesis and production of a variety of smart polymers and textiles, which are based on a good understanding of properties and structures of smart materials. It has become clear that new materials are needed, which are: (a) tailored perfectly to the needs of the applications that they are intended for, (b) created fundamentally for required devices with intelligence and (c) environmentally friendly. Modern macromolecular materials may be tailor-made materials that suit particular needs where we have experience with and believe to be important for future generations of materials and products, particularly textiles.

Smart textile products, such as those for skin care, may be produced by combining novel adaptive polymers and traditional textiles. These skin-care products can moisturise, whiten, brighten, or even prevent aging by firming the skin. For some given products, skin care technology may provide special care for acne, rosacea, eczema and more. Therefore, smart textiles with temperature-controlled release and antibacterial activity have great potential in the cosmetics and medial fields.

Many applications of such materials exist or are envisaged, including associative thickeners in paints, in personal care products, micelle systems for controlled drug release in medicine or for pollutant uptake in environmental applications. Many challenges are faced by researchers today in adaptive and functional polymers and textiles. A clear understanding of the structure-property relationship is essential for a rational design of new functional adaptive polymers that in turn, will enable further development of new technologies.

References

1. Galaev, I. Y. and Mattiasson, B., 'Smart'polymers and what they could do in biotechnology and medicine. *Trends in biotechnology*, 1999. 17(8): pp. 335-340.
2. Hoffman, A. S. and Stayton, P. S. *Bioconjugates of smart polymers and proteins: synthesis and applications*. 2004: WILEY-VCH Verlag Weinheim.
3. Luzinov, I., Minko, S., and Tsukruk, V. V., Adaptive and responsive surfaces through controlled reorganization of interfacial polymer layers. *Progress in Polymer Science*, 2004. 29(7): pp. 635-698.
4. Gil, E. S. and Hudson, S. M., Stimuli-responsive polymers and their bioconjugates. *Progress in Polymer Science*, 2004. 29(12): pp. 1173-1222.
5. Jeong, B. and Gutowska, A., Lessons from nature: stimuli-responsive polymers and their biomedical applications. *Trends in Biotechnology*, 2002. 20(7): pp. 305-311.
6. Kaneko, Y., Nakamura, S., Sakai, K., Aoyagi, T., Kikuchi, A., Sakurai, Y., and Okano, T., Rapid deswelling response of poly (*N*-isopropylacrylamide) hydrogels by the formation of water release channels using poly (ethylene oxide) graft chains. *Macromolecules*, 1998. 31(18): pp. 6099-6105.
7. Schild, H. G., Poly (*N*-isopropylacrylamide): experimental and application. *Prog. Polymer Sci*, 1992. 17: pp. 163-249.
8. Buschle-Diller, G., McClain, A., Walsh, W. K., Lin, W., and Hudson, S., Intelligent, Stimuli-Sensitive Fibers and Fabrics. *National Textile Center Research Briefs: June 2001*.
9. Roy, I. and Gupta, M. N., Smart polymeric materials: emerging biochemical applications. *Chemistry & Biology*, 2003. 10(12): pp. 1161-1171.
10. Lendlein, A., Jiang, H., Jünger, O., and Langer, R., Light-induced shape-memory polymers. *Nature*, 2005. 434: pp. 879-882.
11. Urayama, K., Ogasawara, S., and Takigawa, T., Pure shear deformation of physical and chemical gels of poly(vinyl alcohol). *Polymer*, 2006. 47(19): pp. 6868-6873.
12. Wei, X. and Yu, X. H., Synthesis and Properties of Sulfonated Polyurethane Ionomers with Anions in the Polyether Soft Segments. *Journal of Polymer Science: Part B: Polymer Physics*, 1997. 35: pp. 225-232.

13. Lendlein, A. and Kelch, S., Shape-memory polymers. *Angewandte Chemie International Edition*, 2002. 41(12): pp. 2034-2057.
14. Kim, B. K., Lee, S. Y., and Xu, M., Polyurethanes having shape memory effects. *Polymer-London*, 1996. 37: pp. 5781-5793.
15. Huang, W. M., Yang, B., An, L., Li, C., and Chan, Y. S., Water-driven programmable polyurethane shape memory polymer: Demonstration and mechanism. *Applied Physics Letters*, 2005. 86: pp. 114105.
16. Cho, J. W., Kim, J. W., Jung, Y. C., and Goo, N. S., Electroactive shape-memory polyurethane composites incorporating carbon nanotubes. *Macromolecular Rapid Communications*, 2005. 26(5): pp. 412-416
17. Liu, M., Bian, F., and Sheng, F., FTIR study on molecular structure of poly(*N*-isopropylacrylamide) in mixed solvent of methanol and water. *European Polymer Journal*, 2005. 41(2): pp. 283-291.
18. Hu, J., Zhu, Y., Lu, J., Yeung, L.-y., and Yeung, K.-w., *Uniqueness of shape memory fibers in comparison with existing man-made fibers*, in *The 9th Asian Textile Conference*. 2007, Federation of Asian Professional Textile Associations: Taiwan.
19. Tobushi, H., Hara, H., Yamada, E., and Hayashi, S., Thermomechanical properties in a thin film of shape memory polymer of polyurethane series. *Smart Materials and Structures*, 1996. 5(4): pp. 483-491.
20. Qiu, Y. and Park, K., Environment-sensitive hydrogels for drug delivery. *Advanced Drug Delivery Reviews*, 2001. 53(3): pp. 321-339.
21. Mahdavinia, G. R., Pourjavadi, A., Hosseinzadeh, H., and Zohuriaan, M. J., Modified chitosan 4. Superabsorbent hydrogels from poly (acrylic acid-co-acrylamide) grafted chitosan with salt- and pH-responsiveness properties. *European Polymer Journal*, 2004. 40(7): pp. 1399-1407.
22. Minko, S., Patil, S., Datsyuk, V., Simon, F., Eichhorn, K. J., Motornov, M., Usov, D., Tokarev, I., and Stamm, M., Synthesis of adaptive polymer brushes via "grafting to" approach from melt. *Langmuir*, 2002. 18(1): pp. 289-296.
23. Lu, Y. and Liu, J., Smart nanomaterials inspired by biology: dynamic assembly of error-free nanomaterials in response to multiple chemical and biological stimuli. *Accounts of chemical research*, 2007. 40(5): pp. 315-324.

24. Ballauff, M. and Lu, Y., "Smart" nanoparticles: Preparation, characterization and applications. *Polymer*, 2007. 48(7): pp. 1815-1823.
25. Buzea, C., Pacheco, I. I., and Robbie, K., Nanomaterials and nanoparticles: Sources and toxicity. *Biointerphases*, 2007. 2(4): pp. MR17-MR71.
26. Jin, W. and Brennan, J. D., Properties and applications of proteins encapsulated within sol–gel derived materials. *Analytica Chimica Acta*, 2002. 461(1): pp. 1-36.
27. Gan, D. and Lyon, L. A., Tunable Swelling Kinetics in Core? Shell Hydrogel Nanoparticles. *Journal of the American Chemical Society* 2001. 123(31): pp. 7511-7517.
28. Kawaguchi, H. and Fujimoto, K., Smart latexes for bioseparation. *Bioseparation*, 1998. 7(4): pp. 253-258.
29. Hegemann, D., Hossain, M. M., and Balazs, D. J., Nanostructured plasma coatings to obtain multifunctional textile surfaces. *Progress in Organic Coatings*, 2007. 58(2-3): pp. 237-240.
30. Ahn, S. H., Kim, S. H., and Lee, S. G., Surface-modified silica nanoparticle-reinforced poly (ethylene 2, 6-naphthalate). *Journal of Applied Polymer Science*, 2004. 94(2): pp. 812-818.
31. Yuranova, T., Laub, D., and Kiwi, J., Synthesis, activity and characterization of textiles showing self-cleaning activity under daylight irradiation. *Catalysis Today*, 2007. 122(1-2): pp. 109-117.
32. Bozzi, A., Yuranova, T., and Kiwi, J., Self-cleaning of wool-polyamide and polyester textiles by TiO_2-rutile modification under daylight irradiation at ambient temperature. *Journal of Photochemistry & Photobiology, A: Chemistry*, 2005. 172(1): pp. 27-34.
33. Wang, R. H., Xin, J. H., and Tao, X. M., Multi-functional protective nano-finishing of textile materials 1: Self-cleaning surfaces. *Abstracts of Papers of the American Chemical Society*, 2005. 230: pp. 273-INOR.
34. Yeqiu, L., Jinlian, H., Yong, Z., and Zhuohong, Y., Surface modification of cotton fabric by grafting of polyurethane. *Carbohydrate Polymers*, 2005. 61(3): pp. 276-280.
35. Rodrigues, S. N., Martins, I. M., Fernandes, I. P., Gomes, P. B., Mata, V. G., Barreiro, M. F., and Rodrigues, A. E., Scentfashion® Microencapsulated perfumes for textile application. *Chemical Engineering Journal*, 2009. 149(1-3): pp. 463-472.

36. Cai, X., Conley, S., and Naash, M., Nanoparticle applications in ocular gene therapy. *Vision Research*, 2008. 48(3): pp. 319-324.
37. Xia, X., Hu, Z., and Marquez, M., Physically bonded nanoparticle networks: a novel drug delivery system. *Journal of Controlled Release*, 2005. 103(1): pp. 21-30.
38. Hosokawa, M., Nogi, K., Naito, M., Yokoyama, T., Masuo, H., Kiyoshi, N., Mario, N., and Toyokaz, U. Y., *Development of new cosmetics based on nanoparticles*, in *Nanoparticle Technology Handbook*. 2008, Amsterdam: Elsevier. pp. 521-527.
39. Gregory, R. V., Samuel, R. J., Hanks, T., and National Textile Center Annual Report (USA). 2001.
40. Brahim, S., Narinesingh, D., and Guiseppi-Elie, A., Bio-smart hydrogels: co-joined molecular recognition and signal transduction in biosensor fabrication and drug delivery. *Biosensors and Bioelectronics*, 2002. 17(11-12): pp. 973-981.
41. Hu, J. L., Zeng, Y. M., and Yan, H. J., Influence of processing conditions on the microstructure and properties of shape memory polyurethane membranes. *Textile Research Journal*, 2003. 73(2): pp. 172.
42. Sarier, N. and Onder, E., The manufacture of microencapsulated phase change materials suitable for the design of thermally enhanced fabrics. *Thermochimica Acta*, 2007. 452(2): pp. 149-160.
43. Agrawal, A. K., Jassal, M., Vishnoi, A., and Save, N. S., Temperature responsive fibers with anisotropic transitional behavior. *Journal of Applied Polymer Science*, 2005. 95(3): pp. 681-688.
44. Greiner, A. and Wendorff, J. H., Electrospinning: a fascinating method for the preparation of ultrathin fibers. *Angewandte Chemie International Edition*, 2007. 46(30): pp. 5633.
45. Gibson, P. W., Factors influencing steady-state heat and water vapor transfer measurements for clothing materials. *Textile Research Journal*, 1993. 63(12): pp. 749.
46. Peppas, N. A. and Kim, B., Stimuli-sensitive protein delivery systems. *Journal of Drug Delivery Science and Technology*, 2006. 16(1): pp. 11-18.
47. Yang, S. Y., Ryu, I., Kim, H. Y., Kim, J. K., Jang, S. K., and Russell, T. P., Nanoporous membranes with ultrahigh selectivity and flux for the filtration of viruses. *Advanced Materials*, 2006. 18(6): pp. 709-712.

48. Ulijn, R. V., Bibi, N., Jayawarna, V., Thornton, P. D., Todd, S. J., Mart, R. J., Smith, A. M., and Gough, J. E., Bioresponsive hydrogels. *Materials Today*, 2007. 10(4): pp. 40-48.
49. Chen, V. J. and Ma, P. X., Nano-fibrous poly (L-lactic acid) scaffolds with interconnected spherical macropores. *Biomaterials*, 2004. 25(11): pp. 2065-2073.
50. Ignatova, M., Starbova, K., Markova, N., Manolova, N., and Rashkov, I., Electrospun nano-fibre mats with antibacterial properties from quaternised chitosan and poly (vinyl alcohol). *Carbohydrate research*, 2006. 341(12): pp. 2098-2107.
51. Yang, J. H., Chun, B. C., Chung, Y. C., and Cho, J. H., Comparison of thermal/mechanical properties and shape memory effect of polyurethane block-copolymers with planar or bent shape of hard segment. *Polymer*, 2003. 44(11): pp. 3251-3258.
52. Tobushi, H., Hayashi, S., Ikai, A., and Hara, H., Thermomechanical properties of shape memory polymers of polyurethane series and their applications. *Journal de Physique IV*, 1996. 6(1): pp. 377-384.
53. Hu, J. L., Ji, F. L., and Wong, Y. W., Dependency of the shape memory properties of a polyurethane upon thermomechanical cyclic conditions. *Polymer International*, 2005. 54(3): pp. 600-605.
54. Meng, Q., Hu, J., Zhu, Y., Lu, J., and Liu, Y., Polycaprolactone-based shape memory segmented polyurethane fiber. *Journal of Applied Polymer Science*, 2007. 106(4): pp. 2515-2523.
55. Hu, J., Lu, J., and Zhu, Y., New Developments in Elastic Fibers. *Polymer Reviews*, 2008. 48(2): pp. 275-301.
56. Liu, X., Hu, J., Murugesh Babu, K., and Wang, S., Elasticity and Shape Memory Effect of Shape Memory Fabrics. *Textile Research Journal*, 2008. 78(12): pp. 1048.

Chapter 2

Shape Memory Polymers

This chapter introduces the principle of shape memory function in polymers and classifies shape memory polymers according to their molecular structures, trigger patterns and applications based on a literature review. One of the novel functional shape memory polymers with substrate bonding antibacterial activity is detailed and a new design of supramolecular shape memory polymers is discussed. Finally, shape memory fibres are introduced in comparison with other existing man-made fibres to clarify the uniqueness of such smart textile materials.

Keywords: shape memory, stimulus sensitive, segmented polyurethane, net point, switch, antibacterial activity, supramolecular

2.1 Introduction

Shape Memory materials can be deformed and fixed in a temporary shape and subsequently recover their original shape only when exposed to an appropriate external stimulus such as heating, lighting or PH value change. This ability opens up numerous application areas in smart textiles, adaptive facilities, sensors and actuators. As far as material categories are concerned, such smart materials can be attributed to shape memory polymers (SMPs), shape memory gel, shape memory alloy and shape memory ceramics. In recent decades, SMPs, as one group of most applicable smart materials, were developed widely in not only the academic area, but industrial application field as well, which is due to the low cost, good processing ability, high shape recoverability, large stretch-ability and wide range of shape recovery temperature in comparison with shape memory alloy.[1,2] It is noteworthy to investigate the principles behind the phenomenon of shape memory function in SMPs. In *Classification of SMPs*, the relations between shape memory effect and micro-structures are introduced according to present research progress. Shape memory polyurethane ionomers with substrate bonding antibacterial activity were originally prepared and investigated systematically. Supramolecular shape memory polymers possessing novel quadruple hydrogen bonding units are also covered in this chapter though it is only studied in the academic stage.[3] For the application of shape memory polymers for textiles, shape memory fibres have been engineered systematically at Hong Kong Shape Memory Textiles Research Centre.[4,5] The uniqueness of such smart fibres is illuminated based on the experimental results.

2.2 Principles of Shape Memory Function in SMPs

Shape memory polymer was one of the key smart materials developed widely in both academia and industry for the low cost, good processing ability, high shape recoverability and larger range of shape recovery temperature compared with shape memory alloys.[6] Meanwhile, shape memory behaviour can be observed in various polymer systems which are significantly different in the molecular structure and morphology. In the investigation of shape memory effect, it is also obvious that the shape memory behaviour is not only related to the molecular structure and morphology, but also dependent on the applied processing and programming.[2,7] Therefore, for the study of shape memory effect of various shape memory polymers, several kinds of methodology, such as cyclic tensile investigation,[8] strain recovery test,[9] bending test for the determination of the shape memory effect,[10] shrinkage determination of heat-shrinkable products,[11] were established so as to satisfy the research and application requirement.

Nowadays, a variety of polymers were reported to possess shape memory properties such as trans-polyisoprene (TPI), poly(styrene-co-butadiene), polynorbornene, crosslinked polycyclooctene and segmented polyurethane.[1,12] And shape memory polymers are elastic macro-molecular networks possessing suitable stimulus sensitive switches and elastic net-points as far as the molecular principle is concerned. As shown in Fig. 2.1 for direct thermal sensitive shape memory polymers, the switches may be composed of crystalline phase, amorphous phase or liquid crystal phase. Correspondingly, the melting point, glass transition point or the transition point from isotropic phase to nematic phase can be used as switches. As far as the net-points are concerned, the physical or chemical crosslink may be used to provide recovery performance. Therefore, in a certain programming condition, shape memory function is much dependent on the stimulus sensitive switches and the elastic netpoints.

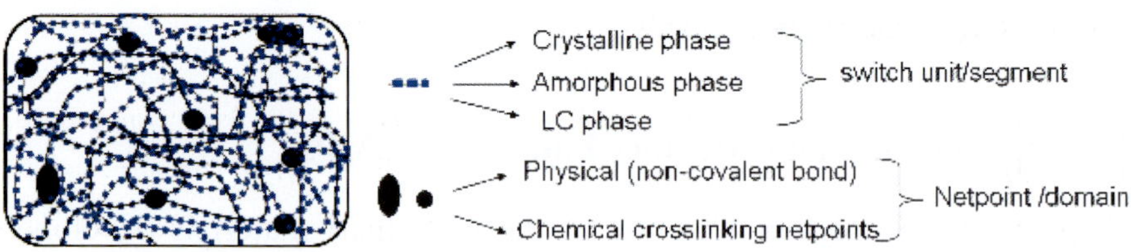

Fig. 2.1 Summary of molecular structures in thermal induced shape memory polymers.

2.3 Classification of SMPs

2.3.1 *Molecular structures of SMPs*

2.3.1.1 *Covalent bond between switches and net-points*

Since the 1960s, polyethylene (PE) covalently crosslinked by ionising radiation has been found to possess shrinkable effect with heating stimulus and has begun to be gradually and widely used in the electric wire and cable area, because it can store large deformation that can be recovered on application of heat.[13,14] In the molecular structure of crosslinked PE, the molecular chains entangle each other and could be categorised into crystalline regions and amorphous regions. Ionising radiation can induce the crosslinking structure and formation of the three-dimensional network, which was believed to occur preferably in amorphous regions. Crystalline regions appear to retain the deformed temporary shape, when the film is cooled down below the melting point of crystalline regions. Heating up the film above the melting point of the crystalline region cleaves the restraint for shape recovery. Then the original shape will be recovered, which is due to the crosslinking structure and the entropy elasticity. 'Memory effect' means that irradiated crosslinking PE is capable of memorising its original irradiated structure however it is deformed. In the late 1970s, amorphous poly-norbornene with the molecular weight of 3×10^6 was initially used as shape memory polymer.[15,16] The poly-norbornene containing 70-80 mol% of trans-linked norbornene units has a glass transition (T_g) at 30-40°C used as switching temperature and physical crosslinking network due to entanglement of the high molecular weight chains, which is regarded as fixed phase. When the stretched sample is cooled down below T_g rapidly enough, the freezing of temporary deformation will be applicable because the relaxation time for disentanglement is much longer than that for the stretching and cooling process. After that, shape recovery can be observed with the heating above the switching temperature. In addition, some co-polymer systems such as poly(styrene-co-butadiene),[17] polyethylene-poly(vinyl acetate) copolymers[18], poly-ethylene/nylon-6-graft copolymer,[19] and ethylene oxide-ethylene tere-phthalate segmented copolymer[20] were also found to have shape memory effect for their specific molecular structure. Even in the natural polymer—'starch' or chemically modified starch 'starch-poly'[21,22]—the significant shape memory function can be found when suitable humidity as external stimulus is applied.

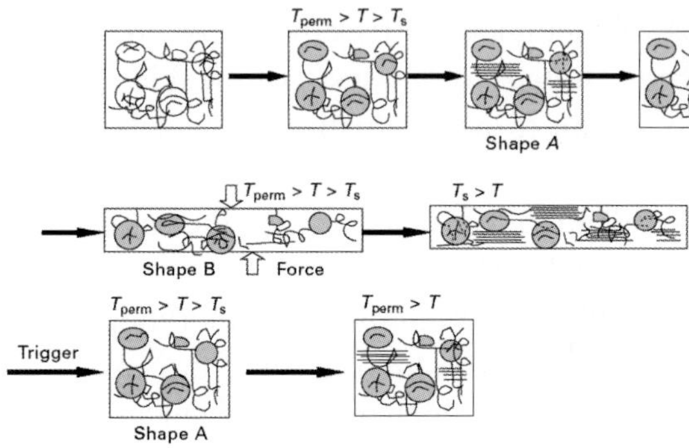

Fig. 2.2 Schematic representation of the molecular mechanism of the thermally sensitive shape memory effect in segmented polyurethanes. (T_s: T_m of soft segments).[23]

As for practical applications, the most readily available is segmented shape memory polyurethane (SMPU) which has aroused much attention.[24] Foremost among the reported microscopic structures on segmented SMPU system is the two-phase heterogeneous structure which consists of rigid fixed as net-points and a soft reversible phase as shown in Fig. 2.2. According to the mechanism of shape memory effect, the reversible phase having a melting or glass transition temperature of the soft segments as the transition temperature (T_{trans}) is used to hold the temporary deformation, whereas the fixed phase is referred to hard segments covalently coupled to the soft segments. Thus, the fixed phase inhibits slippage between the molecular chains by having physical crosslinkage points that are responsible for memorising the permanent shape.[8-10,25,26] In the linear segmented polyurethane (PU) system, the strong inter-molecular force among hard segments results from their possessing high polarity and hydrogen bonding among hard segments due to the presence of urethane and urea units. The hard segment phase with the highest thermal transition temperature (T_{perm}) acts as the physical crosslink and controls the permanent shape. Heating above the highest thermal transition point, the physical crosslinks among hard segments will be destroyed. The whole molecular chains, therefore, melt and can be processed with the traditional processing techniques, such as extrusion, injection or blow molding, by which the permanent shape, such as products with specific figures, can be formed easily by cooling. Therefore, the process and recycle usage of shape memory polyurethane can be realised easily.

2.3.1.2 *Non-covalent bond between switches and net-points*

From the above molecular structures of shape memory polymers, it can be seen that the stimuli sensitive switches are all chemically bonded with elastic net-points, whether the interaction between net-points is physical or chemical. Nevertheless, some polymer blends, semi-interpenetrating network and interpenetrating network (IPN) can also be engineered to realise shape memory function. The novel styrene-butadiene-styrene tri-block copolymer (SBS) and poly(ε-caprolactone) (PCL) blend were introduced for its shape memory properties.[27] It was found that the two immiscible components of the blend separates contributed respectively to shape memory performances. The elastomer was used to provide the stretching and recovery performances and the crystallisable polymer provided the fixing and unfixing performances. In addition, the miscible interpenetrating network (IPN) composed of polyesterurethane / poly-(ethylene glycol) dimethacrylate was reported to show the obvious shape memory function. The strain fixity rate and strain recovery rate were above 93%.[28] Simultaneously, T_gs, wettability, mechanical properties of the IPNs is easily controlled by composition. For semi-IPN, Liu and Peng *et al.* reported that a novel kind of PMMA-PEG (poly(methyl methacrylate)(PMMA)-poly(ethylene oxide)(PEO)) semi-IPN can show excel shape memory behaviour at two transition temperatures, the T_m of the PEG crystal and the T_g of the semi-IPN. Based on a reversible order-disorder transition of crystals below and above the T_m of PEG, and the large difference in storage modulus below and above the T_g of the semi-IPN, the polymer can demonstrate the recovery ratio of 91 and 99% repectively.[29]

In these shape memory systems with non-covalent bond between switches and net-points, the switches and net-points can be synthesised separately. Subsequently, the blend, semi-IPN or IPN can be formed with further techniques. Therefore, the composition is conveniently adjusted to satisfy various applications.

2.3.2 *Trigger patterns of SMPs*

2.3.2.1 *Direct Thermal sensitive shape memory effect*

In the aforementioned part about the principle of direct thermal-sensitive shape memory polymer, the different types of thermal transition including glass transition, melting points of crystallisation and the transition from nematic phase to isotropic phase can be used as switches in shape memory function.

There are a large number of studies on this type of trigger pattern for the shape memory function. Polymer analogous partial reduction of polyketones with $NaBH_4$-THF can be used to prepare poly(ketone-*co*-alcohol)s with shape memory functionality.[30] The degree of reduction of the poly(ketone-*co*-alcohol) was controlled by the amount of $NaBH_4$-THF and was found to affect T_g and mechanical properties of the polymer. Increasing the ratio of [OH]/[CO] functions allows continuous change of the T_g of the resulting polymers from below the room temperature to about 75°C. The maximum recovery ratio can be 95%. Covalently crosslinked polymer networks with shape memory function were prepared by crosslinking linear or branched polymers, in which the successful crosslinking polymers include irradiating polyethylene, PE copolymers with ionising radiation (γ-radiation, neutrons) and crosslinked poly(cyclooctene).[12,14,31,32] The shape memory function depends on the melting of the crystallites. For the pure polycyclooctene with 81 wt% *trans*-vinylene content, a melting temperature of 60°C was determined. Shape recovery of these materials can be observed within 0.7s at 70°C.

As an innovation of chemical structures in direct thermal sensitive shape memory polymers, a triple-shape polymer was designed to change from a first shape (A) to a second shape (B) and from there to a third shape(C).[33] Shape B and shape C are recalled by subsequent temperature increases. Therefore shape A and shape B are fixed by physical crosslinks and shape C is defined by covalent crossliniks established during network formation where the first polymer network system, named MACL containing poly(ε-caprolactone) and poly(cyclohexyl meth-acrylate), and the second network system, named CLEG containing PEG segments introduced as side chains having one dangling end, were used to form the polymers with crystalline PCL domains (T_m=50-60 °C) and PEG domains (T_m=17-39 °C). The entire polymer systems are prepared by photoinduced copolymerisation of a methacrylate-monomer and poly(ε-caprolactone) dimethacrylate (PCLDMA) as crosslinker, to form the elastic networks. In this system, the two transitions can be adjusted to the demands of a specific application by variation of the molecular weight of PCLDMA and PEG monomethylether-monomethacrylate.

Liquid crystalline polymer networks can be formed by the reaction of liquid crystalline moieties with start-shaped reactive precursors.[34] The shape memory function was stimulated by the thermal transition of the liquid crystalline domains. Main chain smectic-C elastomers were composed of coupling tetra-functional silanes as net-points, with oligomeric silanes as spacers, to which two distinct benzoate-based mesogenic groups had been attached. The crosslinking structure defined the permanent shape. In the shape fixity process, the heated sample from liquid crystalline domains to the isotropic state was stretched or twisted and finally cooled below the clearing transition of the smectic-C domains. When the sample is reheated, the permanent shape will be recovered.

2.3.2.2 *Indirect Thermal sensitive shape memory effect*

As for the trigger pattern, except for the commonly used direct heating to the switching temperature, some researchers have explored a variety of new kinds of shape memory polymer, which can respond to different external stimulus, such as water immersion,[35-38] electricity,[39] remote electromagnetic activation[40] and infrared irradiation.[41] In general, only two strategies can be summarised among the above studies.

One method is to lower the T_{trans} by diffusing small molecules to plasticise the polymer bulk, triggering the thermal sensitive shape memory effect. An example of shape memory polymers whose T_{trans} can be lowered by the diffusion of water into the polymer are polyurethanes.[42] Moisture absorption in the polyurethane is the main reason for the decrease of T_g

that results in the loss of shape fixity at room temperature. Moisture absorbed by the polyurethane can be roughly divided into two portions. One is easy to evaporate below 140°C, but has a great effect on T_g. Therefore, a stable T_g can be obtained either by keeping the material dry all time or immersing it in water for a long time (400h, in this experimental case) at room temperature until the saturation of the moisture is reached. A similar mechanism can be found in the solution-driven shape memory polymer.[43] In the study, the immersion in DMF (*N,N'*-Dimethylformamide) decreases the T_g and then causes direct shape recovery. Thus, shape recovery of shape memory polymer can be triggered not only by applying energy, but also by non-energy.

The other method is to heat the sample indirectly by irradiation in which the heat transfer by irradiation was enhanced by incorporation of conductive filler or other fillers sensitive to an alternating magnetic field. In this case, the particle size, distribution and the molecular structure of the particles have to be considered to reach an admirable photo-thermal/radiation-thermal effect. The permanent shape of the photo/radiation-sensitive shape memory polymers is determined by net-points, which are physical or chemical crosslinks of an amorphous permanent polymer network. One of the characteristics of such materials is non-contact triggering. In contrast, the traditional heating stimulus depends on the heat transfer by fluid or gaseous medium to heat samples above a specific switching temperature. The remote triggering may be realised by radiation or alternating magnetic field. One example was reported with a polyetherurethane (TFX) and a biodegradable multiblock copolymer (PDC) with poly(p-dioxanone) as hard segments and poly(ε-caprolactone) as soft segments in the matrix component.[44] Nanoparticles consisting of an iron(III)oxide core in a silica matrix could be processed into both polymers. Compounds with a homogeneous particle distribution have suitable elastic and thermal properties for the shape memory effect. TFX compounds were elongated at increased temperature, and subsequent elongation can be fixed by cooling under constant stress. Cold-drawing of PDC compounds at 25°C resulted in a temporary fixation of the mechanical deformation by 50-60%. The shape memory effect of both composite systems could be induced by indirect heating in an alternating magnetic field (f=258 kHz; H=30 $kA \cdot m^{-1}$). The maximum temperatures achievable by indirect heating in a specific magnetic field depend on sample geometry and nanoparticle content. Shape recovery rates of composites triggered by magnetic field are comparable to those obtained by increasing the environmental temperature. Schmidt reported that the incorporation of surface-modified superparamagnetic nanoparticles into shape memory polymers can remotely actuate the shape transitions by electromagnetic fields. The composite thermosets of oligo(E-cap-rolactone)dimethacrylate / butyl acrylate contain between 2 and 12 wt% magnetite nanoparticles serving as nanoantennas for magnetic heating.[40] Particles are dispersed homogenously within the matrix and the basic thermal and mechanical properties of the polymer matrix are maintained. The specific loss power of the particles is determined to be 30 $W.g^{-1}$ at 300 kHz and 5.0 W. During the shape transition at 43°C, no further temperature increase is observed. The experiments demonstrate that it is possible to stimulate shape memory effect by a touchless and highly selective electromagnetic stimulus. The novel materials are of interest for medical applications as well as in sensor and actuator systems.

2.3.2.3 *Other external stimulus*

2.3.2.3.1 Light as external stimulus

Responsiveness towards light can be introduced into shape memory polymers by different kinds of groups, including azobenzene, reacting molecular switches or photochemical phase transition.[45-48] The mechanism of light sensitive shape memory effect is based on photomechanical effects, light-stimulated phase transitions or photo sensitive chemical bond.

It is well known that azobenzene can show the reversible *trans-cis* isomerisation with the significant change in molecular length from about 9.0Å in the *trans* form to 5.5Å in the *cis* form, when it is exposed in alternating irradiation of UV and visible light. Therefore, the introduction of azobenzene chromophores into polymer backbones or side chains can be used to produce reversible contraction and expansion, under the stimulus of light.[48] The disadvantage is the too small deformation in the solid film for practical applications.

As an example of light stimulated phase transitions, azobenzene LCEs (Liquid Crystal Elastomer) showing large photo-contraction were investigated by Finkelmann and the coworkers.[49] The driving force for the deformation is the variation of alignment order caused by the photo-chemical phase transition. Due to UV irradiation, the azobenzene liquid crystals present a reduction in alignment order. It is because of the photo-isomerisation of the azobenzene moieties in which the rod-like *trans*-azobenzene moieties stabilise the liquid crystal alignment, but the bent *cis* formation decreases the order parameter.

UV (ultraviolet)[2,46] was used to trigger shape memory function in two series of photo-responsive polymer networks in which a grafted polymer with HEA(hydroxyethyl methacrylate)-CA(cinnamic acid) molecules and the polymer network dopped with SCAA(start-poly(ethylene glycol) containing CAA(cinnamylidene acetic acid) terminal groups) molecules were used. The programming process to investigate photo responsive polymers was similar to that for thermally induced shape memory polymers. After the deformation caused by external force, the exposure to UV longer than 260 nm led to the fixity of the deformation due to photo-induced cyclo-addition reaction. The deformation can be fixed for quite a long time after the external force was removed until the irradiation with UV shorter than 260 nm at ambient temperature was used to cleave the photo induced crosslinks. Then the original shape will be recovered. The potential applications which need different pre-determined temporary shapes and remote activation without the temperature limitations can be expected by using such materials.

2.3.2.3.2 Multi-external stimulus in Hydrogels

In addition to the aforementioned bulk (solvent free) shape memory polymers, another class of external stimulus sensitive polymer networks processing shape memory function are shape memory gels. Osada's work has presented that these materials are more flexible and softer than the commonly used shape memory polymers.[50] The typical shape memory gels with crosslinking structures possessing a hydrophilic fraction can be swelled in water and hydrophobic sections with reversible order-disorder parts, which can be controlled by temperature and other external stimulus. In comparison with shape memory alloys and shape memory polymers, shape memory gels have some advantageous properties, such as very large deformation, and sensitivity to many external stimuli.[51] Therefore, they are highlighted in this section. Random copolymers composed of poly(acrylic acid) and poly(vinyl sulfonic acid) were prepared and their swelling ratios at various temperature and PH, and their deformation ratio in an electrical field were measured by Kim and coworkers.[52] The polymer hydrogels exhibiting contraction as response to the external stimulus were, therefore, developed as shape change hydrogels. It can be utilised as biosensors and as artificial muscle in the application field. More details are given in Chapter 3.

2.3.3 *Applications of SMPs*

Having the physical property change during transition points, SMPs could be applied in various product areas. End-users and engineers are evaluating applications as diverse as space appliances, toys, implantable medical devices, building and construction products, recreation goods and industrial materials. Because of the intrinsic shape change property, expedite setup and remolding, SMP is even being proposed for tooling. The underlying idea is to use the change in the physical properties (young modulus shape, and dielectric constant, etc.) when the polymer interacts with the changing environment.

Concerning the application of shape memory polymers, some progress has been made in the past decades. The first shape memory polymer of commercial application was covalently crosslinking polyethylene. Since the 1960s, crosslinking polyethylenes prepared by ionising radiation have been used in the broad application of packaging films and heat shrinkable tubing, especially for insulation of electric wires or as protection against corrosion of pipe lines.[13,14] The mechanism behind the apparent phenomenon is thermal-induced shape memory function. The permanent shape is fixed by covalent crosslinks. The reversible shape is controlled by the polyethylene crystallites. At the same time, a number of other polymers were developed as heat shrinkage plastics, such as crosslinking polyvinylchloride (PVC), polyamide and polytetrafluoroethylene (PTFE). Among various shape memory polymers, the shape memory polyurethanes (PUs) made by Diaplex Company have made great progress because of the following advantages: the forming processes used for other thermoplastic polymers could be utilised directly; the shape recovery temperature could be set at any value within 50K around the room temperature; there are large differences of mechanical properties, optical property and water vapor permeability at the temperatures above and below the designed T_g. Based on these advantages, the shape memory PU has been expected to be used as self-repairing, and smart materials or biomaterials.

2.3.3.1 *SMPs used for bio-medical areas*

Shape memory polymers have already been used in biology and medicine especially for bio-medical devices, which may permit some novel medical therapies or operations.[53] In addition to shape memory function, a number of other functions, such as biodegradability, antibacterial activity and controllable recovery stress can be introduced into the polymer systems through proper molecular designs and processing technologies. It broadens the application scope of such smart materials tremendously.

2.3.3.1.1 Fixity and orthodontic application

The shape memory function, such as the fixity ability for temporary deformation, makes SMPU quite suitable for immobilisation/rehabilitation for patients.[54] Hu and coworkers prepared a series of PCL-based shape memory polyurethanes and investigated their thermal, mechanical properties, shape memory properties, softening and hardening processes, in comparison with a commercially available orthotic material.[55] It was found that in order to obtain shape memory polyurethanes with excellent low-temperature thermoplastic properties for orthopaedical-surgical use, the hard segment content should not be above 22 wt%. In their experiments, a prototype wrist orthosis was easily fabricated at 60 °C with hand using shape memory polyurethanes with 16 wt% hard segment content. Cytotoxicity tests indicate that the wrist orthotic material is not cytotoxic. On the other hand, the substrate bonded antibacterial activity was introduced into shape memory polyurethanes by using ionic groups are detailed in the following sections of this chapter. The resulting shape memory polymer systems can provide rapid shape fixity and admirable biocidal activity at the same time. The neglectable diffusion of antibacterial agents makes such materials more environmentally friendly and avoids bringing toxicity to customers.

Nakasima and coworkers reported that polynorbornene, as a shape memory plastic with a glass transitional point of 35 °C, can generate force during the elastic phase to displace human teeth.[56] It was found that the shape memory plastic wire of 1 mm in diameter stretched to two to three times of its original length at a temperature of 50 °C and a speed of 0.5 mm/sec would exert a relatively stable continuous light force of 119-156 g to move the teeth. In comparison with conventional elastic modules used in orthodontic therapy, it exhibited a lesser degree of force degradation at a body temperature of 37 °C for a long time.

2.3.3.1.2 Smart implant materials

In the biomedical material area, shape memory function around human body temperature range together with biocompatibility offer tremendous application potential for use in minimally invasive surgery.[2,57] Through a small incision in the human body, degradable implants can be inserted in a smaller compressed (temporary) shape. When they are placed at the correct position, they obtain their application relevant shape after being warmed up to body temperature. After a defined time period, the implant is degraded and becomes desorbed. In this case a follow-on surgery to remove the implant is not necessary.[58] A group of biomaterials potentially used as implant, medical device with shape memory property has been described and developed for biomedical applications.[59,60] In summary, it can be found that the advantages of shape memory polymers used as smart implant materials include the wide range of tunable stiffness, fast actuation, tailorable transition temperature, large shape deformation and complete recovery. Particularly for a fabricated self-tightenable biodegradable suture, the programmed sutures can be loosely stitched through incised rat belly tissue and the abdominal muscle.[61] When heated to 41 °C, the recovery force will tighten the suture. When the suture is stretched to 200%, the impressive force generated during shape recovery will be 1.6N. Such smart suture will provide flexibility during connecting the suture and simplify ligation.

2.3.3.1.3 Micro-actuator and others in biomedical areas

In the application about biomedical micro-actuators, shape memory polymers have been proposed for stent usage. The stent can be designed to activate at body temperature, leading to a natural deployment without a need for auxiliary devices.[62] Shape memory polymer networks were synthesised by using photopolymerisation of tert-butyl acrylate and poly(ethylene glycol) dimethacrylate. The free recovery response of the polymer stents at body temperature was studied. The glass transition temperature (T_g), crosslink density, geometrical perforation, and deformation temperature can be independently adjusted.

It is potentially used in developing the minimally invasive cardio-vascular devices. The other example of shape memory polymers as biomedical micro-actuator is a novel shape memory polymer micro-actuators for treating stroke.[63] The permanent shape of the micro-actuator was a cone-shape coil, made in injection molded thermoset poly-urethanes. Before the surgery, the straight shape was fixed to be easily delivered distally into the occlusion through a catheter. Thereafter, the shape recovery triggered by optical heating will restore the original coil shape to remove the thrombus from the vessel, restoring the blood flow.

2.3.3.2 *SMPs applied to textiles*

In the textile area, the shape memory polyurethane was reported to be used in coating on the fabric surface to offer temperature dependence of water vapour permeability to improve the comfort sense of fabric.[64] The smart water vapour permeability dependent on temperature arises from the increased free volume in soft segment domains, which allow water vapour molecules with an average diameter of 3.5Å to be easily transmitted through the polymer thin film.[65]

Since 2002, the Shape Memory Textile Centre at The Hong Kong Polytechnic University has applied shape memory polyurethane in fibre spinning, fabric finishing and garment finishing with various techniques so as to impart shape memory function into fibre and garment products.[4,66-71] In these ways, the crease retention and flat appearance of fabric were expected to be improved simultaneously.

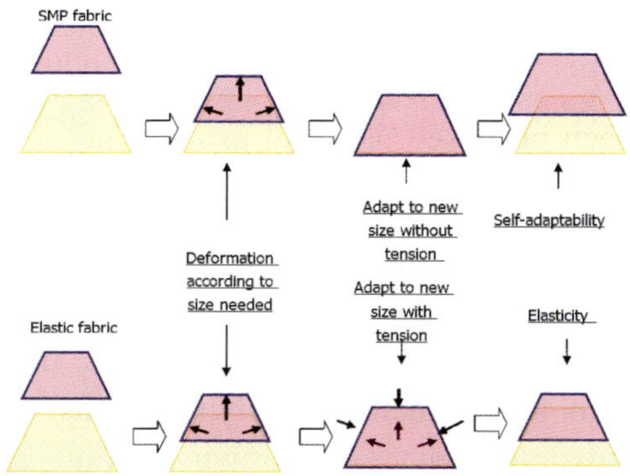

Fig. 2.3 SMPU fabric, garment fits different wearer's figure without tension.

The fabric composed of shape memory fibres was designed to possess the following advantages: when the garment with given size is enlarged to fit different wearers' figures, the fixity ability to temporary deformation caused in the wearing process was expected to diminish most pressure sensation to wearers as shown in Fig. 2.3. After wearing, heating in the washing or drying process will give rise to the recovery of the original size. The mentioned mechanism stems from the shape memory function of SMPU fibres, which was expected to be imparted into fabric and garment in this study. The relevant experimental results have been reported previously.[5,72] In Chapter 6, shape memory textiles are detailed to clarify the specific properties of such adaptive products and the related applications.

2.3.3.3 *Usage in thermal recording, sensors, actuators and others*

Other potential applications for shape memory polymers can be in almost all areas of daily life: from switches to sensors, from intelligent packing to tools, from self-repairing paint to different utensils, from self-adaptive garment to temperature-sensitive cosmetics. Different kinds of product design have brought shape memory polymers into the gradually enlarged practical scopes, such as shape memory polymers used to impart shape memory function on human hair,[73,74] the releasable fastener system composed of shape memory polymer hook elements.[75] Nanometer scale imprints in shape memory polymers are reported in the application of recoverable nanometer scale indents.[76] It was found that the temperature-dependent recovery of atomic force microscope tip-formed indentations in a thermo-set shape memory polymer can be observed. The indentations were made at room temperature and 69°C. Subsequently, the recovery at temperatures between 40 and 70°C can happen automatically. These experiments show shape memory behaviour in the scale of 1 - 100 nm. For larger scale of shaped protrusion arrays, the significant shape recovery ability in shape memory polymers was utilised to produce micro-sized protrusion arrays. As shown in Fig. 2.4, there are two methods to produce micro-sized protrusion arrays including laser heating and indentation. Through the latter method, i.e. an indentation-polishing-heating process, different protrusive bumps can be made. In comparison with laser heating, using shape memory polymers is more convenient and powerful.[77]

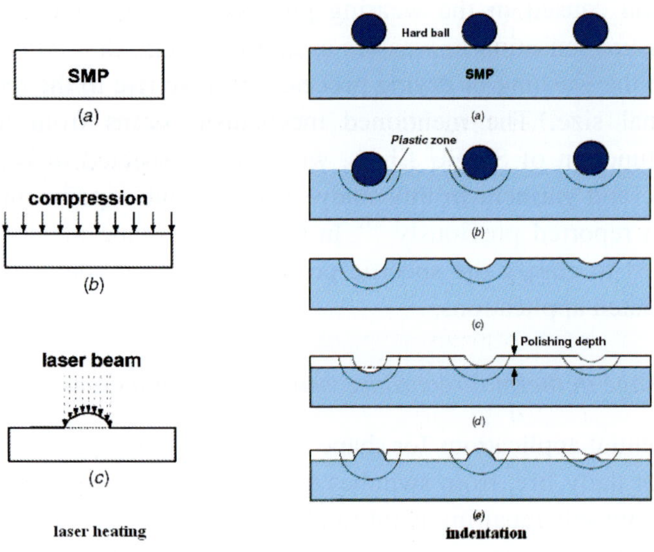

Fig. 2.4 Left graphs: Illustration of forming a protrusive bump atop a shape memory polymer using laser heating. (a) Original sample; (b) Pre-compressed; (c) After laser heating, a bump is formed. Right graphs: Illustration of forming a protrusive bump. (a) Original sample; (b) Indented to different depths using hard balls; (c) After indentation; (d) after polishing; (e) After heating, different shaped bumps can be formed.[77]

The applications of shape memory polymer are based on the design about molecular structures and functions, in which shape memory polymers are able to fix the determined deformation by the programming process, subsequently recover the original shape upon the suitable external stimulus such as heating and lighting. The shape recovery process is dependant on the random coil conformation of networks. In light of practical requirements, both thermoplastic and thermosetting types of shape memory polymers having crystalline or amorphous reversible phase can be chosen. For example, if the better recoverability and reproducibility of shape memory functions are required, the crosslinked shape memory polymers would be the good candidates because of their inherent lower creep characteristics. On the other hand, if better transparency was expected, the amorphous reversible phase would be more suitable in comparison with the crystallisable one. For remote sensing systems, radiation-sensitive shape memory polymers open up the new fields of applications. Even in the satellite, shape memory polymers provide potentials to the deployable thin film structures in which the folded shape memory parts can save storage space and be opened to the full span by suitable stimulus.[53] It is therefore believed that more and more novel shape memory polymers in future will be extensively explored and appear in enlarged usage areas.

2.4 SMPs with Substrate Bonded Antibacterial Activity

Bacterial infection remains one of the most serious complications associated with the use of medical devices. An estimated 45% of hospital infections are related to implants and medical devices.[78] Bacteria can adhere to the surface of biomaterials as well as the proteins absorbed to that surface and form the colony or bio-film. Then the bacteria might further the hematogenous spread and colonisation. Biocidal coatings are widely used to prevent the growth of microorganisms on the surface of materials. At present, the protection is achieved by leaching of bioactive molecules from the coating.

However, these molecules are highly toxic to the environment and the protection is short-lived due to the difficulty of controlling the rate of diffusion. One effective strategy was developed to kill the microorganism in contact with the substratum, which has given to the deep study of polymer materials containing immobilised biocides.

2.4.1 *Molecular structure of SMP with antibacterial activity*

In polyurethane ionomers, a variety of polyurethane anionomers and cationomers have been found to possess substrate bonding antibacterial activity with suitable ionic groups and the counter-ion agent. For polyurethane anionomers, for instance, Kim and the associates reported that the waterborne polyurethane-urea anionomers composed of isophorone diisocyanate(IPDI), poly(tetramethylene oxide)(PTMO), dimethylol propionic acid(DMPA) and ethylene diamine with the 28:1 mol/mol ammonium hydroxide/cupric hydroxide as counter-ion, has a strong antibacterial halo, which was attributed to the bactericidal power of Cu^{2+} cation itself.[79] In the study of antibacterial activity of cationomers, the polymer bearing quaternary ammonium salts (QAS) has been studied for QAS possessing at least one alkyl substituent of eight carbon or more atoms are able to kill microorganisms such as bacteria, fungi and moulds by interacting with the cell membrane.[80] Kawabata and Nishigushi observed that the soluble pyridinium-type polymer cationomers showed strong antibacterial activity against gram-positive bacteria, whereas it was less active against gram-negative bacterial.[81] Cooper and the co-workers synthesise and characterise non-leaching biocidal polyurethane with 1-iodooctane and 1-iodooctadecane as neutralisation agent, N, N-bis(2-hydroxyethyl)isonicotinamide (BIN) as chain extender, PTMO as the soft segment.[82] The pyridine ring in BIN was quaternised with a variety of alkyl halides to form cationic polyurethanes possessing the biocidal activities against Staphylococcus aureus and Escherichia coli as well as the good mechanical properties in the dry state.

Table 2.1 Formulation of SMPU ionomers and non-ionomers.[83]

Sample Code	PCL[a] (wt%)	BIN[a] (wt%)	PCL[a] (mole)	BIN[a] (mole)	BDO[a] (mole)	MDI[a] (mole)
75-0	75	0	1	0	9.07	10.07
BIN75-6N	75	6	1	3.81	3.91	8.71
BIN75-11N	75	10.56	1	6.7	0	7.7
BIN75-6-C8	75	6	1	3.81	3.91	8.71
BIN75-11-C8	75	10.56	1	6.7	0	7.7

Sample Code	C8I[a] (mole)	BIN[a] (mol%)	Pyridinium content[b] (mol%)	Mw[c] ($\times 10^{-4}$)	$\frac{Mw}{Mn}$
75-0		0	0	9.8	5.02
BIN75-6N		0	0	11.3	8.96
BIN75-11N		0	0	10.8	9.11
BIN75-6-C8	3.81	21.86	6.72	9.7	6.14
BIN75-11-C8	6.7	43.51	29.55	17.6	6.72

[a]: feeding composition of SMPU ionomer and non-ionomer.
[b]: real composition of pyridinium content determined by using NMR.
[c]: based on a polystyrene relative calibration in N,N-dimethylformamide by using GPC.

In our design of antibacterial SMPU ionomers, BIN was incorporated into segmented PCL based polyurethane as chain extender. The soft segment length and hard segment content were fixed so as to investigate the ionic group content on this series of segmented polymer ionomers. The series of polyurethane copolymers were nomenclated by the abbreviation of ion chain-extender and three numbers. As shown in Table 2.1, the first two numbers denote the soft segment weight content. The third number represents the BIN weight content. For the corresponding PU non-ionomer, the alphabet 'N' in the end indicates the sample without neutralisation. For the ionomers, '-C8' following numbers means that the neutralisation agent is C8I. For example, the sample BIN75-6-C8 contains 75 wt% of soft segment, 6 wt% of BIN and it is neutralised with stoichiometric C8I. Its non-ionomer is BIN75-6N. The sample 75-0 as the control sample contains no BIN but only MDI, BDO and PCL. The used testing cultures for the antibacterial activity are limited in *Staphylococcus aureus* and *Klebsiella pneumoniae*.

2.4.2 *Characterisation of substrate bonded antibacterial activity*

Antibacterial activity against both Gram-positive (*Staphylococcus aureus*) and Gram-negative bacteria (*Klebsiella pneumoniae*) was tested by using American Association of Textile Chemist and Colourist (AATCC147) test, because it is a relatively quick and easily executed qualitative method to determine antibacterial activity of diffusible antimicrobial agents on treated surface. A zone of inhibition occurs as a result of the diffusion of an antimicrobial agent from the specimen. After incubation, a clear area of interrupted growth underneath and along the sides of the test material indicates antibacterial activity of the specimens. Table 2.2 and Fig.2.5 show the results for 75-0, BIN75-6-C8 and BIN75-11-C8. In all samples, no clear zone of inhibition can be seen with the naked eye, suggesting that there are few diffusible antibacterial agents in the samples. However, it can be observed that the colony can be formed on the surface of 75-0. Therefore, the comment for this sample is 'not acceptable'. For BIN75-6-C8 and BIN75-11-C8, the colony on the surface cannot exist after incubation, because biocidal quaternised pyridine moieties incorporated into hard segments through the chain extender of SMPU ionomers can kill the bacteria with the long alkyl chain neutralisation agent.

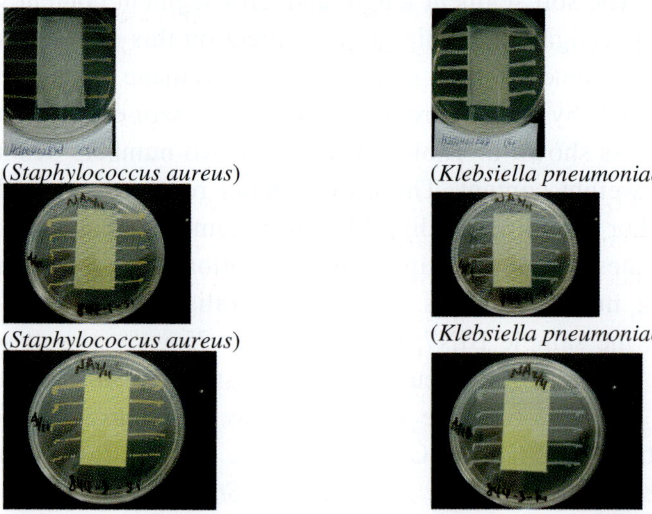

75-0: (*Staphylococcus aureus*) (*Klebsiella pneumoniae*)

BIN75-6-C8: (*Staphylococcus aureus*) (*Klebsiella pneumoniae*)

BIN75-11-C8: (*Staphylococcus aureus*) (*Klebsiella pneumoniae*)

Fig. 2.5 Zone of inhibition and bacterial growth under specimens in AATCC 147.[83]

Table 2.2 Antibacterial activity testing result according to AATCC 147.[83]

Sample code	Test microorganism	Culture type	Width of clear zone of inhibition (mm)	Colony forming on specimen surface	Comments
75-0	Staphylococcus aureus	Gram positive	0	Yes	Not acceptable
	Klebsiella pneumoniae	Gram negative	0	Yes	Not acceptable
BIN 75-6-C8	Staphylococcus aureus	Gram positive	0	No	Acceptable, but not significant
	Klebsiella pneumoniae	Gram negative	0	No	Acceptable, but not significant
BIN 75-11-C8	Staphylococcus aureus	Gram positive	0	No	Acceptable, but not significant
	Klebsiella pneumoniae	Gram negative	0	No	Acceptable, but not significant

Considering the one test method of AATCC 147 that are directly dependent on diffusion of antimicrobial agent are not sufficient for evaluating immobilised antibacterial activity; the other testing method by using ASTM E2149 was chosen to quantitatively determine the antimicrobial activity of SMPU ionomers. ASTM E2149 is designed to evaluate the resistance of non-leaching antimicrobial treated specimens to the growth of microbes under dynamic contact conditions. In this way, the antimicrobial activity of substrate bound antimicrobial is dependent on direct contact of microbes with specimens. The samples were shaken in concentrated bacterial suspension for a one-hour contact time and the percent reduction is calculated based on initial counts. The results are listed in Table 2.3. It shows that for the control sample 75-0, the antibacterial activity to Gram-positive bacteria (*Staphylococcus aureus*) is not acceptable and that to Gram-negative bacteria (*Klebsiella pneumoniae*) is insignificant; for SMPU ionomers (BIN75-6-C8 and BIN75-11-C8), the antibacterial activity to Gram-positive bacteria (*Staphylococcus aureus*) is significant and the reduction rate of bacteria even is 100%; the antibacterial activity to Gram-negative bacteria (*Klebsiella pneumoniae*) is lower than that to Gram-positive bacteria (*Staphylococcus aureus*) in SMPU ionomers studied, which is quite consistent with the result reported by Cooper *et al.*[82] The reduction of bacteria of BIN75-6-C8 to Gram-negative bacteria (*Klebsiella pneumoniae*) is 83.6%. In Cooper's interpretation for the lower antibacterial activity of quaternary ammonium to Gram-negative bacteria compared to the activity to Gram-positive bacteria, the difference in the structure of the cell envelope of the organisms is a predominant factor. The substrate bound biocide in SMPU ionomers destroys the cell cytoplasmic membrane in Gram-positive bacteria (*Staphylococcus aureus*) with simple envelope structure. However, for the multilayered Gram-negative bacteria (*Klebsiella pneumoniae*), this effect is lowered. Therefore, the comment based on our testing result for the antibacterial activity of SMPU ionomers to Gram-negative bacteria (*Klebsiella pneumoniae*) is only acceptable. In addition, it can be observed that the antibacterial activity of SMPU ionomers to Gram-negative bacteria (*Klebsiella pneumoniae*) increased with the increase of BIN content. When the BIN content is 10.56 wt% in BIN75-11-C8, the reduction rate of bacteria is 90.7%.

Table 2.3 Antibacterial activity testing result according to ASTM E2149.[83]

Sample code	Standard	Test microorganism	Culture type	Percent reduction of bacteria(%)	Comments
75-0	ASTM E2149	*Staphylococcus aureus*	Gram positive	0	Not acceptable
		Klebsiella pneumoniae	Gram negative	42.5%	Insignificant
BIN 75-6-C8	ASTM E2149	*Staphylococcus aureus*	Gram positive	100%	Acceptable and significant
		Klebsiella pneumoniae	Gram negative	83.6%	Acceptable
BIN 75-11-C8	ASTM E2149	*Staphylococcus aureus*	Gram positive	100%	Acceptable and significant
		Klebsiella pneumoniae	Gram negative	90.7%	Acceptable

2.4.3 *Shape memory function*

The parameters such as shape fixity ratio and shape recovery ratio for this testing have already been reported previously.[2,84] In cycle tensile testing, the stress-strain curves of the segmented polyurethane films tend to be identical after the first several cycles and without significant variation in further cycles as reported before.[7,8,85-87] It was explained with the destruction of weak net-points in the initial cycles in deformation, and the subsequent formation of ideal elastic network.[2] Therefore, the fixed strain (ε_u) and irreversible strain (ε_p) in the first two tensile cycles were used to assess shape memory effect in this study.

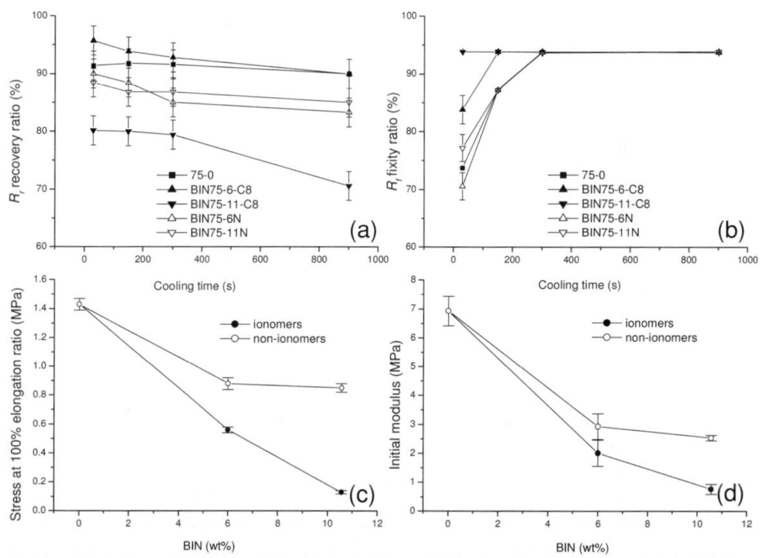

Fig. 2.6 Dependence of recovery ratio (R_r) (a), shape fixity ratio (R_f) (b) on cooling time and the dependence of the stress at 100% elongation (c), the initial modulus (d) on BIN wt% in ionomers and non-ionomers.[83]

$$R_r = \frac{\varepsilon_m - \varepsilon_\mathrm{p}}{\varepsilon_m} \times 100\% \qquad\qquad R_f = \frac{\varepsilon_\mathrm{u}}{\varepsilon_m} \times 100\%$$

In a further study, the effect of cooling time on shape memory function was investigated. Based on previous research,[8,9,26] the crystallisation of soft segments in segmented polyurethane was regarded to fix the temporary deformation at T_{low} in shape memory function. It is therefore supposed that the slow crystallisation of soft segment will be detrimental to the fixity to temporary deformation in short fixing time. In this study, the T_{low} is around 24 °C, when the temporary deformation was fixed. Fig. 2.6(c) and Fig. 2.6(d) show that the stress at 100% elongation and initial modulus of the second tensile cycle are much dependent on the BIN content. For each individual specimen, the fixity ratio increases with the increase of cooling time, suggesting that the crystallisation time of soft segments has substantial influence on the shape fixity in shape memory function. After sufficient cooling time, the fixity ratio of all specimens can reach a high value (≈ 95%). The recovery ratio of all samples decreases slightly with the extension of cooling time. For BIN75-11-C8, an interesting result can be observed: the fixity ratio can achieve quite a high value (93.8%) even in a very short period, 30s, though the crystallisation rate of BIN75-11-C8 tested in isothermal crystallisation investigation is the lowest one among samples according to the $t(0.5)$ at 24 °C. Instead, the shape fixity ratio of the control sample 75-0 is only 73.7% after cooling for 30s. Moreover, the recovery ratio of BIN75-11-C8 is much lower than other samples and monotonously decreases with the increase of cooling time. On the other hand, from Fig. 2.6 (c) and Fig. 2.6 (d), it can be observed that the stress and initial modulus during stretching at T_{high} ($T_{high} > T_{ms} > T_{low}$) drop significantly in BIN75-6-C8 and BIN75-11-C8. This indicates that physical crosslinks among hard segments were disrupted greatly, especially in BIN75-11-C8, which is because the hard segments are totally responsible for the load bearing capability over T_{ms}.[87] It can be concluded that the physical crosslinks, offering the recovery force from temporary deformation to the original shape, were weakening with the presence of pyridinium within hard segments. It is postulated that a part of crystallisation of soft segments in BIN75-11-C8 can withstand the internal stress stored in deformed polymer network. Subsequently better shape fixity can be achieved in a short time period (30s). In that, when the physical crosslinks were much weakening, the crystallisation rate of soft segments has less effect on shape fixity in short cooling time (30s).

2.5 Supramolecular SMPs

2.5.1 *Molecular design of supramolecular SMPs*

Beyond molecular chemistry based on the covalent bond, supramolecular chemistry has developed as the chemistry of entities generated via intermolecular non-covalent interactions.[88,89] Due to the liability of non-covalent interactions, the supramolecular system tends to adapt their architectures in response to environmental conditions, such as medium (solvent), presence of interacting species (protons, metal ions, substrate molecules, etc.) or physical factors (temperature, pressure, electric or magnetic fields, etc.).[90] It is noteworthy that the quadruple hydrogen-bonded supramolecular entities possess high dimerisation constant ($K_{dim}=10^6$ M^{-1} in $CHCl_3$), reversibly bonding/debonding ability in solution ($CHCl_3$) and even in solid state. Such self-complementary arrays of four cooperative hydrogen bonds (ureidopyrimidinone (UPy)) are highly reversible and strongly temperature-dependent.[91,92] The temperature-dependent rheology has been revealed in poly (styrene)(PS), poly(isoprene)(PI), Poly(butylenes terephthalate) and Poly(butylenes isophthalate) containing UPy end groups.[93,94] One of the applicable advantages arising from the quadruple hydrogen bonding (QHB) end groups in engineering polymers is the improvement of mechanical properties, such as the elongation at break and impact strength. The mechanical properties of UPy terminated polymers with lower molecular weight can be comparable to those with higher molecular weight. Meanwhile, the UPy terminated polymers retain excellent flow characteristics in melt status. The flow property is comparable to those engineering polymers with lower molecular weight. This characteristic is due to the aggregation of QHB entities which reinforce the polymer network at room temperature and completely dissociate above 80 °C in processing.[94] Instead, the polymer with higher molecular weight can only improve the mechanical properties at the expense of the loss of admirable process ability. This is because the high molecular weight gives rise to higher melt viscosity at processing temperature.

In our study, QHB entities were introduced into soft segments of thermoplastic segmented polyurethane as pendant groups. The soft segments were synthesised with 2,2-Bis(hydroxymethyl)propionic acid

(DMPA), Poly(propylene oxide)diol (PPG) and 1,6-diisocyanatohexane (HDI). The hard segments were from the reaction of 4, 4'-Methylenebis (phenyl isocyanate) (MDI) and 1,4-butanediol (BDO). The sufficient hard segments as a prerequisite to comprise fixed phases for shape memory function were prepared by 27.9 wt% MDI and BDO, so that the stable hard domains acting as physical crosslinks can be formed.[9] The 29.8 wt% PPG was used to control the relative composition of soft segments. Carboxyl groups with designed contents were introduced into soft segments by DMPA. Then stoichiometric GL (Glycidol) reacted with carboxyl to covalently bind pendant hydroxyl groups in soft segments. As shown in Fig. 2.7, the control sample with the code "PU" was prepared. Based on the obtained control sample, the (2(6-isocyanatohexylaminocarbonyl-amino)-6-methyl-4[1H]pyrimidinone) (MIMIS) was added into PU solution to synthesise segmented polyurethane with UPy pendant groups. The prepared segmented polyurethanes with UPy pendent groups are denoted by 'PUMIMIS' as shown in Fig. 2.7.

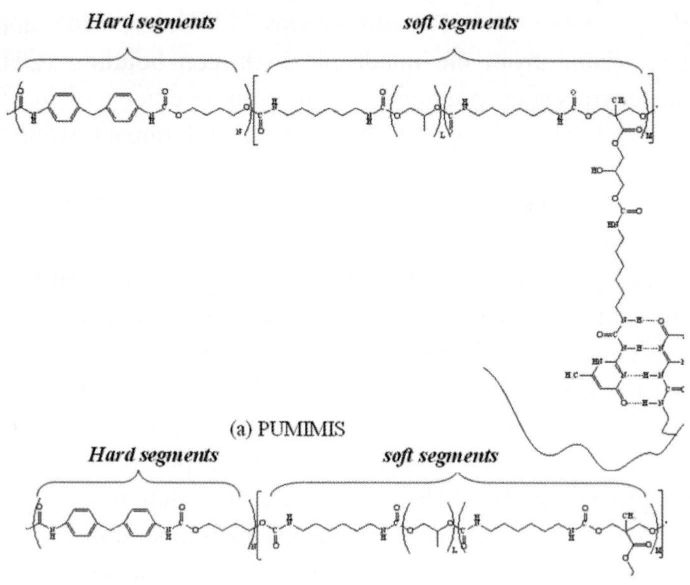

Fig. 2.7 Scheme of shape memory polyurethane with self-complementary supramolecular switch.[95]

2.5.2 *Characterisation of supramolecular SMPs with AFM, DMA*

Tapping-mode AFM was used on thin films to obtain height and phase imaging data simultaneously on a SII AFM from Seiko Instruments Inc. Micro-fabricated cantilevers (Nanosensors) with 5.0±1um thickness, 225±10um length and 33±7.5um width were used at their fundamental resonance frequencies of 96~175kHz. As shown in the phase images of Fig. 2.8, the darker areas are indicative of softer material and the lighter areas represent harder material.[96-98] The phase-separated morphology in PU and PUMIMIS can be observed. The darker spots and region were dispersed simultaneously in lighter areas, suggesting that there are various soft segments in the samples, which might be caused by the rigidity differences between different parts in soft segments composed of PPG, DMPA and HDI.

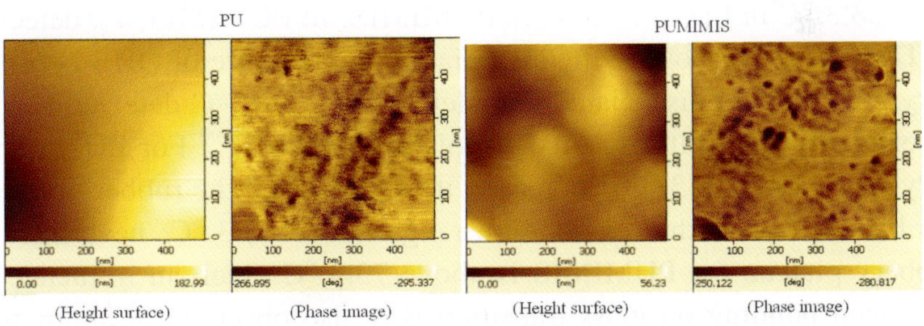

Fig. 2.8 The morphology of PU and PUMIMIS observed by AFM.[95]

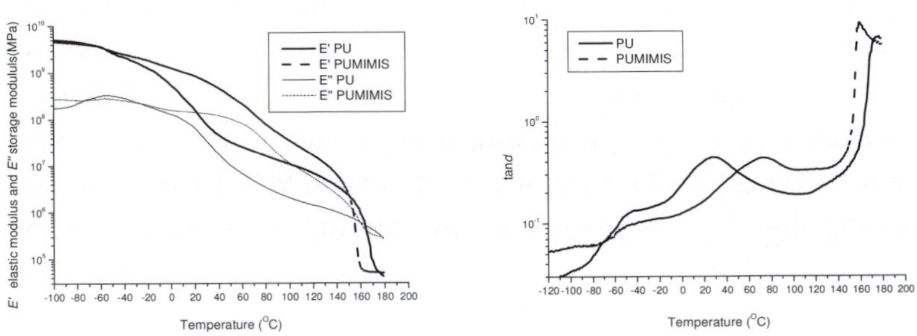

Fig. 2.9 Dynamic mechanical analysis of the control sample PU and PUMIMIS.[95]

Figure 2.9 shows the dynamic mechanical properties of PU and PUMIMIS. In the heating range from zero to 100°C, there is at least one order magnitude of decrease of E' (elastic modulus) for both PU and PUMIMIS. From the WAXD profiles for PU and PUMIMIS (detailed in reference[99]), there are only broad, amorphous scattering haloes between 10° and 30° of 2θ. The results of DSC (detailed in reference[99]) for PU and PUMIMIS also suggest that there are not any crystallisation/melt peaks from -40 °C to 180 °C in 20 °C/min heating scan. Therefore, the transition located at 0-100 °C should belong to glass transition (T_{gs}) of soft segments of segmented thermoplastic polyurethane. However, the sharp decreasing range for PU is mostly located in 0~40 °C, whereas the range for PUMIMIS is 40~80 °C. Also, such a trend can be observed from the peak point of Tanδ curve. When UPy was bonded with soft segments as pendent groups, T_{gs} was raised from 28.3 °C in PU to 73.3 °C in PUMIMIS. In PUMIMIS, T_{gs} detected by DMA is quite close to the dissociation temperature of QHB between UPy units, which dissociated completely around 80 °C detected by the investigation of melt viscosity of polymer network.[100] For PU without UPy, T_{gs} is only 28.3 °C. Therefore, it can be found the rubbery plateau for PU located in the temperature range above 28.3 °C; nevertheless, rubbery plateau for PUMIMIS is above 73.3 °C. Such an influence of hydrogen bonding on glass transition was also observed in alternating copolymer of styrene and maleimide and reported by Lange and Meijer.[101] In their copolymers, a three-dimensional hydrogen bonding network disrupts the imide-imide interaction, giving rise to the decrease of glass transition. Figure 2.9 also illustrates that the transition around -50 °C can be observed from the peaks of E'' (loss modulus) curve or the shoulders of Tanδ(loss tangent) curve for both PU and PUMIMIS. They are attributed to the secondary transition related to PPG segments. The location of this transition point for PU or PUMIMIS is quite similar, suggesting that PPG segments are not significantly influenced by the introduction of UPy, which is because the PPG molecular chain was not bonded covalently with pendent UPy groups and possesses similar activities in PU and PUMIMIS. In higher temperature scope around 160 °C, the dissociation of hard domains can be observed from the sharp decrease of E' and the peaks in Tanδ curve. The introduction of UPy leads to a slight decrease of dissociation temperatures of hard domains (T_{hs}) from 177 °C (in PU) to 159.5 °C (in PUMIMIS). It might be due to the combination of both end-to-end hydrogen bonding (UPy) and lateral hydrogen-bonding (urethane) functionalities in PUMIMIS.[102]

2.5.3 *Shape memory function of supramolecular SMP*

Figure 2.10 demonstrates the routine of cyclic tensile test. Considering that T_{gs} in PU is quite close to T_l (25 °C), V_r is set to be 100, 40, 1 mm/min respectively to investigate the effect of programming on SME. Also, one pair of stretched samples (PU and PUMIMS) was relaxed at 25 °C for 24 hours after the first tensile cycle and then used in strain recovery test as shown in Fig. 2.11(a). Figure 2.12 shows R_f of PUMIMIS increase with the decrease of V_r, but R_f of PU shows a reverse trend. It is because the quadruple hydrogen bonding between UPy in PUMIMIS constrains the activity of soft segments and freezes the deformation. For PUMIMIS, the temporary shape fixity is facilitated by the slow unloading procedure, because the time of stress executed on deformed samples was extended in the slow unloading procedure. In addition, R_f can be kept in a relatively high value for PUMIMIS, especially when V_r=1mm/min, T_l=25°C, R_f is 95.9%. Even when T_l=46 °C, the R_f can be 93%, suggesting that UPy pendent groups within soft segments is strong enough to withstand the entropy elasticity and internal stress stored in deformed samples. Nevertheless, R_f of PU decreased with decrease of V_r. It is only 67% around when T_l=46 °C, V_r =40mm/min. Therefore, it can be concluded that the unloading procedure can significantly affect R_f of PU. As far as the recovery ability in PU and PUMIMS is concerned, the R_r decreases slightly and R_r' can gradually increase to 95.8% with the increase of tensile cycle number. The increase of R_r' with cycle number is due to the formation of perfect elastic network after the disruption of weak net-points in the initial tensile cycles.[2] It also cannot be neglected that the introduction of UPy into soft segments slightly affects the aggregation of hard segments. Accordingly, the R_r of PUMIMIS slightly is lower than PU.

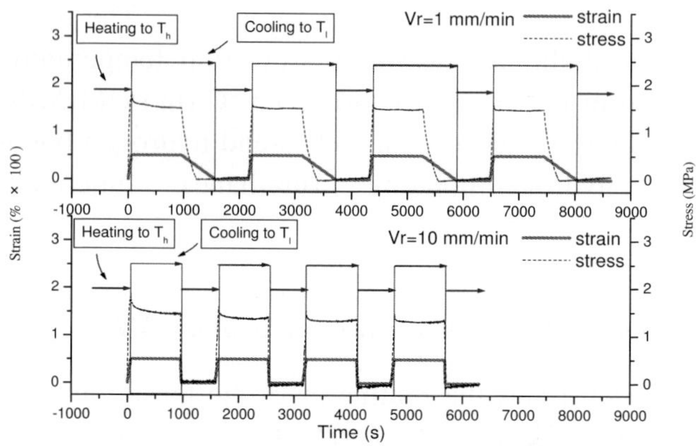

Fig. 2.10 Routine of cyclic tensile testing.[95]

In the following study, after the unloading procedure in the first cycle of cyclic tensile test (V_r=40mm/min, T_l=25°C), 24-hour relaxation was performed to the stretched films. It was found that the fixed lengths of PU and PUMIMIS films are 21.0 and 26.0mm respectively. The original length of films before cyclic tensile testing is 20mm. Therefore, the fixity ratios for stretched PU and PUMIMIS after 24-hour relaxation are 10% and 60%, suggesting that only 10% deformation for PU can be fixed, but for PUMIMIS, 60% deformation was fixed. Figure 2.11(a) demonstrates the strain recovery behaviour of PU and PUMIMIS strained films. A sharp shape recovery behaviour was observed only in PUMIMIS. Thus, it can be concluded that sample PU is mostly elastic in used thermal dynamic conditions. The deformation cannot be fixed effectively, and therefore the shape recovery phenomenon cannot be observed. However, for PUMIMIS, self-complementary QHB can constrain the activity of soft segments and cause the increase of T_{gs} of soft segments from 28.3 °C to 73.3 °C. During the used tensile cycle, most deformation can be fixed for a relatively long time. Subsequently the sharp shape recovery phenomenon is significant. The shape recovery temperature range is located between 40 and 80 °C. To better compare the shape memory function for PU and PUMIMIS, the integrated shape memory programming cycles shown in Fig. 2.11(b) and (c) are illustrated by using 3-D graphs (point A to G) for PU and PUMIMIS.

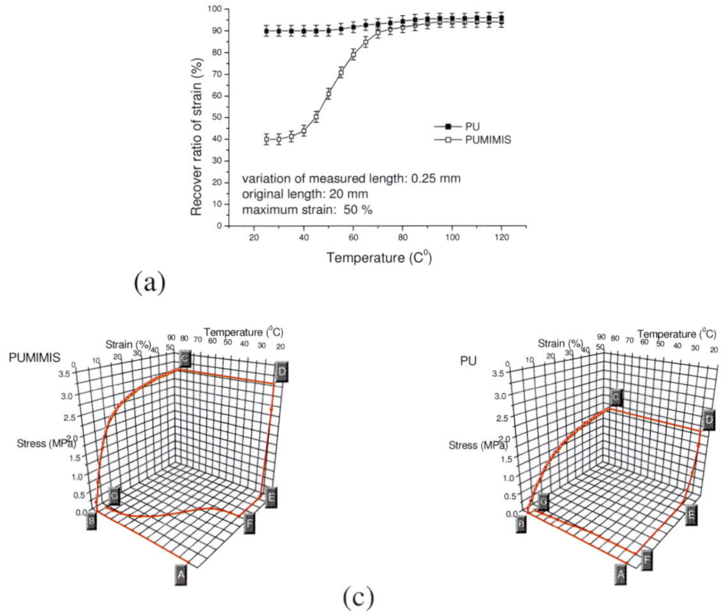

Fig. 2.11 (a): Strain recovery at 5 °C/min heating for stretched PU and PUMIMIS films after 24 hours relaxation; (b) and (c): integrated shape memory programming process (A→B: heating from 25 to 86 °C in 600s; B→C: stretching to 50% strain at 10mm/min at 86 °C; C→D: cooling the strained sample at fixed 50% strain from 86 to 25 °C in 900s; D→E: unloading the strained sample with 40mm/min at 25 °C; E→F: relaxing the strained sample for 24 hour at 25°C; F→G: reheating the strained sample from 25 to 86 °C at 5 °C /min.)[95]: cooling to T_l (C→D), relaxing the deformation at T_l (D→E→F) and shape recovering in reheating from T_l to T_h (F→G).

It was found that UPy units into elastic PU network can facilitate both immediate shape fixity and shape fixity after 24-hour relaxation. After an integrated tensile cycle, the length of sample film is almost recovered totally. The thermal reversible QHB can adjust the T_{gs} and be used to control shape memory function. It is also expected that other stimulus, which is able to affect QHB, such as solvent or humidity, can probably be used to stimulate the shape memory function. It might help exploit a new area of multi-responsive shape memory polymer to satisfy the booming demand for smart materials in various application areas.

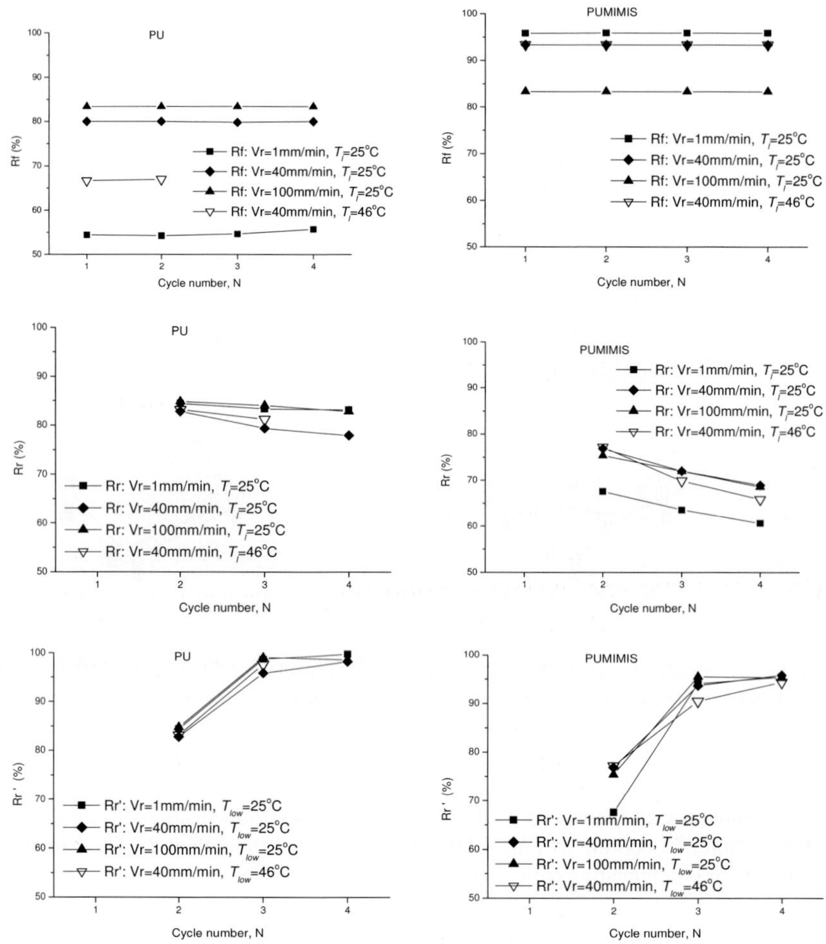

Fig. 2.12 Routine in cyclic tensile test and the result of R_f, R_r and R_r.[95]

Therefore, it can be concluded that with the introduction of quadruple hydrogen bonding unit into soft segments as pendent groups in elastic thermoplastic polyurethane network, T_{gs} can be increased from 28.3°C to 73.3°C. Accordingly, the shape memory function can be realised in commonly used programming conditions. The immediate shape recovery ratio and shape fixity ratio can be 95.8% and 95.9%. After 24-hour relaxation for the deformed film, the corresponding shape recovery ratio and shape fixity ratio can be 94% and 60%. Shape recovery testing suggests that sharp shape recovery does exist in the temperature range between 40 and 80 °C.

2.6 Shape Memory Fibres

2.6.1 *One way shape memory fibre*

As for the application of shape memory polymer used in fibre spinning to impart the smart function into fabric, limited studies in the literature about shape memory polyurethane fibre were found. Nevertheless, the sparse application and production of these kinds of functional fibre has aroused much attention. In the patent, Hayashi *et al.*[103] reported that a woven fabric of shape memory polymer is formed by weaving yarn of shape memory polymer fibres along or by weaving said yarn or ordinary natural or synthetic fibres wherein the shape memory polymer fibres are made of a polyurethane elastomer having a shaped memory property. Chun *et al.*[104] studied the shape memory polyurethane composed of PTMG, BDO, MDI, with the switching temperature ranging from -15ºC to 1.5ºC. The shape memory fibre made of their shape memory polyurethane can be prepared with electrospinning with the use of the mixed solvent of DMF(N,N-dimethylformamide) and THF(tetrahydrofuran).[105] The electrospun polyurethane nonwovens with hard segment concentration of 40 and 50 wt% were found to have the shape recovery of more than 80%. Hu *et al.*[4,106] have studied the shape memory effect of shape memory polyurethane fibre composed of PBA (Poly (buthylene-adipate)) or PEA (Poly (ethylene adipate)), MDI (4, 4'-Methylenebis (phenyl isocyanate)), BDO (1, 4-Butanediol) with the transition temperature at around room temperature (29~64°C). With the usage of various processing technology, the shape memory polymer has been successfully engineered into the series of fibres and yarn with tailormade shape memory functions, applicable tenacity (6~14cN/tex) and maximum strain (35~204%). The shape memory function of the fibres can be demonstrated by the applicable fixity ability (70 %~100%) for stretching at room temperature and the admirable recovery ability after heating the stretched fibre beyond the transition temperature.

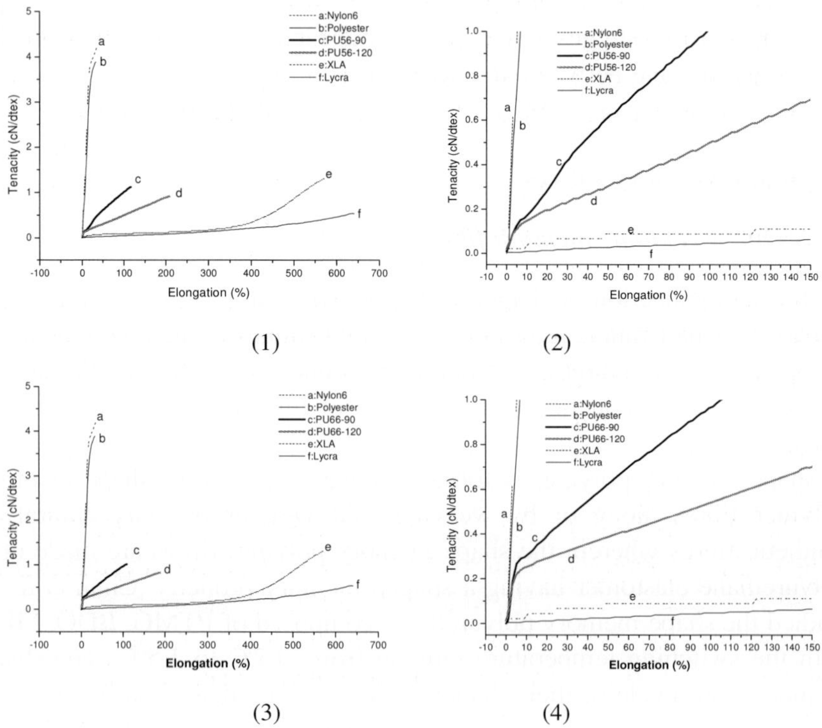

Fig. 2.13 Stress-strain curve of various fibres.[4]

Usually, man-made filaments might be attributed to two species: one is of high tensile strength (tenacity: 5~7 g/den for Nylon 6) and modulus, but low elongation at break (25~40% for Nylon 6) and high elastic recovery with only minor deformation such as Nylon6;[107] the other one is of very low initial modulus, very high elongation ratio at break (500~600%) and nearly complete instant elastic recovery after stretching (90~100%) such as Lycra.[108] For shape memory polyurethane filament, the heating responsive shape memory effect was expected to be introduced into the multi-filaments, namely, to cause the deformation to be recovered with stimulus of heating above the transition temperature. The glassy or crystallisable soft segment at the temperature below transition temperature (T_g or T_m) can impart the fibre with relatively high initial modulus, applicable elongation ratio and strength at break, among which the last factor is usually considered mostly for the kitting process.[4] As shown in Fig. 2.13, the stress-strain curve of SMPU fibres (PU56-90, PU56-120, PU66-90, PU66-120) is located between the high modulus fibre such as nylon and the high elasticity fibre such as Lycra; for stress-strain behaviour, the influence of thermal setting is found to be greater than the hard segment content in the range of hard segments content used in this study; the thermal setting with higher temperature will give rise to the lower modulus and tenacity and the higher maximum elongation ratio in these two series of shape memory polyurethane fibres. Furthermore, the yield points of shape memory polyurethane fibre can be observed at the strain at around 5%; the higher the hard segment content in SMPU fibre, the higher will be the yield strength.

With the investigation of DMA, the elastic modulus (E') in the normal using temperature range about a variety of man-made fibres is demonstrated in Fig. 2.14. The resultant data show that the main difference between SMPU fibre and conventional man-made fibres is the variation of E' in the normal using temperature range. For SMPU fibres such as PU56-120 and PU66-120, the variation of E' is very significant. When the temperature increases above the transition temperature, the E' will sharply decrease and the rubbery state platform will appear and be extended to above 180ºC. However, for other types of man-made fibres such as polyester and Lycra, though in the entire heating scan range, there are some transition areas of E' such as -40ºC for Lycra,[108] 100~105 for Polyester fibres and yarn,[109] the elastic modulus is almost constant, and it changes little with the increase of temperature in room temperature range. Therefore, this transition in SMPU imparts the heating responsive shape memory properties to the SMPU fibres in normal using temperature.

For common elastic fibres, elasticity should be defined as instant recoverability of the length on release of deforming stress. The recoverability in shape memory fibres should be the recovery ability of deformed fibres with external stimulus such as heat or chemicals. In this case, the external stimulus is a must.

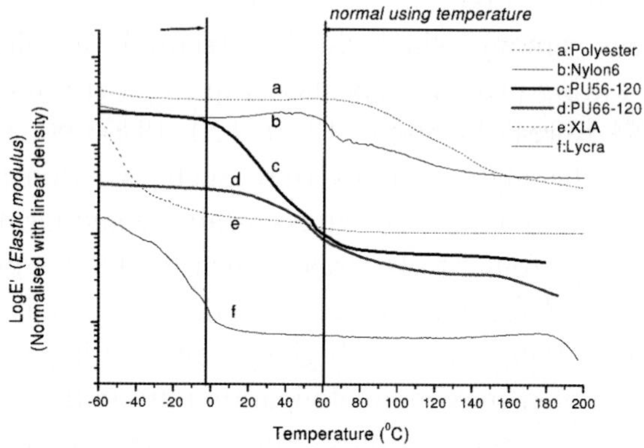

Fig. 2.14 Comparison of elastic modulus between SMPU fibre and various man-made fibres.[4]

2.6.2 *Two way shape memory fibres*

To date, most research on shape memory polymer focuses on shape memory effect based on the programming comprising deforming with external force and automatically recovering from the deformed status with external stimulus, which are usually called 'one-way' shape memory effect in the study of shape memory materials.[110,111] Recently, Terentjev *et al.* developed new thermal-plastic liquid-crystalline elastomers synthesised by using the telechelic principle of microphase separation in triblock copolymers where the large central block is made of a main-chain nematic polymer renowned for its large spontaneous elongation along the nematic director, and the crosslinking is established by small terminal blocks formed of terphenyl moieties.[112] The thin well-aligned fibre, spun from the melt under heating condition at the temperature above the nematic-isotropic transition, shows significant two-way shape memory effect, which is characterised by the thermal actuator behaviour-reversible contraction of heating and elongation on cooling caused by the nematic director. In the spinning process, the nematic ordering and the telechelic crosslinking were formed simultaneously. The amplitude of actuation strain within the studied temperature range (40°C~110°C) can even reach 500%. The transition point located at around 100°C depends on the nematic-isotropic transitions temperature. Presently, although the transition point is quite high for ordinary apparel application, this two-way shape memory fibre can help inspire some novel practice of shape memory fibre in smart and functional textile design.

References

1. Wei, X. and Yu, X. H., Synthesis and Properties of Sulfonated Polyurethane Ionomers with Anions in the Polyether Soft Segments. *Journal of Polymer Science: Part B: Polymer Physics*, 1997. 35: pp. 225-232.
2. Lendlein, A. and Kelch, S., Shape-Memory Polymers. *Angewandte Chemie-International Edition*, 2002. 41(12): pp. 2034 - 2057.
3. Zhu, Y., Hu, J., and Liu, Y. Thermo-reversible Self-complementary Quadruple Hydrogen Bonding to Control Shape Recovery and Shape Fixity in Polyurethanes. in *IUPAC 42nd World Polymer Congress (Macro 2008): Polymers at Frontiers of Science and Technology*. 2008. Taipei, Taiwan.
4. Zhu, Y., Hu, J. L., Yeung, L. Y., Liu, Y., Ji, F. L., and Yeung, K. W., Development of Shape Memory Polyurethane Fiber with Complete Shape Recoverability. *Smart Materials & Structures,* 2006. 15(5): pp. 1385-1394.
5. Hu, J., Zhu, Y., Lu, J., Yeung, L.-y., and Yeung, K.-w., Uniqueness of shape memory fibers in comparison with existing man-made fibers, in *The 9th Asian Textile Conference*. 2007, Federation of Asian Professional Textile Associations: Taiwan.
6. Wei, Z. G., Sandstrom, R., and Miyazaki, S., Review: Shape-memory materials and hybrid composites for smart systems. *Journal of Materials Science,* 1998. 33: pp. 3743-3762.
7. Hu, J. L., Ji, F. L., and Wong, Y. W., Dependency of the shape memory properties of a polyurethane upon thermomechanical cyclic conditions. *Polymer International,* 2005. 54(3): pp. 600-605.
8. Kim, B. K., Lee, S. Y., and Xu, M., Polyurethanes having shape memory effects. *Polymer,* 1996. 37(26): pp. 5781-5793.
9. Li, F. K., Zhang, X., Hou, J. N., Xu, M., Luo, X. L., Ma, D. Z., and Kim B. K., Studies on Thermally Stimulated Shape Memory Effect of Segmented Polyurethanes. *Journal of Applied Polymer Science,* 1997. 64(8): pp. 1511-1516.

10. Lin, J. R. and Chen, L. W., Study on Shape-Memory Behavior of Polyether-Based Polyurethanes. I. Influence of the Hard-Segment Content. *Journal of Applied Polymer Science,* 1998. 69: pp. 1563-1574.
11. Chowdhury, S. R., Mishra, J. K., and Das, C. K., Shrinkability and microstructural properties of composites based on low-density polyethylene (LDPE) and polyurethane (PU) rubber. *Journal of Thermoplastic Composite Materials*, 2000. 13(5): pp. 400-416.
12. Liu, C., Chun, S. B., Mather, P. T., Zheng, L., Haley, E. H., and Coughlin, E. B., Chemically cross-linked polycyclooctene: Synthesis, characterization, and shape memory behavior. *Macromolecules*, 2002. 35(27): pp. 9868-9874.
13. Ota, S., Current Status of Irradiated Heat-Shrinkable Tubing in Japan. *Radiation Physics and Chemistry*, 1981. 18(1-2): pp. 81-87.
14. Chen, W., Xing, K., and Sun, L., The Heat Shrinking Mechanism of Polyethylene Film. *Radiation Physics and Chemistry*, 1983. 22(3-5): pp. 593-601.
15. Sakurai, K. and Takahashi, T., Strain-Induced Crystallization in Polynorbornene. *Journal of Applied Polymer Science*, 1989. 38(6): pp. 1191-1194.
16. Sakurai, K., Kashiwagi, T., and Takahashi, T., Crystal structure of polynorbornene. *Journal of Applied Polymer Science*, 1993. 47(5): pp. 937-940.
17. Sakurai, K., Shirakawa, Y., Kashiwagi, T., and Takahashi, T., Crystal Transformation of Styrene-Butadiene Block-Copolymer. *Polymer*, 1994. 35(19): pp. 4238-4239.
18. Li, F. K., Zhu, W., Zhang, X., Zhao, C. T., and Xu, M., Shape Memory Effect of Ethylene-Vinyl Acetate Copolymers. *Journal of Applied Polymer Science*, 1999. 71: pp. 1063-1070.
19. Li, F. K., Chen, Y., Zhu, W., Zhang, X., and Xu, M., Shape memory effect of polyethylene/nylon6 graft copolymers. *Polymer*, 1998. 39: pp. 6929-6934.
20. Luo, X. L., Zhang, X. Y., Wang, M. T., Ma, D. Z., Xu, M., and Li, F. K., Thermally Stimulated Shape-Memory Behavior of Ethylene Oxide-Ethylene Terephthalate Segmented Copolymer. *Journal of Applied Polymer Science*, 1997. 64: pp. 2433-2440.
21. Chaunier, L. and Lourdin, D., The Shape Memory of Starch. *Starch-Starke,* 2009. 61(2): pp. 116-118.

22. Willett, J., Humidity-responsive starch-poly(methyl acrylate) films. *Macromolecular Chemistry and Physics*, 2008. 209(7): pp. 764-772.
23. Hu, J. L., Shape memory textiles, in Shape Memory Polymers and Textiles. 2007, Cambridge: Woodhead Publishing Limited.
24. Liang, C., Rogers, C. A., and Malafeew, E., Investigation of shape memory polymers and their hybrid composites (Reprinted from Proceedings of the Second Joint Japan/US Conference on Adaptive Structures, Nov, pg 789-802). *Journal of Intelligent Material Systems and Structures*, 1997. 8(4): pp. 380-386.
25. Lin, J. R. and Chen, L. W., Study on Shape-Memory Behavior of Polyether-Based Polyurethanes. II. Influence of Soft-Segment Molecular Weight. *Journal of Applied Polymer Science,* 1998. 69: pp. 1575-1586.
26. Hou, J. A., Ma, X. Q., Luo, X. L., Ma, D. Z., Zhang, X., Zhu, W., and Xu, M., Thermally Simulated Shape Recovery Effect of Segmented Polyurethane, in *The International Symposium on Polymer Alloys and Composite*s, C. L. Choy and F. G. Shin, Editors. 1992, Hong Kong Polytechnic University: Hong Kong Polytechnic University, Hong Kong. p. 211.
27. Zhang, H., Wang, H. T., Zhong, W., and Du, Q. G., A novel type of shape memory polymer blend and the shape memory mechanism. *Polymer*, 2009. 50(6): pp. 1596-1601.
28. Zhang, S. F., Feng, Y. K., Zhang, L., Sun, J. F., Xu, X. K., and Xu, Y. S., Novel interpenetrating networks with shape-memory properties. *Journal of Polymer Science Part a-Polymer Chemistry*, 2007. 45(5): pp. 768-775.
29. Liu, G., Ding, X., Cao, Y., Zheng, Z., and Peng, Y., Novel Shape-Memory Polymer with Two Transition Temperature. *Macromolecular Rapid Communications*, 2005. 26(8): pp. 649-652.
30. Perez-Foullerat, D., Hild, S., Mucke, A., and Rieger, B., Synthesis and properties of poly (ketone-co-alcohol) materials: Shape memory thermoplastic elastomers by control of the glass transition process. *Macromolecular Chemistry and Physics*, 2004. 205(3): pp. 374-382.
31. Kumar, S. and Pandya, M. V., Thermally recoverable crosslinked polyethylene. *Journal of Applied Polymer Science*, 1997. 64(5): pp. 823-829.

32. Chung, T., Rorno-Uribe, A., and Mather, P. T., Two-way reversible shape memory in a semicrystalline network. *Macromolecules*, 2008. 41(1): pp. 184-192.
33. Bellin, I., Kelch, S., Langer, R., and Lendlein, A., Polymeric triple-shape materials. *Proceedings of the National Academy of Sciences of the United States of America*, 2006. 103(48): pp. 18043-18047.
34. Rousseau, I. A. and Mather, P. T., Shape memory effect exhibited by smectic-c liquid crystalline elastomers. *Journal of the American Chemical Society*, 2003. 125(50): pp. 15300-15301.
35. Yang, B., Huang, W. M., Li, C., and Chor, J. H., Effects of moisture on the glass transition temperature of polyurethane shape memory polymer filled with nano-carbon powder. *European Polymer Journal*, 2005. 41: pp. 1123-1128.
36. Huang, W. M., Yang, B., An, L., Li, C., and Chan, Y. S., Water-driven programmable polyurethane shape memory polymer: Demonstration and mechanism. *Applied Physics Letters*, 2005. 86: pp. 114105-1~114105-3.
37. Yang, B., Huang, W. M., Li, C., and Li, L., Effects of moisture on the thermomechanical properties of a polyurethane shape memory polymer. *Polymer*, 2006. 47(4): pp. 1348-1356.
38. Jung, Y. C., So, H. H., and Cho, J. W., Water-Responsive Shape Memory Polyurethane Block Copolymer Modified with Polyhedral Oligomeric Silsesquioxane. *Journal of Macromolecular Science. Part B: Physics*, 2006. 45: pp. 453-461.
39. Cho, J. W., Kim, J. W., Jung, Y. C., and Goo, N. S., Electroactive Shape-Memory Polyurethane Composites Incorporating Carbon Nanotubes. *Macromolecular Rapid Communications*, 2005. 26: pp. 412-416.
40. Schmidt, A. M., Electromagnetic Activation of Shape Memory Polymer Networks Containing Magnetic Nanoparticles. *Macromolecular Rapid Communications*, 2006. 27: pp. 1168-1172.
41. Koerner, H., Price, G., Pearce, N. A., Alexander, M., and Vaia, R. A., Remotely actuated polymer nanocomposites-stress-recovery of carbon-nanotube-filled thermoplastic elastomers. *Nature Materials,* 2004. 25: pp. 115-120.
42. Yang, B., Huang, W. M., Li, C., Lee, C. M., and Li, L., On the effects of moisture in a polyurethane shape memory polymer. *Smart Materials & Structures*, 2004. 13(1): pp. 191-195.

43. Lv, H. B., Liu, Y. J., Zhang, D. X., Leng, J. S., and Du, S. Y. Solution-responsive shape-memory polymer driven by forming hydrogen bonding. in *International Conference on Multifunctional Materials and Structures. 2008.* Hong Kong, PEOPLES R CHINA.
44. Mohr, R., Kratz, K., Weigel, T., Lucka-Gabor, M., Moneke, M., and Lendlein, A., Initiation of shape-memory effect by inductive heating of magnetic nanoparticles in thermoplastic polymers. *Proceedings of the National Academy of Sciences of the United States of America,* 2006. 103(10): pp. 3540-3545.
45. Yu, Y. L. and Ikeda, T., Photodeformable polymers: A new kind of promising smart material for micro- and nano-applications. *Macromolecular Chemistry and Physics*, 2005. 206(17): pp. 1705-1708.
46. Lendlein, A., Jiang, H., Junger, O., and Langer, R., Light-induced shape-memory polymers. *Nature,* 2005. 434: pp. 879-882.
47. Hogan, P. M., Tajbakhsh, A. R., and Terentjev, E. M., UV manipulation of order and macroscopic shape in nematic elastomers. *Physical Review E*, 2002. 65(4): p. 041720 (10 pages).
48. Kinoshita, T., Photoresponsive membrane systems. *Journal of Photochemistry and Photobiology B-Biology*, 1998. 42(1): pp. 12-19.
49. Finkelmann, H., Nishikawa, E., Pereira, G. G., and Warner, M., A new opto-mechanical effect in solids. *Physical Review Letters*, 2001. 8701(1): p. 015501 (4 pages).
50. Osada, Y. and Matsuda, A., Shape-Memory in Hydrogels. *Nature*, 1995. 376(6537): pp. 219-219.
51. Li, Y., Hu, Z. B., and Chen, Y. Y., Shape memory gels made by the modulated gel technology. *Journal of Applied Polymer Science,* 1997. 63(9): pp. 1173-1178.
52. Kim, S. J., Kim, H. I. I., Park, S. J., and Kim, S. I., Shape change characteristics of polymer hydrogel based on polyacrylic acid/poly(vinyl sulfonic acid) in electric fields. *Sensors and Actuators a-Physical*, 2004. 115(1): pp. 146-150.
53. Ratna, D. and Karger-Kocsis, J., Recent advances in shape memory polymers and composites: a review. *Journal of Materials Science,* 2008. 43(1): pp. 254-269.
54. Hu, J., Zhu, Y., and Ye, G. D., Antibacterial Shape Memory Polyurethane Plate for Orthopedics, Fixing and Rehabilitation, and Preparation Method Thereof. 2005: WO2006/102842 A1, China.

55. Meng, Q. G., Hu, J. L., and Zhu, Y., Properties of shape memory polyurethane used as a low-temperature thermoplastic biomedical orthotic material: influence of hard segment content. *Journal of Biomaterials Science-Polymer Edition*, 2008. 19(11): pp. 1437-1454.
56. Nakasima, A., Hu, J. R., Ichinose, M., and Shimada, H., Potential Application of Shape Memory Plastic as Elastic-Material in Clinical Orthodontics. *European Journal of Orthodontics,* 1991. 13(3): pp. 179-186.
57. Langer, R. and Tirrell, D. A., Designing materials for biology and medicine. *Nature,* 2004. 428(6982): pp. 487-492.
58. Alteheld, A., Feng, Y., Kelch, S., and Lendlein, A., Biodegradable, Amorphous Copolyester-urethane Networks Having Shape-Memory Properties. *Angew Chem International Edition*, 2005. 44: pp. 1188-1192.
59. Kusy, R. P. and Whitley, J. Q., Thermal characterization of shape memory polymer blends for biomedical implantations. *Thermochimica Acta*, 1994. 243: pp. 253-263.
60. Baer, G., Wilson, T. S., Matthews, D. L., and Maitland, D. J., Shape-Memory Behavior of Thermally Stimulated Polyurethane for Medical Application. *Journal of Applied Polymer Science*, 2007. 103: pp. 3882-3892.
61. Lendlein, A. and Langer, R., Biodegradable, Elastic Shape-Memory Polymers for Potential Biomedical Applications. *Science,* 2002. 296(5573): pp. 1673-1676.
62. Yakacki, C. M., Shandas, R., Lanning, C., Rech, B., Eckstein, A., and Gall, K., Unconstrained recovery characterization of shape-memory polymer networks for cardiovascular applications. *Biomaterials,* 2007. 28(14): pp. 2255-2263.
63. Maitland, D. J., Metzger, M. F., Schumann, D., Lee, A., and Wilson, T. S., Photothermal properties of shape memory polymer micro-actuators for treating stroke. *Lasers in Surgery and Medicine*, 2002. 30(1): pp. 1-11.
64. Hayashi, S., Properties and applications of polyurethane-series shape memory polymer. *International Progress in Urethanes*, 1993. 6: pp. 90-115.
65. Hu, J. L., Ding, X. M., Tao, X. M., and Yu, J. M., Shape Memory Polymers and Their Applications to Smart Textile Products. *Journal of Dong Hua University (Eng Ed)*, 2002. 19(3): pp. 89.

66. Zeng, Y. M., Hu, J. L., and Yan, H. J., Temperature Dependency of Water Vapor Permeability of Shape Memory Polyurethane. *Journal of Dong Hua University (Eng Ed),* 2002. 19(3): pp. 53.
67. Hu, J. L., Zeng, Y. M., and Yan, H. J., Influence of Processing Conditions on the Microstructure and Properties of Shape Memory Polyurethane Membranes. *Textile Research Journal*, 2003. 73(2): pp. 172-178.
68. Ding, X. M., Hu, J. L., and Tao, X. M., Effect of Crystal Melting on Water Vapor Permeability of Shape-Memory Polyurethane Film. *Textile Research Journal,* 2004. 74(1): pp. 39-43.
69. Hu, J. L. and Mondal, S., Structural characterization and mass transfer properties of segmented polyurethane: influence of block length of hydrophilic segments. *Polymer International,* 2005. 54(5): pp. 764-771.
70. Mondal, S. and Hu, J. L., Segmented shape memory polyurethane and its water vapor transport properties. *Designed Monomers & Polymers*, 2006. 9(6): pp. 527-550.
71. Hu, J., Lu, J., and Zhu, Y., New developments in elastic fibers. *Polymer Reviews*, 2008. 48(2): pp. 275-301.
72. Liu, Y., Chung, A., Hu, J., and Lu, J., Shape memory behavior of SMPU knitted fabric. *Journal of Zhejiang University Science A,* 2007. 8(5): pp. 830-834.
73. Hu, J., Zhu, Y., Chen, S., Luo, H., and Lu, J., Methods for Making Hairs to Have Shape Memory Effect. 2009, in application.
74. Lendlein, A., Kalbfleisch, Duchscherer, Beyer, Graner, Hasson, A., Krause, T., Kratz, Jiang, H.-Y., Moersheim, S., and Ghazlan, Y., Method for hair treatment with shape memory polymers. 2007 Mnemoscience GmbH, US 2007/0275020 A1: United States.
75. Barvosa-Carter, W., Browne, A. L., Johnson, N. L., Momoda, L. A., and Stanford, T. B., Releasable fastener systems and processes. 2007, General Motors Corporation, US 7,308,738 B2: United States.
76. Nelson, B. A., King, W. P., and Gall, K., Shape recovery of nanoscale imprints in a thermoset "shape memory" polymer. *Applied Physics Letters*, 2005. 86(10): p. 103108 (3 pages).
77. Liu, N., Xie, Q., Huang, W. M., Phee, S. J., and Guo, N. Q., Formation of micro protrusion arrays atop shape memory polymer. *Journal of Micromechanics and Microengineering*, 2008. 18(2): p. 027001.

78. Arciola, C. R., Radina, L., Alvergna, P., Cennib, E., and Pizzoferrato, A., Heparin surface treatment of poly-(methylmethacrylate) alters adhesion of a Staphylococcus aureus strain: utility of bacterial fatty acid analysis. *Biomaterials* 1993. 14(15): pp. 1161-1164.
79. Yang, J.E., Kong, J.S., Park, S.-W., Lee, D.-J., and Kim, H.-D., Preparation and Properties of Waterborne Polyurethane-Urea Anionomers. I. The Influence of the Degree of Neutralization and Counterion. *Journal of Applied Polymer Science*, 2002. 86: pp. 2375-2383.
80. Hazziza-Laskar, J., Nurdin, N., Helary, G., and Sauvet, G., Biocidal Polymers Active by Contact. I. Synthesis of Polybutadiene with Pendant Quaternary Ammonium Groups. *Journal of Applied Polymer Science*, 1993. 50: pp. 651-662.
81. Kawabata, N. and Nishiguchi, M., Antibacterial Activity of Soluble Pyridinium-Type Polymers. *Applied and Environmental Micro-biology*, 1988. 54(10): pp. 2532-2535.
82. Grapski, J. A. and Cooper, S. L., Synthesis and characterization of non-leaching biocidal polyurethanes. *Biomaterials*, 2001. 22(16): pp. 2239-2246.
83. Zhu, Y., Hu, J., and Yeung, K.-w., Effect of Soft-segment Crystallization, Hard-segment Physical Crosslink on Shape memory Function in Antibacterial Segmented PU Ionomers. *Acta Biomaterialia*, 2009, in press.
84. Zhu, Y., Hu, J. L., Yeung, K. W., Choi, K. F., Liu, Y. Q., and Liem, H. M., Effect of Cationic Group Content on Shape Memory Effect in Segmented Polyurethane Cationomer Journal of Applied Polymer Science, 2007. 103(1): pp. 545-556.
85. Jeong, H. M., Ahn, B. K., and Kim, B. K., Temperature sensitive water vapour permeability and shape memory effect of polyurethane with crystalline reversible phase and hydrophilic segments. *Polymer International*, 2000. 49(12): pp. 1714-1721.
86. Hu, J. L., Yang, Z. H., Yeung, L. Y., Ji, F. L., and Liu, Y. Q., Crosslinked polyurethanes with shape memory properties. *Polymer International*, 2005. 54(5): pp. 854-859.
87. Kim, B. K., Lee, S. Y., Lee, J. S., Baek, S. H., Choi, Y. J., Lee, J. O., and Xu, M., Polyurethane ionomers having shape memory effects. *Polymer*, 1998. 39(13): pp. 2803-2808.

88. Lehn, J. M., Supramolecular Chemistry. *Science*, 1993. 260(5115): pp. 1762-1763.
89. Lehn, J. M., Supramolecular polymer chemistry- scope and perspectives. *Polymer International*, 2002. 51(10): pp. 825-839.
90. Lehn, J. M., From supramolecular chemistry towards constitutional dynamic chemistry and adaptive chemistry. *Chemical Society Reviews*, 2007. 36(2): pp. 151-160.
91. Sijbesma, R. P., Beijer, F. H., Brunsveld, L., Folmer, B. J. B., Hirschberg, J. H. K. K., Lange, R. F. M., Lowe, J. K. L., and Meijer, E. W., Reversible polymers formed from self-complementary monomers using quadruple hydrogen bonding. *Science*, 1997. 278(5343): pp. 1601-1604.
92. Folmer, B. J. B., Sijbesma, R. P., Versteegen, R. M., Rijt, J. A. J. v. d., and Meijer, E. W., Supramolecular Polymer Materials: Chain Extension of Telechelic Polymers Using a Reactive Hydrogen-Bonding Synthon. *Advanced Materials,* 2000. 12(12): pp. 874-878.
93. Yamauchi, K., Lizotte, J. R., and Long, T. E., Synthesis and characterization of novel complementary multiple-hydrogen bonded (CMHB) macromolecules via a Michael addition. *Macromolecules*, 2002. 35(23): pp. 8745-8750.
94. Yamauchi, K., Kanomata, A., Inoue, T., and Long, T. E., Thermoreversible polyesters consisting of multiple hydrogen bonding (MHB). *Macromolecules*, 2004. 37(10): pp. 3519-3522.
95. Zhu, Y., Hu, J. L., and Liu, Y. J., Shape memory effect of thermoplastic segmented polyurethanes with self-complementary quadruple hydrogen bonding in soft segments. *European Physical Journal E*, 2009. 28(1): pp. 3-10.
96. Garrett, J., Siedlecki, C., and Runt, J., Microdomain morphology of poly(urethane urea) multiblock copolymers. *Macromolecules*, 2001. 34(20): pp. 7066-7070.
97. O'Sickey, M., Lawrey, B., and Wilkes, G., Structure-property relationships of poly(urethane urea)s with ultra-low monol content poly(propylene glycol) soft segments. I. Influence of soft segment molecular weight and hard segment content. *Journal of Applied Polymer Science*, 2002. 84(2): pp. 229-243.
98. O'Sickey, M., Lawrey, B., and Wilkes, G., Structure-property relationships of poly(urethane-urea)s with ultra-low monol content poly(propylene glycol) soft segments. Part II. Influence of low

molecular weight polyol components. *Polymer*, 2002. 43(26): pp. 7399-7408.

99. Zhu, Y., Hu, J., and Liu, Y., Shape Memory Effect of Thermoplastic Segmented Polyurethanes with Self-complementary Quadruple Hydrogen Bonding in Soft Segments. *European Physical Journal E*, 2008. 28: pp. 3-10.
100. Yamauchi, K., Lizotte, J. R., and Long, T. E., Thermoreversible poly(alkyl acrylates) consisting of self-complementary multiple hydrogen bonding. *Macromolecules*, 2003. 36(4): pp. 1083-1088.
101. Lange, R. F. M. and Meijer, E. W., Supramolecular Polymer Interactions Based on the Alternating Copolymer of Styrene and Maleimide. *Macromolecules*, 1995. 28(3): pp. 782-783.
102. Kautz, H., van Beek, D. J. M., Sijbesma, R. P., and Meijer, E. W., Cooperative end-to-end and lateral hydrogen-bonding motifs in supramolecular thermoplastic elastomers. *Macromolecules,* 2006. 39(13): pp. 4265-4267.
103. Kobayashi, Kazuyuki, Hayashi, and Shunichi, Woven fabric made of shape memory polymer. 1992, Mitsubishi Jukogyo Kabushiki Kaisha (Tokyo, JP): United States Patent 5, 128, 197.
104. Lee, H. Y., Jeong, H. M., Lee, J. S., and Kim, B. K., Study on the shape memory polyamides. Synthesis and thermomechanical properties of polycaprolactone-polyamide block copolymer. *Polymer Journal*, 2000. 32(1): pp. 23-28.
105. Cha, D. I., Kim, H. Y., Lee, K. H., June, Y. C., Cho, J. W., and Chun, B. C., Electrospun Nonwovens of Shape-Memory Polyurethane Block Copolymers. *Journal of Applied Polymer Science*, 2005. 96: pp. 460-465.
106. Ji, F. L., Zhu, Y., Hu, J. L., Liu, Y., Yeung, L. Y., and Ye, G. D., Smart polymer fibers with shape memory effect. *Smart Materials & Structures*, 2006. 15(6): pp. 1547-1554.
107. Mark, H. F., Atlas, S. M., and Cernia, E., Man-Made Fibers. Polymer Engineering and Technology, ed. D. V. Rosato, R. B. Akin, H. F. Mark, J. J. Scavuzzo, S. S. Stivala, and L. J. Zukor. Vol. 2. 1968: John Wiley & Sons. 277.
108. Meyer, R. V., Haug, E., and Spilgies, G., Sources of stretch. *Textile Asia*, 1994: pp. 40-45.
109. Manich, A. M., Bosch, T., Carilla, J., Ussman, M. H., Maillo, J., and Gacen, J., Thermal Analysis and Differential Solubility of

Polyester Fiber and Yarns. *Textile Research Journal,* 2003. 73(4): pp. 333-338.

110. Moore, C. L. and Bruck, H. A., A fundamental investigation into large strain recovery of one-way shape memory alloy wires embedded in flexible polyurethanes. *Smart Materials & Structures*, 2002. 11: pp. 130-139.
111. Kneissl, A. C., Unterweger, E., Lojen, G., and Anzel, I., Microstructure and Properties of Shape Memory Alloys. *Microsc Microanal*, 2005. 11: pp. 1704.
112. Ahir, S. V., Tajbakhsh, A. R., and Terentjev, E. M., Self-Assembled Shape-Memory Fibers of Triblock Liquid-Crystal Polymer. *Advance Functional Materials*, 2006. 16: pp. 556-560.

Chapter 3

Adaptive Polymeric Gels and Applications

The definition and classification of polymer gels will be first provided in this chapter. Then, the molecular structure and gelation theory of adaptive polymeric gels are systemically reviewed. The molecular design and synthesis methods of adaptive polymeric gels will then be discussed. Various preparation methods of adaptive polymeric gels, including grating, plasma treatment, *r*-ray and UV irradiation and free radical polymerisation methods are introduced as well. In the last section of the chapter, the properties of adaptive polymeric gels and their applications are summarised.

Keywords: adaptive polymeric gels; hydrogel; volume phase transition; swelling/deswelling; drug delivery; grafting

3.1 Introduction

This section tracks the concepts and development of polymeric gels. The design of novel hydrogel networks for future applications is also stated. In addition to describing the mechanisms of gelating, the fabrication and modeling of several intelligently designed hydrogel systems for various applications are discussed. These are of great interest for the research of hydrogels, both medically and commercially. Conventional hydrogels change little in swelling with environmental condition changes while stimuli-responsive hydrogels may swell or deswell sharply with relatively small changes in the environment.

3.1.1 *Introduction of Gels*

The *Polymer Science Dictionary* defines 'gel' as 'polymers and its swollen matters with three-dimensional network structures that are insoluble in any solvents'.[1] Gels are liquids whose rheological properties could be altered by the addition of certain gelling agents (gellants) and as a result, their behaviour resembles that of solids. The definition of a gel is not straightforward. D. Jordan Lloyd is quoted by Hermans [2] who wrote in her survey on the problem of gel structures in 1926, that '(t)he colloidal condition, the gel, is one which is easier to recognise than to define'. Bungenberg de Jong [3] defined a gel as a '... system of solid character, in which the colloidal particles somehow constitute a coherent structure'.

Hermans[2] used three propositions to define gels, including that: '(a) they are coherent colloid systems of at least two components; (b) they exhibit mechanical properties characteristic of a solid; (c) both the dispersed component and the dispersion medium extend themselves continuously throughout the whole system'.

Most gels (natural or synthetic) are composed of water, which are called hydrogels. Hydrogels are the most widely and deeply researched among polymer gels. Therefore, when speaking of gels, people generally refer to water-medium gels, or the so-called hydrogels.

There are many polymer gels that show sharp changes in response to external stimuli, such as temperature,[4] pH values,[5] solvents, light intensity/wavelength,[6,7] ionic strength,[8] magnetic field,[9] electrical field,[10] and pressure.[11] Hence, they are called smart or intelligent gels. Adaptive polymeric gels are a type of intelligent gel. According to our explanation, it is a type of gel which can display different responses in accordance to different external stimuli and change properties to adapt to environmental variations. Adaptive gels were first revealed by Zrínyi Miklós.[12] Figure 3.1 is a schematic diagram of the different types of responses of environmentally sensitive gels to external stimuli.

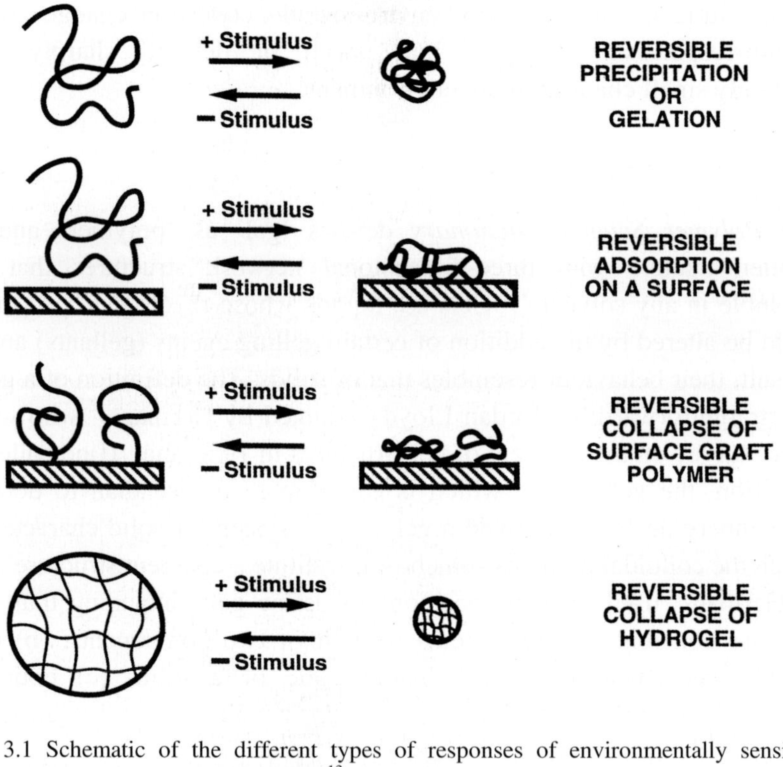

Fig. 3.1 Schematic of the different types of responses of environmentally sensitive polymer gels to environmental stimuli.[13]

3.1.2. *Introduction of polymeric hydrogels*

Hydrogels are polymeric networks which can absorb large quantities of water while remaining insoluble in aqueous solutions due to chemical or physical crosslinking of individual polymer chains. Hydrogels are water swollen polymer networks also. In Wikipedia's definition[14], a hydrogel is defined as a network of polymer chains that are water-soluble, sometimes found as a colloidal gel in which water is the dispersion medium. Hydrogels are superabsorbent (they may contain as much as 99% water) natural or synthetic polymers.

Synthetic hydrogels were firstly established by Wichterle and Lim in 1954,[15] and the growth of hydrogel technologies has developed in many fields ranging from food additives to pharmaceuticals to biomedical implants since then. The developments of an ever-increasing spectrum of functional monomers and polymers also continue to widen the versatilities of hydrogel applications.

Generally speaking, hydrogels are three-dimensional macromolecular networks that contain a large fraction of water within their structure, do not dissolve, and are soft and pliable.[16] These properties are similar to the natural tissue, therefore, hydrogels are particularly useful in biomedical and pharmaceutical applications.[17] Hydrogels have been developed to respond to a wide variety of stimuli. Among the various simulative conditions of sensitive hydrogel, temperature and pH, responsive hydrogels are the most extensively studied experimentally and theoretically due to the importance of these variables in the typical physiological, biological and chemical systems. Temperature/pH-sensitive hydrogels contain both temperature and/or pH-sensitive components. Some hydrogels can reversibly swell or shrink up to 1,000 times in volume in response to a variety of external stimuli.[18]

Temperature responsive hydrogels, such as poly(*N*-isopropylacrylamide) (PNIPAAm) are a well-studied class of systems in drug delivery research.[19,20] Specifically, PNIPAAm undergoes a volume phase transition at a lower critical solution temperature (LCST) of 34°C[21] which is very close to the human body temperature. Below the LCST, it is hydrophilic and swells in water, while above the LCST, it becomes hydrophobic and expels water, collapsing into a smaller volume.

Hydrogels can be prepared from natural or synthetic polymers. Hydrogels made from natural polymers offer several advantageous properties such as inherent biocompatibility, biodegradability and so on. Synthetic polymers usually have well-defined structures that can be modified to yield tailorable degradability and functionality. Table 3.1 lists natural polymers as well as synthetic monomers that are commonly used in hydrogel fabrication.

Table 3.1 Natural polymers and synthetic monomers used in hydrogel fabrication.[22]

Natural polymer	Synthetic monomer
Chitosan	Hydroxyethyl methacrylate (HEMA)
Alginate	*N*-(2-hydroxypropyl) methacrylate (HPMA)
Fibrin	*N*-vinyl-2-pyrrolidone (NVP)
Collagen	*N*-isopropyl acrylamide (NIPAAm)
Gelatin	Vinyl acetate (VAc)
Hyaluronic acid	Acrylic acid (AA)
Dextran	Methacrylic acid (MAA)
	Polyethylene glycol acrylate/methacrylate (PEGA/PEGMA)
	Polyethylene glycol diacrylate/dimethacrylate (PEGDA/PEGDMA)

Polymeric hydrogels have many excellent properties, which make them appropriate for wide applications. Stimuli-sensitive or '*smart*' polymeric hydrogels are hydrogels that may overcome dramatic property changes in response to small changes in the environment. The most important systems are those sensitive to pH or temperature as shown in Table 3.2.[13] Beside the temperature sensitive polymer hydrogel, the pH sensitive hydrogel is another most important type of responsive hydrogel.

Table 3.2. Some types of temperature sensitive polymers.[13]

Ether group	Poly(ethylene oxide), Poly(EO/PO) random copolymers, PEO-PPO-PEO triblock surfactants, Alkyl-PEO block surfactants, Poly(vinly methyl ether)
Alcohol groups	Hydroxypropyl acrylate Hydroxyproyls methylcellulose Hydroxylpropyl cellulose Methylcellulose Poly(vinyl alcohol) derivatives Poly(methacrlic acid)
Substituted amide group	Poly(N-substituted acrylamides) Poly(N-acrylol pyrrolidine) Poly)acryl-L-amino acid amides)
Others	pH sensitive materials polyelectrolytic gels

3.2 Classification and Molecular Structure of Polymeric Gels

3.2.1. Classification of polymeric gels

Polymer gel can be classified based on the types of crosslinking netpoints which create 3D networks, shape and size of the gel configuration, and types of solvents.[23] According to the method of crosslinking foundation, the gels can be divided into covalent and non-covalent bonding gels. According to the configuration size, the gels can be classified into micro and macro-gels. The gel is called aerogel/xerogel, hydrogel or organic gel when the solvent of gels is based on air, water or oil respectively. Because the polymer gel is composited by polymer networks and solvents, it has both the characteristics of a solid and liquid simultaneously. Flory classified gels on the basis of structural criteria into four types: '(a) well-ordered lamellar structures, including gel mesophases; (b) completely disordered covalent polymeric networks; (c) predominantly disordered, but with regions of local order, polymer networks formed with physical aggregation and (d) particulate disordered structures.'.[24]

Polymer gels can be conventionally categorised into two groups according to the types of crosslinking, i.e. chemical and physical gels. The classification of crosslink formation can be generally divided into those formed by a chemical reaction (chemical gels) and those formed by aggregation caused by hydrogen or ionic bonding, and by the physical entanglement of polymer chains (physical gels). Generally, gels formed by chemical bonding cannot be dissolved again, and are thus called irreversible gels. Physical gels, on the other hand, often create gels in a reversible way, by changes in temperature, composition of solvent, and pH and are thus called reversible gels.

According to the source of the material, the environmentally sensitive hydrogels can be generally classified as synthesised gel or natural gel. Synthesised polymer gels include: poly(hydroxyethyl methacrylate), polyacrylamide, poly(*N*-isoproterenol), poly(*N*-Vinyl-2-pyrrolidone), poly(ethylene glycol)-g-poly(propylene glycol)-g-Poly(ethylene glycol) and so on. The natural gels include polysaccharide, protein and peptide, *etc.* The chemical and physical gels are expected to differ significantly in mechanical properties as well as network structure. Multifunctional structural units (monomers) react consecutively and grow like tree branches. When the branches fill the entire space, gelation has been achieved.

In the creation of network structures by chemical bonding(covalent bonding), there are methods: (1) crosslinking at the same time as polymerisation;, and (2) crosslinking by chemical reaction after linear polymer chains have been synthesised.[23]

It is possible to crosslink by intermolecular physical bonding, such as hydrogen, static, coordination, hydrophobic and van der Waals bonding. In general, these interactions form junction zones such as microcrystal helices, ion complexes and micelles as shown in Fig. 3.2.

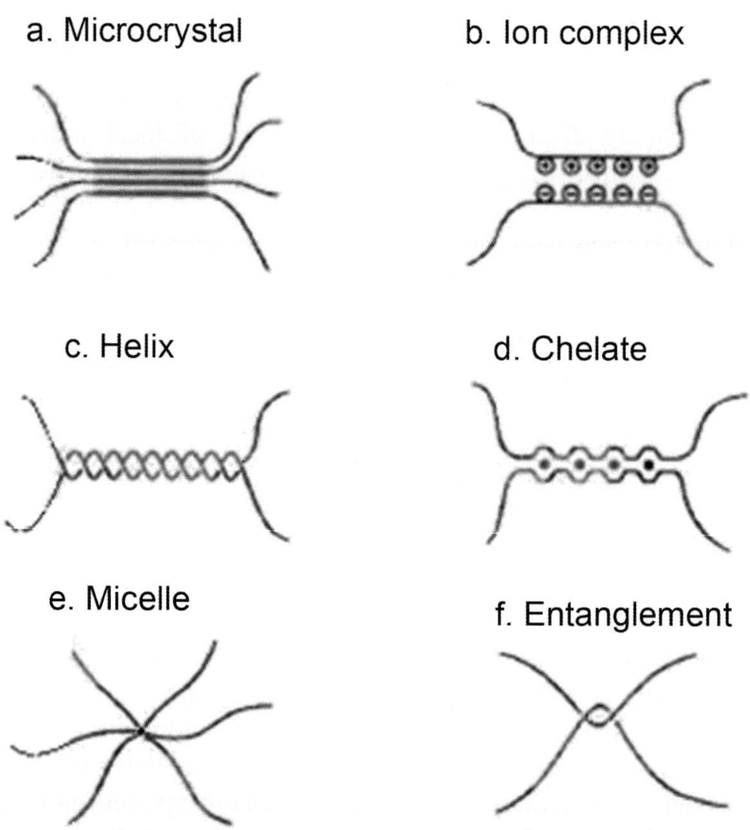

Fig. 3.2 Formation of crosslinking by intermolecular linkage.[23]

Biodegradable hydrogels

Biodegradable hydrogels are favoured over non-degradable gels since they degrade in clinically relevant time scales under relatively mild conditions, which have been studied extensively for their biomedical applications. The application of biodegradable hydrogels depends on the properties of the hydrogels. The most desirable biodegradable hydrogels would be the one which leaves no residuals when applied.

Smart hydrogels

'Smart' or stimuli-sensitive hydrogels can 'sense' the changes in environmental properties such as pH and temperature and respond by increasing or decreasing their degree of swelling. The volume-changing behaviour of 'smart' hydrogels is particularly useful in drug delivery applications as drug release can be triggered upon environmental changes.

Biomimetic hydrogels

One of the principal goals of the biomedical engineering is to design biomaterials capable of replacing, substituting or repairing natural tissues. A drawback of using synthetic and some natural hydrogels for in vivo applications is that they do not reproduce the functional and biological properties of natural tissues. Biomimetic hydrogels are novel biomaterials which are designed to mimic the behaviours of natural tissue. For this reason, adaptive hydrogels are attractive because they can provide a highly swollen 3D environment similar to soft tissues.

3.2.2. ***Gels formation by intermolecular bonding***

The first gel model was suggested by Flory and Stockmayer that the gel is a network by covalent bonding.[25] In this model, in the physical gel, the portion with physical bonding forms clusters in the physical gel. Figure 3.3 shows the common gel network-building mechanisms.

Point (covalent) cross-linking	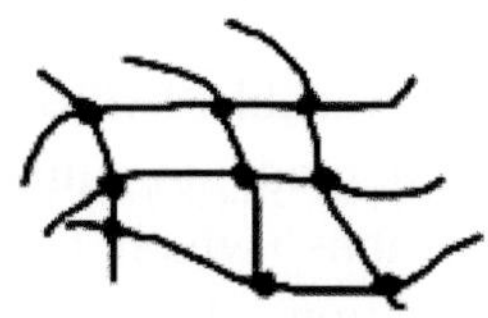
Extended junction zone formation (physical association networks)	
Particle association;	
Spinmodal decomposition	

Fig. 3.3. Common gel network-building mechanism. (a) point(covalent) crosslinking; (b) extended junction zone formation (physical association networks); (c) particle association; (d) spinmodal decomposition.[26]

3.2.3. *Gels design criteria for hydrogels*

Material selection and network fabrication govern the mode of the hydrogels. Table 3.3 lists some important criteria and variables for designing adaptive hydrogels. The transport properties of the hydrogels are affected by their network size. Physical properties of the hydrogel also affect their properties. For example, polymer molecular weights, composition, and polymer/initiator concentrations influence hydrogel swelling and degradation. Finally, the stimuli-responsiveness of a hydrogel network can also influence their physical and chemical properties.

Table 3.3 Design criteria for hydrogels in formulations.[22]

Design criteria	Design variables
Transport properties	• Molecular weight and size of protein
Molecule diffusion	• Molecular weight of polymer
	• Crosslinking density
	• Polymer–protein interactions
	• Hydrogel degradation rate
	• Additional functionalities
Physical properties	• Polymer/crosslinker/initiator concentrations
Gelling mechanisms /conditions	• Temperature, pH, ionic strength
Structural properties	• Molecular weight of polymer
Biodegradability	• Mechanical strength
Stimuli-responsiveness	• Concentration of degradable groups
	• Concentration of responsive groups
Biological properties	• Cytotoxicity of the hydrogel
Biocompatibility	• Capsule formation

Even if a hydrogel formulation is designed with the appropriate physical and transport properties, its special characteristics may still be affected by the relative solvent. Proper material selection, fabrication process and surface texture of the hydrogels are therefore always critical in designing an appropriate hydrogel for different applications.

3.2.4. *Theory and principles of gelation*

In recent years, functional materials have been increasingly investigated at the molecular and/or atomic level. Synthesis and assemblies of various functional polymers have also been researched. Researches on stimuli sensitive hydrogels focus on the synthesis, fabrication and application. With respect to synthesis, the focal point is on stimuli including light, heat or temperature, pH or ion, electric field and chemicals. With respect to fabrication, attention is given to two types of methods: chemical, such as grafting and hybridisation and physical, such as blends, microgels, films, fibres and composites. With respect to industrial applications, many attempts have been made in recent years on the released control of chemicals.[27]

3.3 Synthesis of Adaptive Polymeric Gels

3.3.1. *Fundamentals of gelation*

A hydrogel can be defined as a gel that exhibits the ability to swell in water and retain a significant fraction of water within its 3D structure. Crosslinking polymer chains form the 3Dl network. Crosslinking appoints can be provided by covalent bonds, hydrogen bonding, van der Waals forces or physical entanglements.[28,29] Although 3D crosslinking structures are usually shown by using a lattice model, actually, hydrogels can be rather spread out or coiled up into a ball, depending on its relationship with the solvent. A conceptual diagram for the crosslinking structures is shown in Fig. 3.4.[23] There is the heterogeneous hydrogel, in which the crosslinking units are locally concentrated, and a homogeneous hydrogel with evenly distributed crosslinking units.

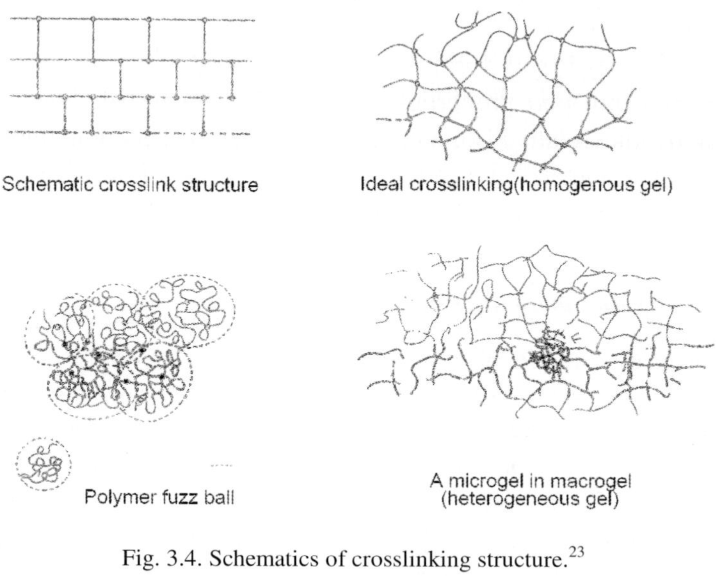

Fig. 3.4. Schematics of crosslinking structure.[23]

3.3.2. *Swelling theory of hydrogels*

The most noted characteristic of hydrogels is that of swelling/deswelling. On responding to external stimuli, hydrogel networks swell by absorbing water or shrink by expelling water until thermodynamic equilibrium is reached. This equilibrium is called swelling equilibrium. Equilibrium is determined by interactions between the hydrogel and water and can respond sensitively to various external conditions, such as temperature, pH and so on. Adaptive polymeric hydrogels may show large discontinuous changes in response to external conditions, such as temperature or solvent composition. Tanaka described the first synthesis of a hydrogel with discontinuous volume phase transition behaviours. The sensitivity of hydrogels towards, for instance, temperature, pH, light, concentration of material and ion, was developed by Tanaka in 1978.[18] The sensitivity of adaptive polymeric gels is based on the reversible and reproducible volume changes by small alterations in the environment. An enormous impact of these so-called stimuli responsive or smart hydrogels on many applications is expected.[30]

3.3.3. *Adaptive sensitive hydrogels*

Hydrogels have been developed in response to a wide variety of stimuli. Among the various simulative conditions of sensitive hydrogels, temperature[18,31-34] and pH responsive hydrogels[35] are the most extensively studied experimentally and theoretically due to the importance of these variables in typical physiological, biological and chemical systems.

A common feature shared by temperature sensitive polymers is that hydrophobic (methyl, ethyl, propyl, *etc.*) and hydrophilic (amide, carboxyl) groups coexist in one macromolecular network simultaneously. Among the temperature sensitive hydrogels, PNIPAAm has attracted considerable attention in the past decades. The main impetus for using PNIPAAm is its novel thermal behaviour in an aqueous media. The transition of the PNIPAAm macromolecules from hydrophilicity to a hydrophilicity structure occurs rather abruptly which is known as the LCST. The LCST of PNIPAAm was first researched by Heskins and Guillet in 1968.[36] Fig. 3.5 is the phase diagram of PNIPAAm in an aqueous solution. Below the LCST, the polymer chain of PNIPAAm is hydrated to an expanded form, which is soluble in water. Above the LCST, PNIPAAm undergoes a reversible phase transition to an insoluble form, and dehydrates to a compact form.

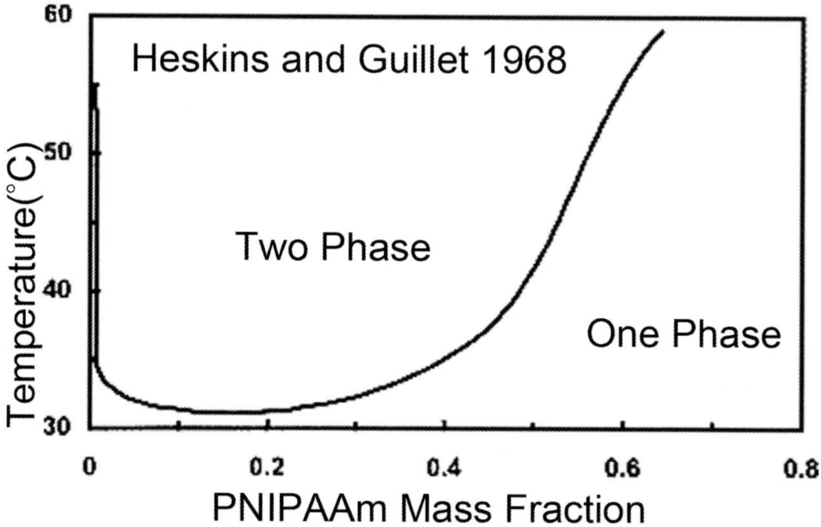

Fig. 3.5 Phase diagram of poly(*N*-isopropylacrylamide)/water.[36]

Recently, an important advancement in PNIPAAm-based hydrogel systems focuses on the mechanistic understanding of phase transition, fine control of the structure-property relationship and novel biomedical applications.[37] By adjusting the PNIPAAm copolymer hydrogel composition and topology, the phase transition could be kinetically and thermodynamically controlled. Hoffman and his colleagues succeeded in demonstrating that the LCST of PNIPAAm can be controlled by incorporating hydrophilic or hydrophobic comonomers.[38] Hydrophilic groups increase the LCST values of copolymers. In contrast, hydrophobic groups reduce the phase transition temperature. Control of deswelling kinetics was also achieved by grafting the copolymer structure.[34] PNIPAAm copolymer hydrogels have been studied for many applications that take advantage of their temperature sensitive properties. Kinetics and thermodynamics of the phase transition might be controlled by well designed molecular parameters. The hydrogel volume is changed ten to 1,000 times by changing temperature, but drastic change is followed by a decrease in the mechanical strength of the hydrogel because of repeated swelling/deswelling and the loose network structure of the hydrogel.[39] Although the stimuli sensitive hydrogel, in particular PNIPAAm, has been widely applied in the medical and biology fields, little attention has been paid to extending them to the textile field.[40]

3.3.4. *Preparation of adaptive hydrogels*

Compared to common hydrophobic polymers, the conditions for fabricating hydrogels are relatively mild. Gels that can be formed at ambient temperatures and organic solvents are rarely required. The fabrication and modelling of adaptive hydrogels are well-developed. In addition to hydrolytically degradable hydrogels, synthetic hydrogels incorporating biological moieties that can be degraded enzymatically are also under investigation. Although hydrogels derived from natural sources can be degraded, many synthetic hydrogels are more favoured because of their tunable physicochemical properties.

3.3.4.1 *Grafting methods onto cellulose based nonwovens*

Cellulose is a most abundant naturally occurring biopolymer and its derivatives have many important applications in fibres, paper and paint industries. Cellulose consists of polydisperse glucose polymer chains, which form supramolecular structures through hydrogen bonding.[41] To control properties, such as hydrophobility, adhesivity,[42] selectivity,[43] drug delivery,[44] wettability,[45] and thermosensitivity,[46] graft copolymerisation of suitable monomers is an indispensable technique for cellulose modification with original excellent properties. Recently, scientists have attempted to develop stimuli-sensitive fabrics or so-called smart textiles by grafting stimuli-sensitive hydrogel on the surface of cellulose. When the stimuli sensitive hydrogel is modified onto the solid surfaces of fabrics, the responsiveness of hydrogels gives a 'smart' surface with varying physical properties that can be controlled by applying external stimuli, such as temperature and pH. Among the hydrogels used for such a purpose, the PNIPAAm hydrogel has attracted considerable attention due to its well-defined LCST in aqueous medium around 32~34°C which is very close to human body temperature.

Among all cellulose sources, cotton cellulose is the most abundant species in nature. Meanwhile, cotton has many appropriate properties for medical applications. The graft copolymerisation of monomers onto cellulose has been carried out by different techniques, such as irradiation with plasma ion beams, gamma rays ultraviolet light and free radical initiation.[47]

3.3.4.2. *Plasma treatment*

Kanamori and his colleagues have prepared a temperature sensitive poly(propylene) (PP) membrane with PNIPAAm(PP-*g*-PNIPAAm) by plasma-induced graft polymerisation.[48] After the formation of radicals on the PP membrane surface by irradiation with argon plasma, the graft polymerisation of a NIPAAm onto a nonwoven PP membrane was performed. The concentration of NIPAAm monomer solution, reaction time of polymerisation, reaction temperature and the average pore size of the PP substrate membrane significantly influence the grafting yield of PNIPAAm. They also studied the relationship between the temperature and water permeability of the PP-*g*-PNIPAAm membrane. The water permeability of PP-*g*-PNIPAAm can be controlled by temperature. When the temperature is below the LCST, the PP-*g*-PNIPAAm membrane exhibits higher water permeability than the original PP substrate membrane. As the temperature increases above the LCST, water permeability gradually decreases. Additionally, the polymer grafting percentage (PGP) of PNIPAAm and the average pore size of the PP substrate influence water permeability. This PP-g-PNIPAAm membrane with unique temperature sensitivity could be useful as a novel device for various applications, such as smart separation membrane.

3.3.4.3. *γ-ray pre-irradiation*

Pre-irradiation has been reported to initiate the graft copolymerisation of NIPAAm hydrogels on cotton fabrics. Liu *et al.* reported the pre-irradiation induced grafting of PNIPAAm onto cellulose fabrics with the LCST being 35.4°C.[49] In Liu's study, temperature sensitive PNIPAAm was grafted onto the surface of cotton cellulose fabrics by γ-ray pre-irradiation. The effects of the absorbed dose, dose rate, grafting temperature and concentration of monomer on the grafting yields were investigated. The main trapped radicals located in the inter-phase between the crystal and amorphous regions, where the grafting reaction occurred, induce the grafting reaction. The results showed that the surfaces of the grafted fabrics have temperature sensitivity as well.

3.3.4.4. *UV-irradiation*

The hydrogel can also be grafted onto fabric surfaces by photo-induced polymerisation. The Kubota group reported the preparation of poly(acrylic acid) and PNIPAAm grafted celluloses by a photo-grafting method using periodic acid oxidation.[50] This binary monomer grafted cellulose might be responsive to pH and temperature changes.

Chen *et al.* even grafted PNIPAAm hydrogels onto nonwoven fabrics by photo-induced graft polymerisation and studied their temperature-responsive characteristics.[51] They grafted temperature sensitive PNIPAAm hydrogels onto plasma-activated polyethylene terephthalate (PET) film and polypropylene (PP) nonwoven fabric surfaces. These grafted gels exhibited an LCST at about 32°C, which showed that the temperature sensitive behaviour of bulk PNIPAAm hydrogels was preserved. The plasma pre-treatment was studied as well. The argon plasma pre-treatment was carried out and subsequent photo-induced surface graft polymerisation was employed to graft NIPAAm. The effects of additives during the grafting were assessed.

The thickness of the stimuli-sensitive hydrogel layer generated by the photo polymerisation or the plasma treatment polymerisation was very thin, which limited some potential applications, such as wound dressing and cosmetics, which need a certain thickness in the hydrogel to release water, nutrients and drugs, and absorb the exudates of a wound. So, another way to obtain a bulky hydrogel is solution polymerisation of NIPAAm using a redox initiator and a crosslinking agent, which must proceed at a temperature below the LCST.[20]

3.3.4.5. *Free radical polymerisation*

NIPAAm can be grafted onto the cotton fibres through ceric-ion-initiated polymerisation in the presence of a crosslinking agent.[52] In order to demonstrate that a 3D polymer network may be grafted onto cotton cellulose, Xie and Hsieh used a two-step initiation and polymerisation process to prepare a cellulose supported temperature sensitive hydrogel.[53] The obtained cellulose supported hydrogels which exhibited thermally responsive swelling and deswelling behaviours. The study demonstrated that cellulose supported hydrogel exhibited lower extents of phase transition over a wider temperature range (28~40°C) than the free PNIPAAm hydrogel (32°C). These findings supported the fact that hydrogels can be used to control the pore structure of cellulose and can be supported with cellulose fibres. A temperature sensitive PNIPAAm hydrogel layer was bonded on the surface of cellulose which was supported by a two-step initiation polymerisation, and the pore sizes of the resulting membrane could be adjusted by the swelling of PNIPAAm in response to temperature variation. *N,N′*-methylenebisacrylamide (MBA), a commercially available reagent, is hydrophilic and water-soluble, and often used as a crosslinking reagent for acrylamide and *N*-isopropyl acrylamide in an aqueous system.[54] MBA is a type of considerable crosslinking agent used in this investigation to improve the mechanical properties without depressing the hydrophilicity of the copolymer hydrogel. The binding extents of hydrogel on the cellulose solids could be controlled by varying in the concentrations of the monomer and crosslinking agent as well as the NIPAAm/MBA solution-to-solid ratios. The cellulose supported hydrogels exhibited a swelling/deswelling behaviour in response to the temperature changes. These initiation and polymerisation procedures offer an approach for reinforcing hydrogels with solid support and modifying cellulose membranes whose pore sizes can be controlled by the swelling of hydrogels.

Gupta *et al.* investigated the grafting of copolymerisation NIPAAm onto cellulose in the presence of ceric (IV) ammonium nitrate (CAN) as the initiator at 25.0 ± 0.1°C in an acidic medium, and grafting parameters were evaluated as a function of experimental conditions.[41] A significant increasing trend below the LCST of PNIPAAm was shown by the grafting of copolymerisation NIPAAm onto cellulose. The PNIPAAm-grafted cellulose displayed improved thermal stability and temperature-dependent swelling degree. Variation in the degree of swelling of PNIPAAm-grafted cellulose as a function of temperature has been used to determine the

LCST. Their study provided a feasible way to test the swelling/deswelling ratio of PNIPAAm grafted cotton cellulose for the future studies.

In our group, we have used ammonium persulfate (APS) as the initiator and TEMED as the accelerator to copolymerise the NIPAAm and polyurethane to obtain a temperature/pH dual sensitive hydrogel that is grafted with nonwoven material. These obtained composites can respond to temperature and pH as well. The morphology of a hydro-gel/nonwoven composite is influenced by a monomer concentration. There may be two mechanisms underlying the modification of a copoly-merisation on the surface of nonwoven fabrics. One is the free radical graft copolymerisation of NIPAAm and PU with polysaccharide initiated by APS.[34] A probable mechanism involved might be as follows:

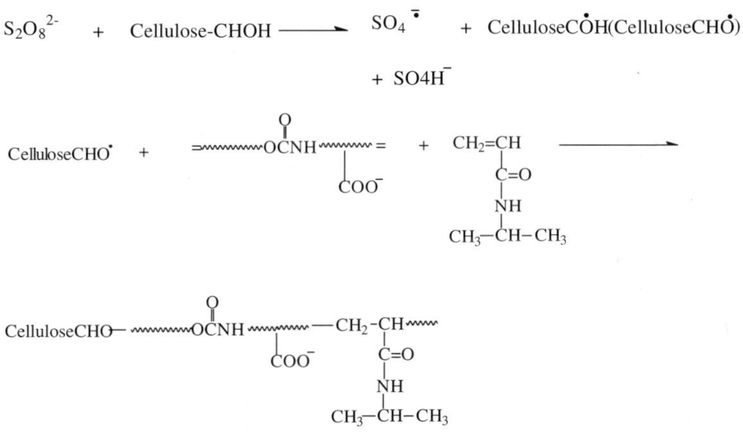

This may not reflect the real reaction procedures due to the random nature of graft copolymerisation aforementioned. Another mechanism involved might be the physical entrapment and anchoring. At a low monomer concentration, physical mechanism is negligible; thus a small amount of ungrafted copolymer is readily washed away. However, at a higher concentration, apart from chemical bonding, more of monomers impenetrate into the inner or the interspaces of fibres, where polymerisation occurred, and consequently, macromolecular chains are entangled and fixed among the fibre networks. With polymerisation processing, a thin or thick layer of hydrogels is generated throughout the fibre network.

3.4 Properties and Applications of Adaptive Polymeric Gels

3.4.1. *Volume phase transition of polymeric gels*

Polymeric gels have many special properties which make them applicable for many fields. One of the most remarkable characteristics of polymer gel is swelling/deswelling. Polymeric gels will swell and expand and the entropy of macromolecule chains among the crosslink points decreases when the solvent is penetrated into the 3D networks. Moreover, the polymeric gels shrink with the increase of elasticity tension between the intermolecules. If there are independent ionised groups among the networks, the difference in ion concentration between the external and internal networks will affect the swelling/deswelling equilibrium. The swelling behaviours of adaptive polymeric gels are related to the properties of the solute and solvent, concentration, pressure and degree of crosslinking.

The volume phase transition (VPT) of polymeric gels is defined as the incontinuous changes in volume of gel in response to various external conditions, such as ion strength, temperature, pH value, light, electric field and solvent components. Polymeric gels can swell or shrink mutually. The volume of a polymeric gel can be changed discontinuously as much as 1,000 times of its original size or more than when surrounding conditions vary continuously, such as solvent component, ionic concentration, pH value and temperature, which is similar to the liquid-gas transition of substances. The smart behaviours of polymeric gels are the basis of their incontinuous VPT. The following is an example used to illustrate this phenomenon. i.e. the volume of poly(acrylamide) gel will vary with the changes in volume fraction of acetone in a water-acetone solvent. It can be seen from Fig. 3.6, that the gels shrink with an increase in acetone concentration. When the concentration of acetone reaches a critical degree, the gels shrink sharply, i.e. swelling changes to deswelling.

The threshold of gels from continuous volume changes to incontinuous volume changes is a critical point. With an increase in acetone concentration, the hydrogen bonds between the water molecules and acylamino are broken, and the hydrophilic affinity among the polymer network and water molecules becomes weak. This volume phase transition is caused by the deference of the solvent constituent.

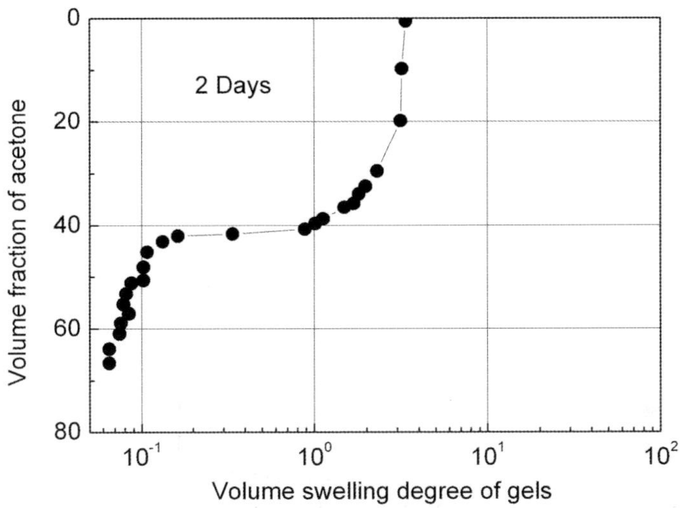

Fig. 3.6 The VPT of poly(acrylamide) gels in the acetone-water solution.

The volume phase transition of poly(*N*-isopropylacrylamide) is a process of balancing the hydrophilic/hydrophobic. In the PNIPAAm gel-water system, the network constringes to a stable state and the entropy decreases with the increase of temperature. This phenomenon is seemingly contravened to the thermodynamics theory. Actually, in the PNIPAAm gel-water system, the gel swells and polymer chains expand in water when the temperature is below LCST. During this process, some water molecules are formed as a hydrogen bond, and some other water molecules are structured around the hydrophobic groups, such as isopropylacryl groups, and cause the macromolecules chains to become hydrophilic. These structured water molecules act cooperatively to form a cage-like structure, which leads to a good solubility of the whole gel system. When the temperature reaches the LCST, these hydrogen bonds are destroyed. Thus the hydrophobic groups are naked, the hydrophobic interactions become strong, and polymer chains begin to dehydrate. When the temperature is above the LCST, the hydrophobic interactions become dominant causing polymer chains to collapse and aggregate abruptly.

3.4.2. *Properties of adaptive hydrogels*

The most unique characteristic of adaptive hydrogel is the reversible swelling/deswelling transition. The high moisture absorption of the hydrogels deems them suitable for use as sanitary products.

Another important property of hydrogels is their ability to exhibit phase transition due to an open structure that interacts with the external environment. Hydrogels can contain energy and information, such as pH, temperature, electric fields and chemical compounds. The open system morphology allows hydrogels to be used in the controlled release of substances. The behaviours of hydrogels can be controlled by using external physical or chemical signals. The microenvironment of a hydrogel may be applied as material exchange, separation[55] or active transfer, as well as for chemical reactions. The applications include electrophoresis substrates, ion exchange resin, and culture substrates based on the characteristics of hydrogels. Pattern formation and nonlinearity of hydrogels are caused by the makeup of a particular microenvironment. Therefore, adsorption, desorption and transport of materials are all possible because of the open system of hydrogel. Table 3.4 summarises the characteristics, function and application examples of hydrogels.

Table 3.4. Characteristics, function and examples of hydrogels.[55]

Characteristics	Function and examples
Water absorption, sustained water	Superabsorbent material, paper diapers, feminine products, oil-water separation materials
Moisture absorption	Drying agents, antifrost agents
Sustained release	Drug substrates, horticultural water holding agents, air fresheners
Absorption of materials, permeation	Impurity removal materials, selective separation membranes, chromatographic column materials, ion exchange resins
Stabilisation of suspension, increased viscosity	Food materials, cosmetics, culture substrates
Transparency, light transmission	Artificial lens, optical lenses, display materials
Biocompatibility	Cell cultures, artificial skins, contact lenses
Chemomechanical responsiveness	Artificial muscles, actuators, switch elements, shape memory materials

Electrical properties	Sensors, electrodes
Vibration damping, sound absorption	Sound barrier walls, insulators, impact-damping materials, artificial joints
Flexibility, plastic deformation, elasticity, floatability, stretching	Shape-deformation materials, slurrying agents for soil, packing materials
Solidification of water at room temperature, freezing point depression	Digestive aids, fertiliser additives, cool pillows, coolants
Swelling	Toys, packing materials, ceiling materials

The properties of hydrogels are related to their structure. In any definition of hydrogels, they have a 3D structure of infinite structural materials that occupy an entire given spaces. Hydrogel formation is the process that forms an infinite structure by intermolecular interaction, such as covalent bonding, i.e. chemical interaction and hydrogen, hydrophobic, and ionic bonding, i.e. physical interactions. In other words, it indicates that intermolecular interaction extends throughout the space. In the process of hydrogel formation, intermolecular interaction spreads throughout the system. However, interaction spreading is not necessarily homogenous. The structure of the hydrogels could be understood for the first time by determining the distribution and bonding states of these domains.

We may understand the structure of hydrogels through the static and dynamic structures. For the static structure, the first classical model of the hydrogel structure that was theorised rigorously using a model by Flory and Stockmayer is formed by covalent bonding.[56] In this model, multifunctional structural units (monomers) react consecutively and grow like tree branches. When the branches fill the entire space, gelation is achieved.[57] The structural models of hydrogels discussed are necessarily random and homogeneous because they are considered statistically and theoretically. In such models, long-range interaction deals with mean-field approximation and will not be the cause of inhomogeneous structures. On the other hand, gelation is incorporated into the general theory of critical phenomenon in which the hydrogel point is regarded as the critical point. The response can be understood well by considering that gelation is the state where the bonding of structural elements, whether chemical or physical, is extended throughout the system. As for the dynamic structure, one of the characteristic properties of hydrogels is its modulus. Therefore, hydrogels can be defined from their mechanical properties. Stated simply, hydrogels are viscoelastic solids that possess the plateau of the real par G_i (storage modulus of complex modulus G^*) in a wide frequency range.[58] Rigorously constructed theoretical models assume a homogeneous structure. Real hydrogels are heterogeneous due to various reasons and their heterogeneity can be understood to some extent by the hierarchical construction of homogenous hydrogel structure models.

Hydrogels can be prepared through copolymerising by two different monomers[59] by forming inter-penetrating polymer networks,[20] or creating networks with microporous structures.[60] They can be made in bulk or particle forms. The bulk gels are easy to handle, but usually have very slow swelling rates. Gel nanoparticles react quickly to external stimuli, but may be too small for some special practical applications. The general preparation methods of hydrogels are plasma treatment, γ-ray pre-irradiation, UV-irradiation and free radical polymerisation technology. Every technology has its merits and shortcomings.

Temperature sensitive hydrogels have been studied most extensively and their unique applications have been reviewed previously.[34] For convenience, temperature sensitive hydrogels can be classified into negatively thermosensitive, positively thermosensitive and thermally reversible hydrogels. If the polymer chains of hydrogels are not covalently crosslinked, temperature sensitive hydrogels may undergo sol-gel phase transitions instead of swelling/deswelling transitions. The thermally reversible hydrogel with inverse temperature dependence become sol at higher temperatures. Temperature sensitive hydrogels can also be made using temperature sensitive crosslinking agents. A hybrid hydrogel system was assembled from water-soluble synthetic polymers and a well-defined.[61] A substitute design of the temperature sensitive hydrogels may be obtained by adding temperature sensitive crosslinking agents.

Our research work aims to develop a smart textile with tunable water absorption in response to the environment. Figure 3.7 displays the relationship between the EWC and temperature. As shown in Fig. 3.7(a), for the original fabrics, and GCE05 and GCE15 without hydrogel, the water absorption is independent of temperature, while at a lower PGP, the EWC of GCE25 and GCE30 shows slight temperature sensitivity. The EWC begins to decrease slightly owing to the shrinkage of PNIPAAm when the temperature is above the phase transition temperature. For

GCE40 with a high content of grafted polymer, prominent temperature responsiveness is observed. When approaching 32°C, the EWC drops dramatically. However, the hydrogel is still in a moderate swelling state. When the temperature decreases (Fig. 3.7(b)), the EWC of the hydrogels grafted nonwoven fabrics increases back to the original state, indicating a reversible temperature sensitiveness.

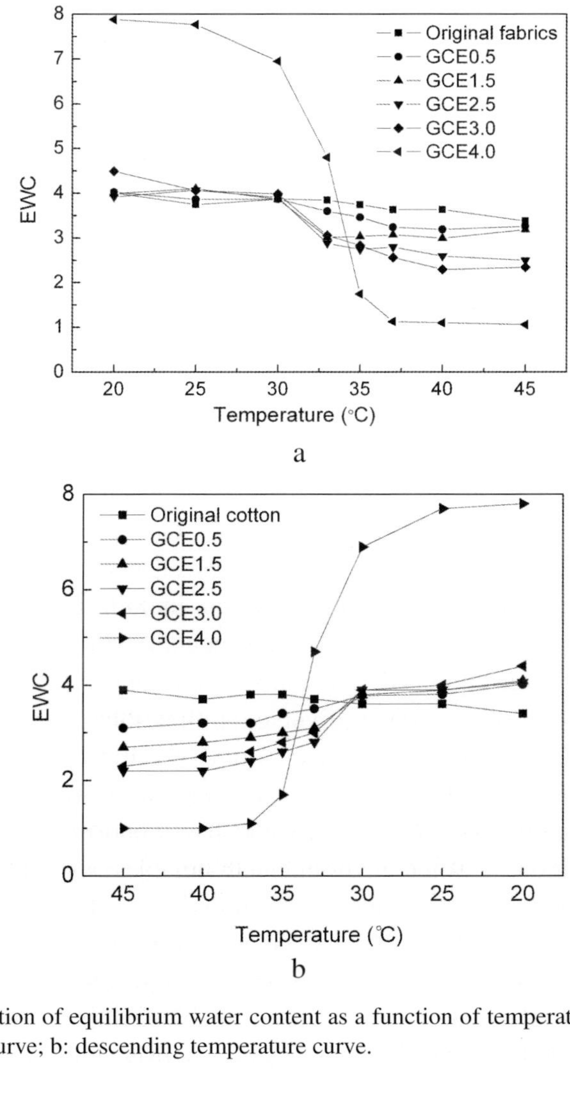

Fig. 3.7 Variation of equilibrium water content as a function of temperature. a: ascending temperature curve; b: descending temperature curve.

3.4.3. *Applications of hydrogels*

Recent research on hydrogels has been conducted broadly at both a fundamental and the applied level. Hydrogel can be grafted onto the substrate surface in many forms, such as polymer chains, polymer loops, networks and so on (See Fig. 3.8).

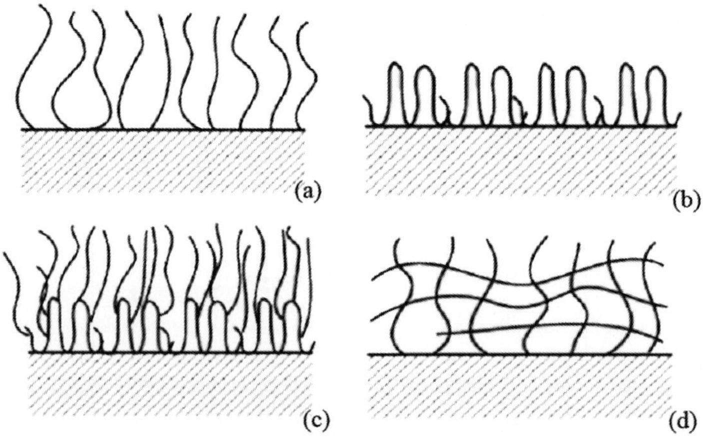

Fig. 3.8 Types of stimuli sensitive polymer chain on surface of substrate (a) Freely polymer chain end grafted. (b) Polymer looped chain grafted. (c) Polymer free end grafted onto loops. (d) Thin hydrogel grafted.[62]

Hydrogel can be applied in many fields, including in medicine.[63,64] Numerous applications of hydrogels, especially for making medical devices, have been attempted by many researchers.[34,36,55,59,65-67] Smart hydrogels have been used in diverse applications, such as making artificial muscles,[68] chemical valves,[10] immobilisation of enzymes[69] and cells,[59] and concentrating dilute solutions in bioseparation.[70]

Environmentally sensitive hydrogels have enormous potential in various applications.[20] Some environmental variables, such as low pH and elevated temperatures, are found in the body. Either pH-sensitive and/or temperature-sensitive hydrogels can be used for site-specific controlled drug delivery.[71] Hydrogels that are responsive to specific molecules, such as glucose[10] or antigens,[72] can be used as biosensors as well as drug delivery systems. Light-sensitive, pressure-responsive and electro sensitive hydrogels also have a potential to be used in drug delivery and bioseparation. Although the concepts of these environmentally sensitive hydrogels are sound, practical applications require significant improvements in hydrogel properties. Usually hydrogels are too fragile and do not have enough mechanical strength in many applications. Synthesis of new polymers and crosslinking agents with more biocompatibility and better biodegradability would be essential for successful applications. Hydrogels have been used in a range of biomedical applications, such as contact lenses. Although such hydrogels are relatively biocompatible, their further use as biomaterials has been restricted because the mechanical properties of these materials are rather poor. The mechanical properties of hydrogels have been the subject of a number of studies.[73]

Development of environmentally sensitive hydrogels with smart properties is a formidable challenge. PNIPAAm copolymer hydrogels are still completely reversible even after more than 30 heat-and-cool cycles.[74] This excellent property adds more incentive for selecting the NIPAAm as one of the polymer hydrogel monomers. By using the temperature-responsive properties of the PNIPAAm, pulsatile drug delivery devices may be triggered intelligently by a minor temperature change. An on-off drug release profile in response to a stepwise temperature change has been observed in PNIPAM and its derivatives.[75]

3.4.4. *Applications of adaptive hydrogels*

The most important temperature sensitive hydrogels researched are negatively thermosensitive reversible gels. Thermosensitive monolithic hydrogels are used to obtain an on-off drug release profile in response to stepwise temperature changes. Hydrophobic comonomers butyl methacrylate (BMA) were introduced into NIPAAm hydrogels to increase the mechanical strength. These hydrogels include crosslinked P(NIPAAm-*co*-BMA),[76] interpenetrating polymer networks (IPN) of PNIPAAm and poly(tetramethyleneether glycol) (PTMEG).

Hydrogels now play a critical role in many tissue engineering scaffolds, biosensor and bio-microelectromechanical system (BioMEMS) devices and drug carriers. A successful drug delivery device not only relies on an intelligent network design, but also on accurate priori mathematical modeling of drug release profiles.

Thermosensitive polymeric hydrogels with LCST close to the human body temperature have physiological value, i.e. PNIPAAm. When an external stimulus is applied, such as temperature, the PNIPAAm hydrogel will give various responses in shape, surface, sol-gel transition or solubility.

Aside from hydrophilic and hydrophobic changes, the graft chains can be acted as a molecular valve to regulate the permeation characteristics in separations field as shown in Fig. 3. 9.

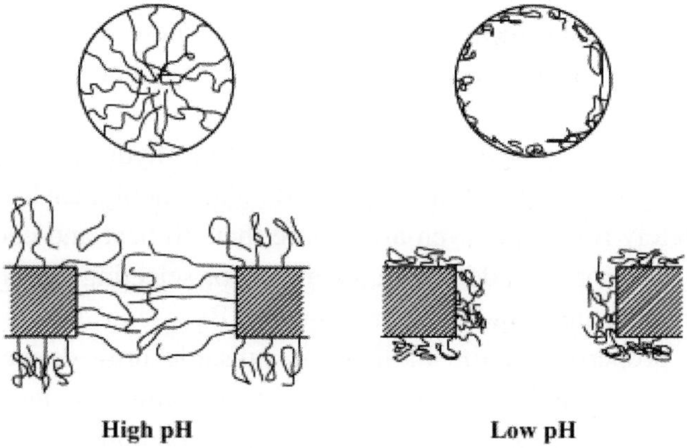

Fig. 3.9. Chemical separation of pH sensitive polymer.[77]

Hydrogels are commonly used as burn and wound dressing materials.[78] They prevent microbial contamination, inhibit the loss of body fluids and provide free flow of oxygen to the wound and generally accelerate the healing process.

Temperature sensitive hydrogels can also be placed inside a rigid capsule containing holes or apertures. As shown in Fig. 3.10, the on-off release is achieved by the reversible volume change of temperature sensitive hydrogels.[79] Such a device is called a squeezing hydrogel device because the drug release is affected by the hydrogel dimension. In addition to temperature, hydrogels can be made to respond to other stimuli, such as pH. In such a system, the drug release rate is proportional to the squeezing rate of the drug-loaded polymer.

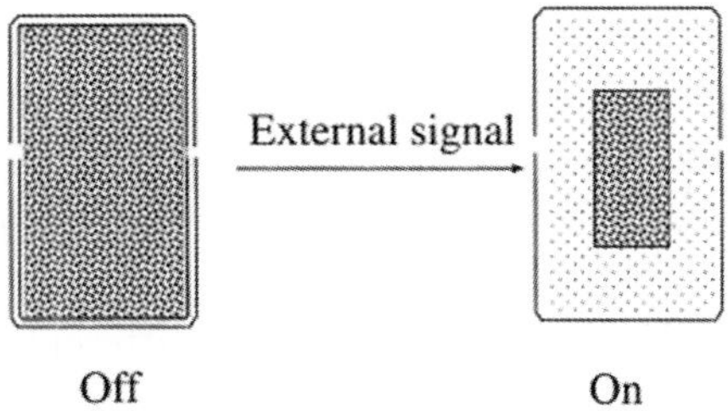

Fig. 3.10 Schematics of on-off release from a squeezing hydrogel device for drug delivery.[80]

Temperature sensitive hydrogels can be secured by placing them into a rigid matrix or grafting them onto the surface of rigid membranes. A composite membrane is prepared by dispersing PNIPAAm hydrogel microparticles into a crosslinked gelation matrix.[81]

Several unique properties that hydrogels possess make them useful in delivering biomolecules. For example, stimuli responsiveness can be easily tailored into hydrogel networks during fabrication. Hydrogels can also be engineered to exhibit bioadhesiveness to facilitate drug targeting, especially through mucus membranes, for non-invasive drug administration. Both natural polymers (*e.g.* chitosan) and synthetic monomers (*e.g.* AA) provide this advantageous property. Some bioadhesive polymers have been used to fabricate hydrogels for oral and buccal drug delivery.

Another prospect of hydrogels is their role as scaffolding materials in tissue engineering applications. An excellent example is cartilage tissue engineering. Controlled release of encapsulated growth factors and other agents from these tissue constructs is critical to providing the necessary cues for cell migration, differentiation and others. As the discussion, adaptive hydrogels have a unique combination of characteristics that make them useful in many applications.

3.5 Summary

The chapter has addressed the fabrication, properties and applications of adaptive polymeric gels. The water of hydrogels is mainly in the 3D network structure. Adaptive polymeric hydrogels can swell/deswell under stimuli. The definition and classification of gels, hydrogels and adaptive gels have been discussed comprehensively. Hydrogel is a colloidal gel in which water is the dispersion medium. Adaptive gel products with certain medical properties and functions have always been indispensable elements in all fields from personal hygiene to intensive care hospital environments. Contrary to such functional products, such as wound dressings, diapers, face masks, etc, hydrogels can be prepared in different technologies. Truly smart materials can combine the function of medical research with the comfort and user-friendliness of apparel products. Combined advancements in different fields, such as putting together adaptive polymers, textiles, advanced materials processing, microelectronics, sensors, telecommunications or informatics, means that more and more adaptive polymeric gels will be produced and the application range of adaptive polymeric hydrogels will be widened accordingly. With the development of crosslinking multi-disciplines, increasingly more achievements in adaptive polymeric gels are possible achieved.

References

1. Alger, M. S. M., ed. *Polymer science dictionary*. 1989, London ; New York : Elsevier Applied Science. 181.
2. Hermans, P. H., *Gels*. Colloid science II, ed. H. R. Kruyt. 1949, Amsterdam: Elsevier. pp. 483-651.
3. Bundenberg de Jong, H. G., *A survey of the study objects in this volume*. *Colloid Science* II, ed. H. R. Kruyt. 1949, Amsterdam: Elsevier Publ. Co., Inc. pp. 2-17.
4. Mukae, K., Bae, Y. h., Okano, T., and Kim, S. W., New thermo-sensitive hydrogel. Poly(ethylene oxide-dimethyl siloxane-ethylene oxide)/poly(N-isopropyl acrylamide) interpenetrating polymer networks I. Synthesis and characterization. *Polymer Journal*, 1990. 22(3): pp. 206-217.
5. Brannon-Peppas, L. and Peppas, N. A., Equilibrium swelling behavior of pH-sensitive hydrogels. *Chemical Engineering Science*, 1991. 46(3): pp. 715-722.
6. He, H., Li, L., and Lee, L. J., Photopolymerization and structure formation of methacrylic acid based hydrogels: The effect of light intensity. *Reactive and Functional Polymers*, 2008. 68(1): pp. 103-113.

7. Stromberg, N. and Hulth, S., A fluorescence ratiometric detection scheme for ammonium ions based on the solvent sensitive dye MC 540. *Sensors and Actuators B: Chemical*, 2003. 90(1-3): pp. 308-318.
8. Rodriguez, R., Alvarez-Lorenzo, C., and Concheiro, A., Cationic cellulose hydrogels: kinetics of the cross-linking process and characterization as pH-/ion-sensitive drug delivery systems. *Journal of Controlled Release*, 2003. 86(2-3): pp. 253-265.
9. Xulu, P. M., Filipcsei, G., and Zrinyi, M., Preparation and responsive properties of magnetically soft poly(*N*-isopropylacrylamide) gels. *Macromolecules*, 2000. 33(5): pp. 1716-1719.
10. Eddington, D. T. and Beebe, D. J., Flow control with hydrogels. *Advanced Drug Delivery Reviews*, 2004. 56(2): pp. 199-210.
11. Meyvis, T. K. L., Stubbe, B. G., Van Steenbergen, M. J., Hennink, W. E., De Smedt, S. C., and Demeester, J., A comparison between the use of dynamic mechanical analysis and oscillatory shear rheometry for the characterisation of hydrogels. *International Journal of Pharmaceutics*, 2002. 244(1-2): pp. 163-168.
12. http://www.kfki.hu/chemonet/hun/teazo/interju/bme/zrinyi.html.
13. Hoffman, A. S., Bioconjugates of intelligent polymers and recognition proteins for use in diagnostics and affinity separations. *Clinical Chemistry*, 2000. 46(Sep.): pp. 1478-1486.
14. Wikipedia, http://en.wikipedia.org/wiki/Hydrogel.
15. Wichterle, O. and Lim, D., Hydrophilic gels for biological use. *Nature*, 1960. 185(4706): pp. 117-118.
16. Huang, G., Gao, J., Hu, Z., St. John, J. V., Ponder, B. C., and Moro, D., Controlled drug release from hydrogel nanoparticle networks. *Journal of Controlled Release*, 2004. 94(2-3): pp. 303-311.
17. Peppas, N. A. and Langer, R., New challenges in biomaterials. *Science*, 1994. 263(5154): pp. 1715-1720.
18. Tanaka, T., Collapse of gels and the critical endpoint. *Physical Review Letters*, 1978. 40(12): pp. 820-823.
19. Hoffman, A. S., Applications of thermally reversible polymers and hydrogels in therapeutics and diagnostics. *Journal of Controlled Release*, 1987. 6(1): pp. 297-305.
20. Qiu, Y. and Park, K., Environment-sensitive hydrogels for drug delivery. *Advanced Drug Delivery Reviews*, 2001. 53(3): pp. 321-339.
21. Hirotsu, S., Hirokawa, Y., and Tanaka, T., Volume-phase transitions of ionized *N*-isopropylacrylamide gels. *Journal of Chemical Physics*, 1987. 87(2): pp. 1392-1395.

22. Lin, C. C. and Metters, A. T., Hydrogels in controlled release formulations: Network design and mathematical modeling. *Advanced Drug Delivery Reviews*, 2006. 58(12-13): pp. 1379-1408.
23. Osada, Y. and Kajiwari, K., eds. *Gels Handbook*. Vol. 1. 2001. 4.
24. Flory, P. J., Gels and gelling processes. *Disc. Faraday Soc*, 1974. 57: pp. 7-18.
25. Flory, P. J., *Principles of Polymer Chemistry*. 1953: Cornell University Press.
26. Schwartzberg, H. G. and Hartel, R. W., *Physical chemistry of foods*. 1992: Marcel Dekker, ISBN 0-08247-8693-9, New York, p. 269.
27. Hirasa, O., Research trends of stimuli-responsive polymer hydrogels in Japan. *Journal of Intelligent Material Systems and Structures*, 1993. 4(4): pp. 538-542.
28. Kamath, K. R. and Park, K., Biodegradable hydrogels in drug delivery. *Advanced Drug Delivery Reviews*, 1993. 11(1-2): pp. 59-84.
29. Barman, S. P., Rajaratnam, J. B. M., Coury, A. J., and Pathak, C. P. *Biodegradable hydrogels for temperature sensitive drug delivery*. in *Transactions of the Annual Meeting of the Society for Biomaterials in conjunction with the International Biomaterials Symposium*. 1996.
30. English, A. E., Edelman, E. R., and Tanaka, T., *Experimental Methods in Polymer Science: Modern Methods in Polymer Research & Technology*. T. Tanaka ed. 2000, New York: Academic Press. 547-589.
31. Amiya, T., Hirokawa, Y., Hirose, Y., Yong, L., and Tanaka, T., Reentrant phase transition of *N*-isopropylacrylamide gels in mixed solvents. *Journal of Chemical Physics*, 1987. 86(4): pp. 2375-2379.
32. Hirokawa, Y. and Tanaka, T., Volume phase transition in a nonionic gel. *The Journal of Chemical Physics*, 1984. 81(12): pp. 6379-6380.
33. Kopecek, J., Smart and genetically engineered biomaterials and drug delivery systems. *European Journal of Pharmaceutical Sciences*, 2003. 20(1): pp. 1-16.
34. Yoshida, R., Design of functional polymer gels and their application to biomimetic materials. *Current Organic Chemistry*, 2005. 9(16): pp. 1617-1641.
35. Tanaka, T., Fillmore, D., Sun, S. T., Nishio, I., Swislow, G., and Shah, A., Phase Transitions in Ionic Gels. *Physical Review Letters*, 1980. 45(20): pp. 1636-1639.

36. Heskins, M. and Guillet, J. E., Solution properties of poly(*N*-isopropylacrylamide). *Journal of Macromolar Science-Chemistry*, 1968. A2(8): pp. 1441-1455.
37. Schild, H. G., Poly(*N*-isopropylacrylamide). experiment, theory and application. *Progress in Polymer Science*, 1992. 17(2): pp. 163-249.
38. Chen, G. and Hoffman, A. S., Graft copolymers that exhibit temperature-induced phase transitions over a wide range of pH. *Nature*, 1995. 373(6509): pp. 49-52.
39. Ilavsky, M., Hrouz, J., and Havlicek, I., Phase transition in swollen gels: 7. Effect of charge concentration on the temperature collapse of poly(*N,N*-diethylacrylamide) networks in water. *Polymer*, 1985. 26(10): pp. 1514-1518.
40. Serra, M., Adaptable skin - hydrogel gives wetsuit protection. *Smart Materials Bulletin*, 2002. 2002(7): pp. 7-8.
41. Gupta, K. C. and Khandekar, K., Temperature-responsive cellulose by ceric(IV)ion-initiated graft copolymerization of *N*-isoproprylacrylamide. *Biomacromolecules*, 2003. 4(3): pp. 758-765.
42. Grohmann, B. A., Wallmersperger, T., and Kroeplin, B., Adaptive Strukturen und gekoppelte Mehrfeldprobleme(Adaptive structures and coupled multi-field problems). *Stahlbau*, 2000. 69(6): p. 9.
43. Urayama, K., Ogasawara, S., and Takigawa, T., Pure shear deformation of physical and chemical gels of poly(vinyl alcohol). *Polymer*, 2006. 47(19): pp. 6868-6873.
44. Ohya, Y., Huang, T. Z., Ouchi, T., Hasegawa, K., Tamura, J., Kadowaki, K., Matsumoto, T., Suzuki, S., and Suzuki, M., Synthesis and antitumor activity of α-1,4-polygalactosamine and *N*-acetyl-α-1,4-polygalactosamine immobilized 5-fluorouracils through hexamethylene spacer groups via urea, urea bonds. *Journal of Controlled Release*, 1991. 17(3): pp. 259-266.
45. Lenz, P., Wetting phenomena on structured surfaces. *Advanced Materials*, 1999. 11(18): pp. 1531-1534.
46. Pan, Y. V., Wesley, R. A., Luginbuhi, R., Dento, D. D., and Ratner, B. D., Plasma polymerized *N*-isopropylacrylamide: synthesis and characterization of a smart thermally responsive coating. *Biomacromolecules*, 2001. 2(1): pp. 32-36.
47. Gupta, K. C. and Sahoo, S., Graft copolymerization of acrylonitrile and ethyl methacrylate comonomers on cellulose using ceric ions *Biomacromolecules*, 2001. 2(1): pp. 239-247.

48. Kim, S. Y., Kanamori, T., and Shinbo, T., Preparation of thermal-responsive poly(propylene) membranes grafted with *N*-isopropylacrylamide by plasma-induced polymerization and Their water permeation. *Journal of Applied Polymer Science*, 2002. 84(6): pp. 1168-1177.
49. Liu, J. Q., Zhai, M. L., and Ha, H. F., Pre-irradiation grafting of temperature sensitive hydrogel on cotton cellulose fabric. *Radiation Physics and Chemistry*, 1999. 55(1): pp. 55-59.
50. Kuwabara, S. and Kubota, H., Water-absorbing characteristics of acrylic acid-grafted carboxymethyl cellulose synthesized by photografting. *Journal of Applied Polymer Science*, 1996. 60(11): pp. 1965-1970.
51. Chen, K.-S., Tsai, J.-C., Chou, C.-W., Yang, M.-R., and Yang, J.-M., Effects of additives on the photo-induced grafting polymerization of N-isopropylacrylamide gel onto PET film and PP nonwoven fabric surface. *Materials Science and Engineering: C*, 2002. 20(1-2): pp. 203-208.
52. Hebeish, Z. H. and El-Hilw, Z. H., Preparation of DEAE cotton-*g*-poly(methacrylic acid) for use as ion exchanger. *Journal of Applied Polymer Science*, 1998. 67(4): pp. 739-745.
53. Xie, J. B. and Hsieh, Y. L., Thermosensitive poly(*N*-isopropylacrylamide) hydrogels bonded on cellulose supports. *Journal of Applied Polymer Science*, 2003. 89(4): pp. 999-1006.
54. Das, M., Zhang, H., and Kumacheva, E., Microgels: old materials with new applications. *Annual Review of Materials Research*, 2006. 36: pp. 117-142.
55. Guven, O., Sen, M., Karadag, E., and Saraydin, D., A review on the radiation synthesis of copolymeric hydrogels for adsorption and separation purposes. *Radiation Physics and Chemistry*, 1999. 56(4): pp. 381-386.
56. Flory, P. J., *Principles of Polymer Chemistry*. 1956, Ithaca, New York: Cornell University Press.
57. Ferri, F., Frisken, B. J., and Cannell, D. S., Structure of silica gels. *Physical Review Letters*, 1991. 67(25): pp. 3626.
58. Garaita, M. G. and Kennedy, J. F., Chemical and physical networks. formation and control of properties in: K. te, Nijenhuis, W.J., Mijs (Ed.). 2001. Wiley.
59. Feil, H., Bae, Y. H., Feijen, J., and Kim, S. W., Effect of comonomer hydrophilicity and ionization on the lower critical

solution temperature of n-isopropylacrylamide copolymers. *Macromolecules*, 1993. 26(10): pp. 2496-2500.
60. Kabra, B. G., Gehrke, S. H., and Spontak, R. J., Microporous responsive HPC gels 1. synthesis and microstructure. *Macromolecules*, 1998. 31(7): pp. 2166-2173.
61. Wang, C., Stewart, R. J., and Kopecek, J., Hybrid hydrogels assembled from synthetic polymers and coiled-coil protein domains. *Nature*, 1999. 397(6718): pp. 417-420.
62. Kikuchi, A. and Okano, T., Intelligent thermoresponsive polymeric stationary phases for aqueous chromatography of biological compounds. *Progress in Polymer Science*, 2002. 27(6): pp. 1165-1193.
63. Kiritoshi, Y. and Ishihara, K., Synthesis of hydrophilic cross-linker having phosphorylcholine-like linkage for improvement of hydrogel properties. *Polymer*, 2004. 45(22): pp. 7499-7504.
64. Hoffman, A. S., Hydrogels for biomedical applications. *Advanced Drug Delivery Reviews*, 2002. 54(1): pp. 3-12.
65. Li, F., Liu, W. G., and Yao, K. D., Preparation of oxidized glucose-crosslinked N-alkylated chitosan membrane and in vitro studies of pH-sensitive drug delivery behaviour. *Biomaterials*, 2002. 23(2): pp. 343-347.
66. Anca, M. Y., Bayramoglu, G., Anca, B., Yalcum, E., Ito, K., and Yagci, Y., Novel Hydrogel Membrane Based on Copoly(hydroxyethyl methacrylate/*p*-vinylbenzyl-poly(ethylene oxide)) for Biomedical Applications: Properties and Drug Release Characteristics. *Macromolecular Bioscience*, 2005. 5(10): pp. 983-992.
67. Peppas, N. A. and Sahlin, J. J., Hydrogels as muco-adhesive and bio-adhesive materials: a review. *Biomaterials*, 1996. 17(16): pp. 1553-1561.
68. Ueoka, Y., Gong, J., and Osada, Y., Chemomechanical polymer gel with fish-like motion. *Journal of Intelligent Material Systems and Structures*, 1997. 8(5): pp. 465-471.
69. Helander, I. M., Latva-Kala, K., and Lounatmaa, K., Permeabilizing action of polyethyleneimine on Salmonella typhimurium involves disruption of the outer membrane and interactions with lipopolysaccharide. *Microbiology*, 1998. 144: pp. 385-390.
70. Jeong, B. and Gutowska, A., Lessons from nature: stimuli-responsive polymers and their biomedical applications. *Trends in Biotechnology*, 2002. 20(7): pp. 305-311.

71. Coughlan, D. C., Quilty, F. P., and Corrigan, O. I., Effect of drug physicochemical properties on swelling/deswelling kinetics and pulsatile drug release from thermoresponsive poly(*N*-isopropylacrylamide) hydrogels. *Journal of Controlled Release*, 2004. 98(1): pp. 97-114.
72. Miyata, T., Uragami, T., and Nakamae, K., Biomolecule-sensitive hydrogels. *Advanced Drug Delivery Reviews*, 2002. 54(1): pp. 79-98.
73. Trevett, A. S., *The mechanical properties of hydrogel polymers*. 1995, Aston University. p. 246.
74. Lin, H. H. and Cheng, Y. L., In-Situ thermoreversible gelation of block and star copolymers of poly(ethylene glycol) and poly(*N*-isopropylacrylamide) of barying architectures. *Macromolecules*, 2001. 34(11): pp. 3710 - 3715.
75. Gutowska, A., Bae, Y. H., Feijen, J., and Kim, S. W., Heparin release from thermosensitive hydrogels. *Journal of Controlled Release*, 1992. 22(2): pp. 95-104.
76. Okano, T., Bae, Y. H., Jacobs, H., and Kim, S. W., Thermally on-off switching polymers for drug permeation and release. *Journal of Controlled Release*, 1990. 11(1-3): pp. 255-265.
77. Kato, K., Uchida, E., Kang, E.-T., Uyama, Y., and Ikada, Y., Polymer surface with graft chains. *Progress in Polymer Science*, 2003. 28(2): pp. 209-259.
78. Rosiak, J. M., Ulanski, P., and Rzeznicki, A., Hydrogels for biomedical purposes. *Nuclear Instruments and Methods in Physics Research Section B: Beam Interactions with Materials and Atoms*, 1995. 105(1-4): pp. 335-339.
79. Dinarvand, R. and D'Emanuele, A., The use of thermoresponsive hydrogels for on-off release of molecules. *Journal of Controlled Release*, 1995. 36(3): pp. 221-227.
80. Gutowska, A., Seok Bark, J., Chan Kwon, I., Han Bae, Y., Cha, Y., and Wan Kim, S., Squeezing hydrogels for controlled oral drug delivery. *Journal of Controlled Release*, 1997. 48(2-3): pp. 141-148.
81. Chun, S. W. and Kim, J. D., A novel hydrogel-dispersed composite membrane of poly(*N*-isopropylacrylamide) in a gelatin matrix and its thermally actuated permeation of 4-acetamidophen. *Journal of Controlled Release*, 1996. 38(1): pp. 39-47.

Chapter 4

Adaptive Polymeric Particles and Applications

This chapter provides a comprehensive review on adaptive polymeric particles and their applications. The definition, classification, structure, theory and processing of fabrication, properties and applications are discussed in sequence. Adaptive polymeric particles exhibit a number of unique properties due to their special forms and large surface area. Adaptive polymeric particles are a type of particle which can express a sharp change within a narrow range of external stimuli, such as light, heat or temperature, pH or ion, electric field, chemicals and so on. Researchers have developed two types of fabrication methods: chemical and physical. The synthesis theories have been suggested as well. Stimuli-responsive polymer particles are very promising in the field of academic studies and applied research. Adaptive particles have a wide range of applications in various fields, such as drug delivery, chemical separation, use in biomaterials, textile finishing, self cleaning and so on.

Keywords: adaptive polymeric particles; nanoparticles; core-shell; self cleaning; drug delivery; textile finishing; cosmetics

4.1 Introduction of Polymeric Particles

A particle is a tiny mass of material or can be defined as a conglomeration of discrete matter with properties distinctively different from its bulk. The particle can be formed by solids or liquids. Most of them are composed in a solid state. The simplest shape of particles is a sphere. Due to their unique forms and high surface area/mass ratio, particles exhibit many special characteristics. People may combine the benefits of traditional and particle materials to create a new generation product that can be seamlessly integrated into a complex system. Commonly, a polymer particle is a spheriform matter and could be prepared by a polymerisation process (Fig. 4.1). In order to serve special applications, different adaptive/functional particles could be obtained by controlling the synthesis conditions, selection of composition, optimisation of technology and so on. The properties of particles include size, morphology, surface group attachment, composition, dispersion, etc.[1]

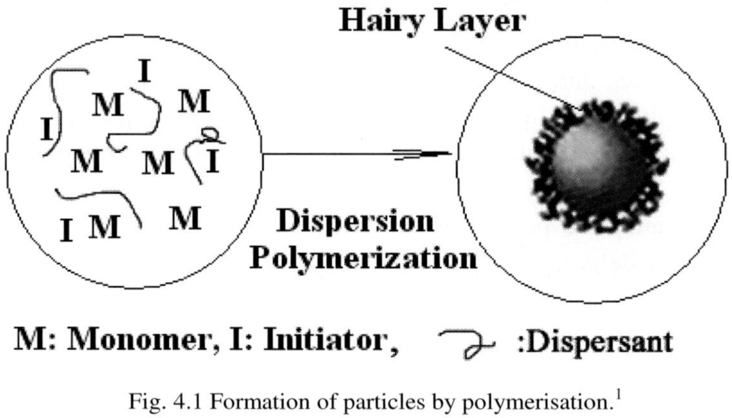

Fig. 4.1 Formation of particles by polymerisation.[1]

Depending on their source and chemical composition, polymeric particles have different sizes, concentrations, properties and functions. The physical and chemical compositions of a particle have strong impacts on its behaviours, properties and applications.

With higher surface area/mass ratio, customised shape and size, special group attachment, mechanical properties (elasticity, conductivity, high surface energy, etc.) and so on, the particles may have many particular properties that differ from those of bulk matter. Many researchers have studied the theory of polymeric particles[2-4], their fabrication,[5,6] properties[7,8] and applications.[3,9,10] Adaptive/functional polymeric particles are widely applied in many areas, including in different fields, such as collided chemistry[11], bio-medicals[12], and environmental,[13] as well as different areas, such as in painting materials,[14] dyeing materials,[10,15] textile and fabrics[16,17] cosmetics,[18] and others.

This chapter will first introduce the classification of adaptive polymeric particles and their properties. Then, the aggregation theory and fabrication technologies are summarised. Some important applications of adaptive polymeric particles are introduced, and the future aspects of these smart particles are provided. As one of the most important particles, nanoparticles are also reviewed in this chapter.

4.2 Classification of Adaptive Polymeric Particles

It is quite difficult to classify the different types of adaptive polymeric particles based on just one concept. The classification of adaptive polymeric particles can be divided into many classes, with respect to size, shape/morphology, distribution/uniformity, composition and external stimuli,[2] as shown in Table 4.1.

In terms of diameter,[2,19-21] particles can be classified as course (40~500 μm), fine (10~100 μm) and nano (1-100 nanometres (nm)). At the small end of the size range, an adaptive polymeric nanoparticle is very important. A nanoparticle is a microscopic particle which is sized between one and 100 nm with one or more dimensions in the order of 100 nm or less. Nanoparticles may or may not exhibit size-related properties which differ significantly from those observed in fine particles or bulk materials. To some degree, nanoparticle matter should be considered a distinct state of matter due to its large surface area and size effects. The fraction of the atoms at the surface in nanoparticles is increased compared to micro-particles or bulk. Compared to bulk matter, nanoparticles have a tremendous surface area.

Nanoparticles can be generally classified based on their dimensionality, morphology, composition, uniformity and agglomeration.[22] With respect to dimensionality, nanoparticles are a type of three-dimensional nanomaterial. Nanoparticles can be classified as spherical, oval, cubic, prism, helical or pillar through morphology. They can be composed of a single constituent or a composite of several materials. Based on their chemistry and electromagnetic properties, nanoparticles can exist as dispersed aerosols, suspensions/colloids or in an agglomerate state (See Fig. 4.2).

Table 4.1 Classification of adaptive polymeric particles.

Classification Factors	Classifications
By Size	Course particles (40~500μm)
	Fine particles (10~100μm)
	Nanoparticles(1-100nm)
By Composite	Microgels
	Micro emulsion
	Polymeric modified particles
	Composite particles
By Distribution	Mono-dispersed particles
	Multi-dispersed particles
By Shape	Spheriform particles
	Dish particles
	Rod particles
	Irregular particles
By Structure	Core-shell particles
	Micro-capsule particles
	Uniform particles
By Stimuli	Temperature sensitive particles
	pH sensitive particles
	Ion sensitive particles
	Light sensitive particles
	Chemical sensitive particles
	Electric sensitive particles
	Duo-sensitive particles
	Multi-sensitive particles

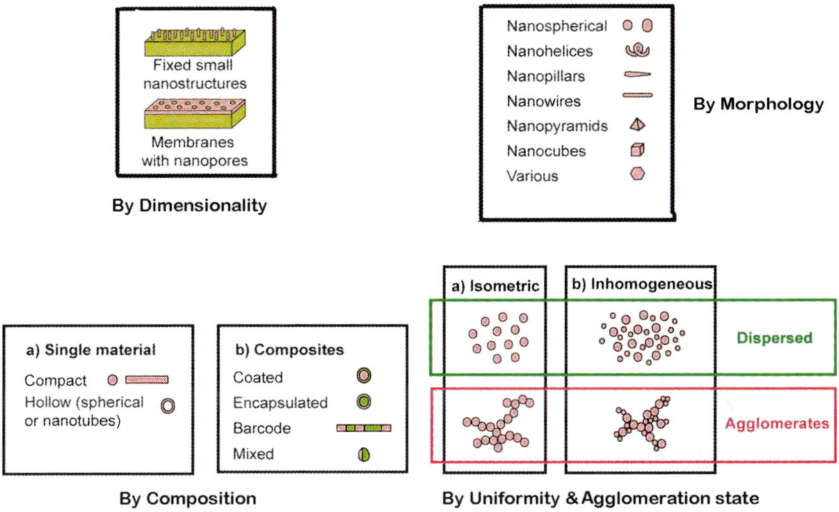

Fig. 4.2 Classification of nanostructured materials with respect to dimension, morphology, composition, uniformity and agglomeration state.[22]

As surface area and size effect are two import aspects of nanoparticles, the aggregation of particles must be considered in applications. If a polymeric particle can express intense changes to external stimuli, such as light,[23] temperature,[24] pH,[20] ions,[5] electric fields, magnetic fields, glucose, etc. or combines two or more of the above,[25] it is called an adaptive polymeric particle.

4.3 Properties of Adaptive Polymeric Particles

There are many unique chemical and physical properties of adaptive polymeric particles, such as size, shape, morphology, light reflection/diffraction, viscosity, solvent ability, absorbency, groups and so on. Many interesting properties of adaptive polymeric particles are the result of their dimension, which lies between atomic and bulk. The surface area and energy of adaptive polymeric particles increase with decreases in their size. Almost all properties are more or less related to the particle size, shape, group attachment and surface properties. Hence, we will introduce four key properties of adaptive polymeric particles: surface chemistry, size, shape and modification of particles.

4.3.1 *Surface properties of polymeric particles*

Surface properties influence the interactions of particles with other matter. The modifications on particles also involve the surface properties of polymeric particles. The surface properties of polymeric particles include hydrophobicity/hydrophilicity, surface activity, van de Vaals force, reflection and diffraction, contact angle, absorbency and so on. The properties of particles are closely related to their distribution, aggregation and dispersant-solvent interactions of particles. The surface energy and interactions of the surface of particles are also important for powder particles. Fowkes and Mostafa[26] focused on particle-dispersant interaction and found that basic polymers adsorb strongly from neutral organic solvents onto acidic sites located on the particles and vice versa.

The characteristics of particles are related to the most fundamental property attributes that are affected by the particle size surface area. As the example shows, dividing a single cubic crystal into 27 cubes of only 1/3 in width increases the surface area of the particle by three times, and a similar increase occurs if we reduce the particle size by as much again. A crude rule of thumb is that surface area increases X times if the average particle size is reduced by 1/X (Fig. 4.3).

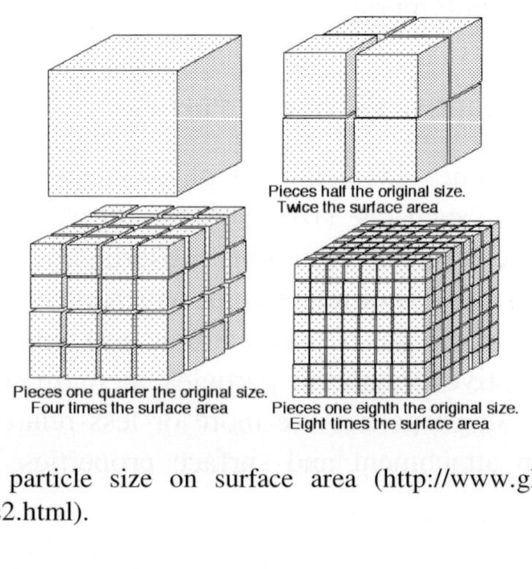

Fig. 4.3 Effect of particle size on surface area (http://www.gly.uga.edu/railsback/-GeologicalDiagrams2.html).

4.3.2 *Size of polymeric particles*

Particle size is the most important property in governing many aspects of particle behaviours. The size of polymeric particles has an important influence on the properties of particles. Bulk material should have constant physical properties regardless of its size, but at nanoscale, this is often not the case. Size-dependent properties are observed, such as quantum confinement in semiconductor particles,[27] surface plasmon resonance[28] in some metal particles and superparamagnetism in magnetic materials.[29] The physical and chemical properties of particles are dependent on relevant aspects of particle behaviour. The linear geometric dimension or equivalent geometric sizes of particles are based on the particle volume or surface area. Particles that are large enough can have their surface or volume considered as continua, but for much smaller particles, their physical nature and particle size needs to be treated differently. The physical, chemical and biological interactions of particles with surroundings need to be reconsidered and the associated particles may change their physical, chemical or even their mechanical properties for aggregation.[30] However, particle size is not a simple attribute. Most particles clump or flocculate into larger chunks called aggregates or agglomerates.

Ford and colleagues stated that drug particle size is important in insoluble drugs, but for soluble drugs, the differences are only noticeable at low levels of hydroxypropyl methyl cellulose (HPMC) and when the drug particle size is large.[31]

The colour of particles in turn, depends on its material size. First among these is the particle size, measured both as (1) the median or average particle size, and (2) the range or distribution of particle sizes. Different particle sizes can be intended for different applications (colouring plastics, ceramics, cements, cosmetics, paints, inks, etc.). There are many properties that are determined by the particle size. Of these, the most important for painting materials are light scattering properties, including the pigment *lightfastness, colour, opacity* and *tinting strength.* Smaller particles present a larger surface area for light to strike, which means that the light scattering effects of the particle surface, and the eroding effects of light absorption, are also increased.

Nanoparticles are of great scientific interest as they are effectively, a bridge between bulk materials and atomic or molecular structures. Bulk material should have constant physical properties regardless of its size.

4.3.3 *Shape of polymeric particles*

Precisely shaped polymeric particles and structures are widely used for applications in photonic materials, biomaterials and self-assembly. The shape of a particle has many influences on its properties. Nonspherical microparticles that use microfluidics have been studied by Patrick's group.[32] Sozzani *et al.* synthesised poly(styrene) (PS) vase-shaped particles through direct replication.[33] Controlled morphogenesis and shape replication are challenges for several rapidly developing fields of materials science. Julie *et al.* studied the effects of particle shape in drug delivery.[34] Xu *et al.* generated monodisperse particles by using microfluidics to control the size, shape and composition of tripro-pyleneglycol diacrylate (TPGDA), particles.[35]

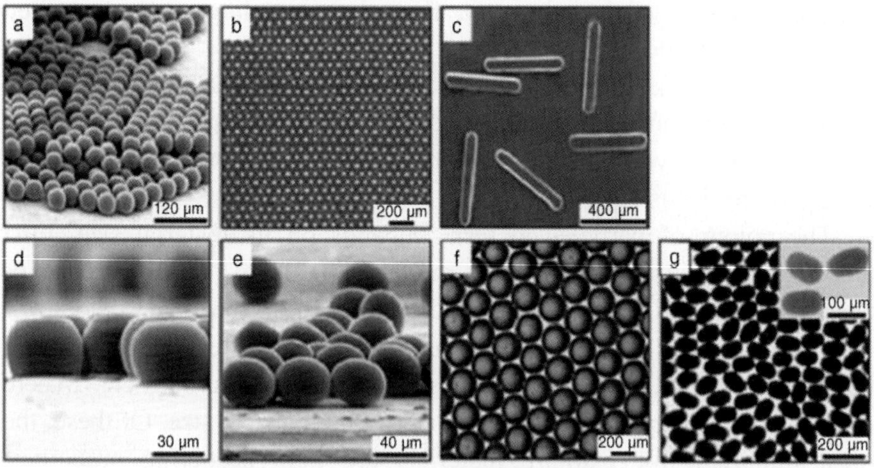

Fig. 4.4. Optical microscopy images of polyTPGDA particles: a) microspheres, b) crystal of microspheres, c) rods, d) disks, and e) ellipsoids. Optical microscopy images of: f) agarose disks and g) bismuth alloy ellipsoids produced using thermal solidification. Inset: micrograph of bismuth alloy ellipsoids at higher magnification.[35]

Exploiting the inhibition of free-radical polymerisation near polydimethylsiloxane (PDMS) surfaces, Dhananjay *et al.* were able to repeatedly pattern and flow rows of particles in less than 0.1 s, affording a throughput of nearly 100 particles per second by using the simplest of device designs.[36]

The second type of fabrication technique uses spherical particles as starting materials and manipulates them into different morphologies. Manoharan *et al.* utilised self-assembly of PS spheres on the surface of an emulsion droplet to form clusters of spheres containing two to 15 particles.[37] Yin *et al.* used template-assisted self-assembly to form clusters and chains of PS spheres by trapping spheres in molded cavities of various shapes and sizes.[38] Alternately, Ho *et al.* stretched spherical PS particles that were embedded in a polymer film to create ellipsoidal particles.[39]

Sun *et al.* recently reported a nanocasting method to make a superhydrophobic PDMS surface as shown in Fig. 4.5.[40] They first made a negative PDMS template using lotus leaf as an original template and then used the negative template to make a positive PDMS template—a replica of the original lotus leaf. The positive PDMS template had the same surface structure and superhydrophobicity as the lotus leaf. Given the difference in composition and consequent surface energy between the lotus leaf (paraffinic wax crystals, –CH2–, 30–32 mN/m) and the PDMS replica (–CH3, 20 mN/m), the similarity of the hydrophobicity obtained is surprising, which is the lotus effect.

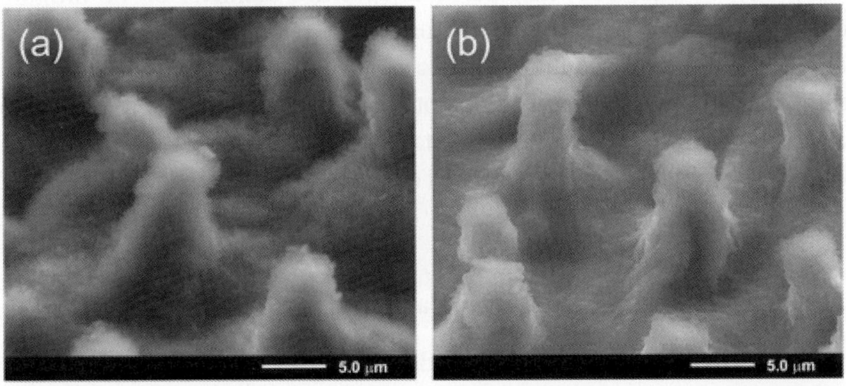

Fig. 4.5 Higher-magnification SEM images of (a) a natural lotus leaf, and (b) its positive replica. They almost have the same surface morphology at both micro- and nanoscale.[40]

The properties of materials change as their size approaches the tiny scale and the percentage of atoms at the surface of a material becomes significant. For bulk materials larger than one micrometre, the percentage of atoms at the surface is minuscule relative to the total number of atoms of the material.

4.3.4 *Modification of polymeric particles*

The properties of polymeric particles can be changed as desired by medication through chemical, physical, mechanical or other techniques. The surface structure, surface energy, electrical properties, surface absorbency and reaction abilities can be modified by medication to serve the requirements of new materials, techniques and technology. The modification of polymeric particles can broaden the application fields of polymeric particles or endorse polymeric particles with novel characteristics. The modification methods include coating, use of a coupling agent, precipitation reaction, surface chemistry, grafting, plasma or ion treatment, micro-encapsulation and many others.

The grafting modification method can combine the advantages of organic particles and polymers to obtain novel functional particles. For example, the surface modified TiO_2 nanoparticles dissolved in toluene were encapsulated in PMMA by *in situ* radical polymerisation of methyl methacrylate initiated by 2,2'-azobisisobutyronitrile.[41] The surface modification of the TiO_2 nanoparticles (average diameter of 4.5 nm) was achieved by the formation of a charge transfer complex between TiO_2 nanoparticles and 6-palmitate ascorbic acid. The glass transition temperature of the polymer was not influenced by the presence of the nanoparticles while the thermal stability was significantly improved. Sometimes, copolymerisation of two or more mononers can obtain special functional particles.

4.4 Manufacturing of Polymeric Particles

Many of polymeric particles are made directly as dry powders, and it is a common myth that these powders will stay in the same state when stored. In fact, they will rapidly aggregate through a solid bridging mechanism in as little time as a few seconds. Whether these aggregates are detrimental will depend entirely on the application of the material. If the particles need to be kept separate, then they must be prepared and stored in special conditions.

4.4.1 *Aggregation theory*

Equilibrium thermodynamics has nothing to say about the rate at which processes occur. It is a fact that much two-phase dispersion appear unchanged over very long periods of time, but the situation is analogous to the thermodynamic instability of diamond with respect to graphite. Two-phase dispersions will always spontaneously change into a smaller number of large particles when given sufficient time.

Most of these systems are based on PNIPAAm or related copolymers[42] and their derivatives.[43] When PNIPAAm-based microgels are functionalised with pH-ionizable, hydrophilic and reactive carboxylic acid groups, the 'smart' microgels can be tuned specifically in order to generate fast and targeted swelling responses to multiple external stimuli, such as temperature and pH.[44,45]

Ngai *et al.* studied the novel emulsions stabilised by pH and temperature sensitive microgels.[44] Surfactant-free oil-in-water emulsions prepared with temperature and pH sensitive PNIPAAm microgel particles offer unprecedented control of emulsion stability.

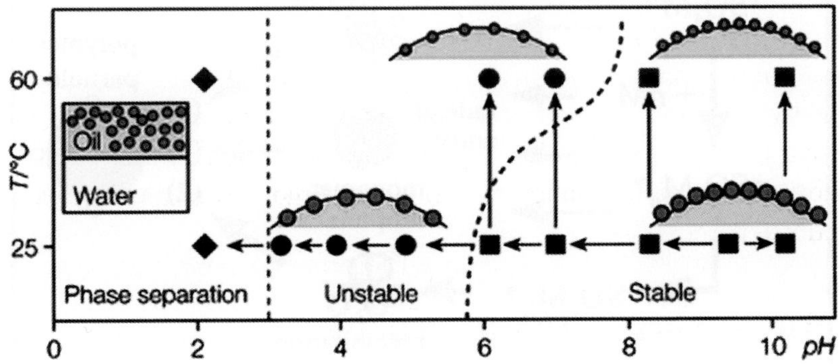

Fig. 4.6 The stabilising efficiency of PNIPAAm microgel particles for octanol-in-water emulsions as a function of pH and temperature. ■: Stable, ●: Unstable, ◆: Phase separation. (Arrows indicate the probed transitions).[44]

4.4.2 *Manufacturing polymeric particles*

There are various chemical and physical events which take place simultaneously during the process of particle formation and growth. Namura *et al.* depicted the generally accepted scheme for the kinetics of emulsion polymerisation.[1] Three major mechanisms for particle formation have been proposed to date. Figure 4.7 shows the proposed scheme for particle formation in emulsion polymerisation initiated by water-soluble initiators.[1] Particle formation is considered to have taken place when (1) a free radical in the aqueous phase enters a monomer-swollen emulsifier micelle and propagation proceeds therein; (2) the chain length of a free radical growing in the aqueous phase exceeds its solubility limit and precipitates to form a particle nucleus or (3) a free radical growing in the aqueous phase enters a monomer droplet and propagation proceeds therein. However, if the resultant polymer particles are not stable enough, the final number of polymer particles produced, regardless of the mechanism of particle formation, is determined by coagulation between the existing particles.

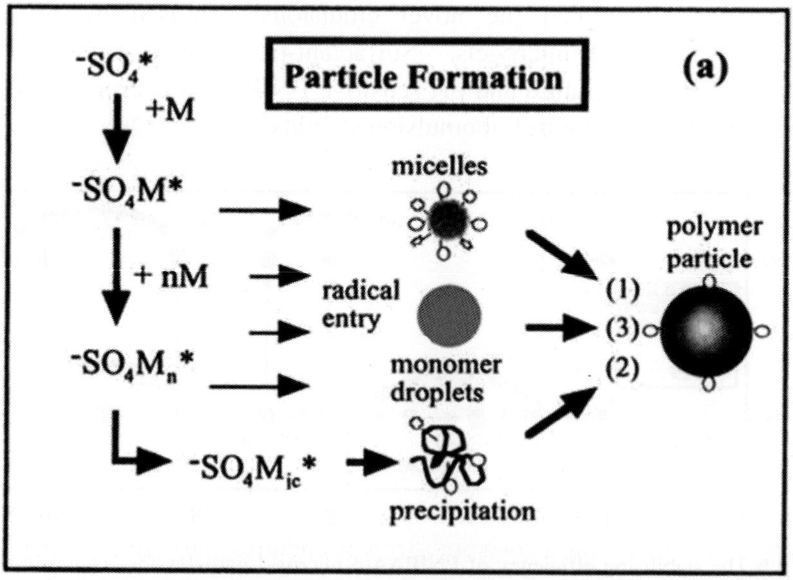

Fig. 4.7 Three major mechanisms of particle formation.[1]

There are several methods for creating particles. In addition, macro or micro scale particles are ground in a ball mill, planetary ball mill or other size reducing mechanism. Lu *et al.*[46] synthesised monodisperse thermosensitive PS-NIPAAm core-shell particles composed of a PS core and a crosslinked PNIPAAm shell successfully by photoemulsion polymersation. Cryo-TEM images clearly indicated the core-shell morphology of the PS-NIPAAm particles where a homogeneous regular PNIPAAm shell was affixed on the spherical PS core. The shell of PNIPAAm networks with different crosslinking densities can shrink and re-swell with temperature and the volume transition temperature is around 32°C in all cases. In the first step, a PS core with 5 mol% NIPAAm was prepared by emulsion polymersation. In the second step, the PS core was covered with a thin layer of photo-initiator HMEM. In the third step, the shell of a crosslinked PNIPA network was formed by photo-emulsion polymersation. Evidently, compared with conventional emulsion polymersation, photo-emulsion polymersation leads to a more homogeneous crosslinked PNIPAAm shell. Also, the shells are tightly bound to the cores. This is to be expected as the shell results from a 'grafting-from' process, that is, from radicals tightly bound to the surface of the core particles.

These particle manufacturing techniques historically come under the title of 'colloid chemistry' and involve classical 'sol-gel' processes or other aggregation processes. These wet chemistry techniques can offer the best quality particle from a number of viewpoints:

A. They produce particles that are already in the form of dispersion, hence high inter-particle forces can be designed to prevent agglomeration.
B. The formation of aggregates can be reduced or eliminated.
C. The particles can be made to be very monodisperse, i.e. all the same size to within small tolerances.
D. The chemical composition and morphology can be closely controlled. The production of particles in liquid carbon dioxide was a most recent development, which offered the promise of controlled conditions with a 'bottom-up' wet chemistry approach, where the dispersant can be removed easily by simply reducing the pressure of the reaction container.

4.4.2.1 *Sol-gel process*

The sol-gel process is a wet-chemical technique that has been recently and widely used in the field of materials science as shown in Fig. 4.8. This method is used primarily for the fabrication of materials starting from a chemical solution (sol, short for solution) which acts as the precursor for an integrated network (or gel) of either discrete particles or network polymers. The sol-gel process is a versatile solution process for making advanced materials. In general, the sol-gel process involves the transition of a solution system from a liquid 'sol' (mostly colloidal) into a solid 'gel' phase. The sol-gel process is possible for fabricating advanced materials in a wide variety of forms, including: ultrafine or spherical shaped powders, fibres, porous or dense materials, and extremely porous aerogel materials.

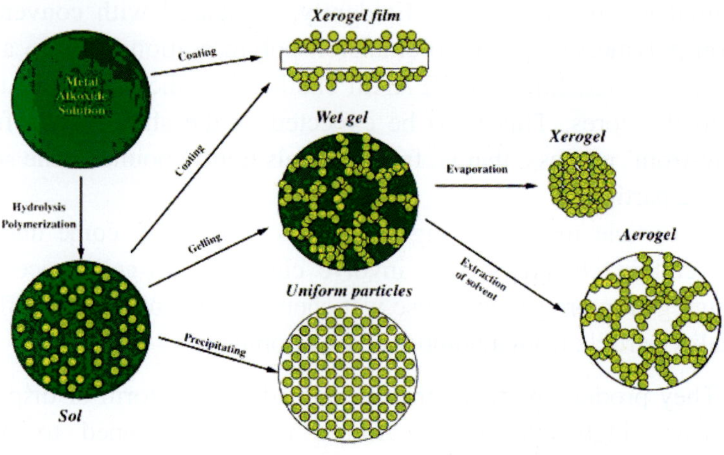

Fig. 4.8 Sol-Gel Technologies and Their Products.[47]

4.4.2.2 *Surface modification of particles*

The coating of particle surfaces can modify the properties of particles. The catalytic activity of particles is inactivated thereby preventing the deterioration of the fragrance, oils and agents that are present together with the particles in cosmetics. Si-H groups are present in this networked silicone, and it is possible here to introduce unsaturated compounds in a hydrosilylation chemical reaction.[48] By introducing glycerin residue, dual purpose foundations with moisture retentivity, that can be used both with and without water, have been developed.[49]

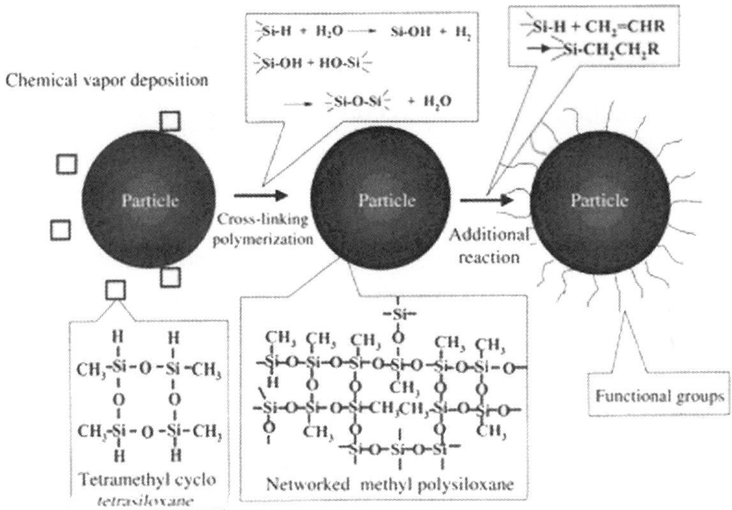

Fig. 4.9 Schematic diagram of functional nanocoating.[49]

4.4.2.3 *Fabrication of core-shell structure polymeric particles*

Core-shell particles have attracted much scientific and industrial interest because their chemical and mechanical properties can be customised for many applications, such as paints, adhesives, paper and textile manufacturing and impact modifiers.[50] Core-shell particles are a type of composite material which consists of a core domain covered by a shell domain. The core and shell domains may be composed of a variety of materials, including polymers, inorganic solids and metals. Core-shell materials are prepared with polymer shells to keep the core dispersed in a solvent or to protect medicines or other materials from dissolution or hydrolysis.[51]

4.4.2.4 *Micro-encapsulation technology*

Micro-encapsulation is a process in which tiny particles or droplets are surrounded by a coating to give small capsules many useful properties. Microcapsule production may be achieved by means of physical or chemical techniques. A microcapsule is a small sphere with a uniform wall around it. The material inside the microcapsule is referred to as the core, internal phase or fill, whereas the wall is sometimes called a shell, coating or membrane. There are many advantages of the micro-encapsulation technique, for example, isolation of material oxidation, retarding evaporation of a volatile core, prevention of chemical attacks, controlling release of drugs in a target area or even simply masking unwanted odors. One of the first industrial applications of micro-encapsulation was invented by B. K. Green (National Cash Register Co.) in 1953.[52] After that, there were increasingly more developments in micro-encapsulation technology. Micro-encapsulation can be applied in medicine, food, painting, light-sensitive materials, printing, textiles, cosmetics and many other fields.

The size of a micro-capsule is about 5-200 μm, and sometimes the size of an encapsulation can be as tiny as 100 nm (nanocapsules). The shapes of micro-capsules are different based on the composite material and fabrication techniques as shown in Fig. 4.10. Most of them are spheriform, floccus precipitate, particles and so on. According the number of core particles, the capsules can be divided into mono-core and multi-core capsules. After being classified by the wall structure, the capsules can be then classified into mono-layer, double-layer or multi layer.

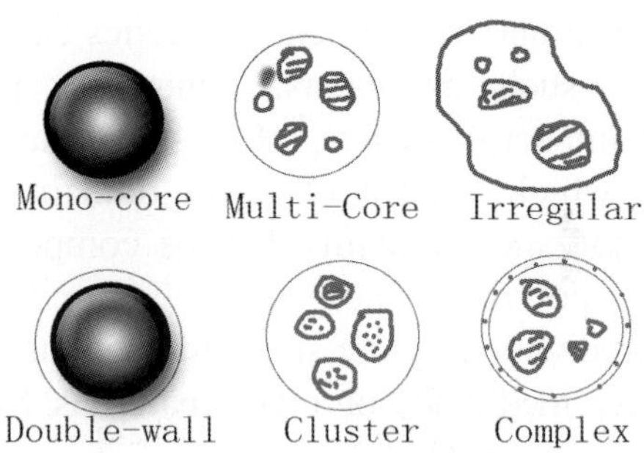

Fig. 4.10 Structure of micro-capsules.

There are mainly three methods that can be applied to obtain micro-capsules, including chemical, physical and physical/chemical. Now there are numerous methods which have been developed based on these three methods, for example, the interfacial solvent exchange technique,[53] *in situ* polymersation, interface polymersation, phase separation, double emulsion/solvent evaporation, layer-by-layer nano self-assembly technique, supercritical fluid technique and so on. Physical methods are mainly spray drying or centrifugal and fluidised bed processes which are inherently incapable of producing microcapsules smaller than 100 μm. The most suitable chemical processes are *in situ* polymersation techniques.[54]

Although there are many methods, there are four procedures in general to prepare micro-capsules (Fig. 4.11).

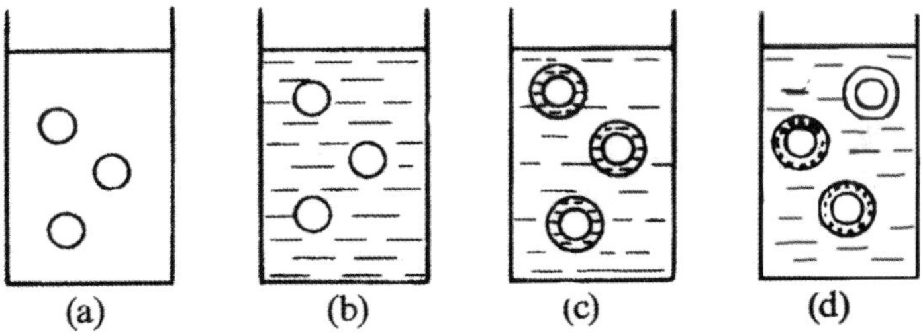

Fig. 4.11 Schematics of micro-encapsulation a. dispersement of core, b. adding shell material, c. precipitating, d. solidification of shell.

The micro-encapsulation method which is based on an *in situ* polymersation technique is quite successful for producing microcapsules (Fig. 4.12) with an enhanced thermal capacity in relation to the phase change material content. Seventy-seven percent of microcapsules are 100 μm in diameter and 90% are smaller than 185 μm by using *in situ* polymersation techniques.[54]

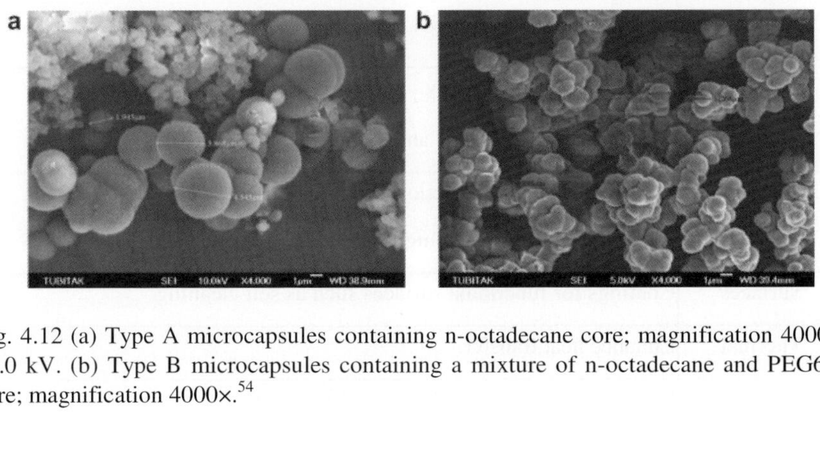

Fig. 4.12 (a) Type A microcapsules containing n-octadecane core; magnification 4000×; 10.0 kV. (b) Type B microcapsules containing a mixture of n-octadecane and PEG600 core; magnification 4000×.[54]

4.5 Applications of Adaptive Particles

Adaptive polymeric particles have wide potential applications as shown in Table 4.2 as they have many advantages, such as stability, ease of synthesis, good control over particle size and easy fictionalisation providing stimulus-responsive behaviour. It becomes clear that this type of material holds great promise for nanotechnology. A recent review given by Nayak and Lyon[42] showed studies of the detailed structure-function relationships in hydrogel particles.

Table 4.2 Properties and current applications of particles.

Property	Current Application
Biomedical	Coating with antibacterial properties. Drug delivery systems.
Electronic	Smaller components with high performance. Enhance conductivity of materials.
Energy	Increase durable and energy for batteries. Hydrogen storage applications. High efficient catalysts for combustion engines.
Environmental	Pollution deduction. High efficient for water filtration.
Magnetic	Storage media. MRI images.
Mechanical	Improved wear durability. New structural materials and composites.
Optical	Coatings with anti-reflection. Act as sensors for detecting cancer diagnosis.
Surfaces	Coatings for functional surfaces such as self cleaning.
Thermal	Enhance heat transfer.

4.5.1 *Application to textiles and fibres*

Textiles and fibres are very important in daily life. The most common uses of textiles are for clothing and containers, such as bags and baskets. Also, they can be used in carpeting, window shades, towers, table padding and so on. For industrial applications, textiles can be applied as medical textiles, agrotextiles, protective textiles, laboratory fabrics and in firefighting. In order to meet these multiple applications, peoples have developed many techniques to modify textiles and fibres. Textiles can be treated by many methods to obtain the desired characteristics.[55] The new techniques have offered many important platforms for the functional modification of textiles to serve special and versatile requirements. These methods include nanoparticle surface modification, sol-gel treatment, chemical vapour disposition, micro-encapsulation and many others. Polymeric particles which are adaptive and functional in many ways will be popular, especially for multiple purposes.

4.5.1.1 *Nanoparticle modification on textiles*

The Textiles Nanotechnology Laboratory at the Cornell University has attached nanoparticles to textile fibres in order to create smart and functional clothing (http://nanotextiles.human.cornell.edu/). They have modified existing textile materials by using electrostatic self-assembly and atomic layer deposition techniques to create multifunctional and customisable textile surfaces by depositing polyelectrolyte onto the conventional textile substrates with a myriad of nanoparticles and non-spherical colloids. They used magnetic and electric fields to achieve precise position control of functional nanoparticles inside and outside polymeric fibres, hence creating novel nanocomposite materials (Fig. 4.13). These nanofibres may be hailed as anti-counterfeiting, antibacterial, anti-odour and self-decontaminating applications. They also include tunable colouration and control of near infrared signatures for active camouflage applications.

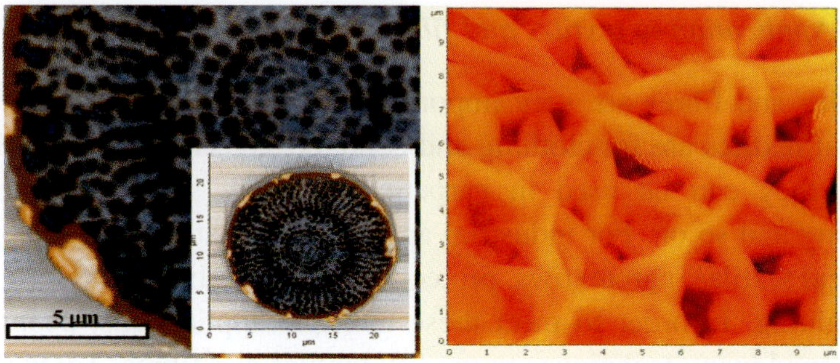

Fig. 4.13 Acoustic force atomic microscopy image of conjugated fibre (Islands on the Sea) containing 1120 nanofibres of polyester in a sea of polyethylene (L). Electrostatic force microscopy of a nanofibre nonwoven web used in an electret filtration media (R). (Permission from the website: http://nanotextiles.human.cornell.edu/).

4.5.1.2 *Self cleaning modification on textiles*

Another important application of a modified nanoparticle; TiO_2, is the self cleaning modification of cotton textiles[56-59] or wool-polyamide and polyester textiles.[60] Meilert *et al.* attached cotton by formation of an ester-bond.[59] The TiO_2 binds to the cotton by chemical means and the textiles present self cleaning properties. The increase of the curing temperature during the esterification of the spacer by the cellulose-OH is shown to increase the amount of spacer that could be bound to the cotton textile, but curing at higher temperatures do not always produce the coated cotton textiles with the best self cleaning properties. The self cleaning cotton could be useful in the event of wine, make-up and other stains. The self cleaning action makes it possible to space at larger intervals, the cleaning of cotton tissues that are used in aircrafts, and clothing and health industries, since self cleaning is only partially effective in completely abating dirt.

4.5.1.3 *Shape memory finishing on textiles*

Biodegradable materials have been widely applied in textiles. Xin JH and his group modified a cotton surface with chitosan particles to obtain multi-functional protective and self cleaning textile materials through nanofinishing.[61] His group had also used chitosan-based polymeric core-shell particles to finish cotton fabric and obtain durable antibacterial ability.[62] In their results, cotton fabric with excellent antibacterial durability was obtained when treated with chitosan containing core-shell particles without any chemical binders. The SEM of chitosan-poly(butyl acrylate) (CTS-polyBA) particles is shown in Fig. 4.14. These particles with antibacterial chitosan shells covalently grafted onto polymer cores were prepared via a surfactant-free emulsion copolymersation in aqueous chitosan. Herein, two core-shell particles, one with poly (n-butyl acrylate) soft core and another with a crosslinked poly(N-isopropylamide) hard core, were synthesised and applied to cotton fabric by a conventional pad-dry-cure process. The testing results showed that the treated fabric has an excellent antibacterial property with bacterial reduction higher than 99%. After 50 times of home laundering, the antibacterial activity is maintained at over 90%.

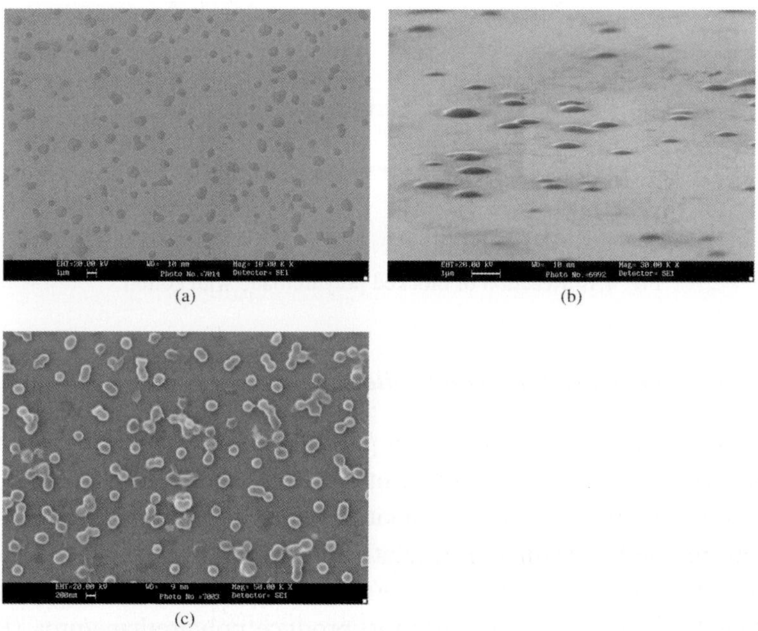

Fig. 4.14 SEM micrographs of CTS-polyBA (a, b) and CTS-polyNIPAM (c) latex particles. (a) CTS-polyBA, top-view; (b) CTS-polyBA, 80ºC side-view; (c) CTS-polyNIPAM, top-view.[62]

Hu's group synthesised the emulsion of shape memory polyurethane particles and used it to finish cotton fabrics.[63] In their study, the blocked shaped memory polyurethane emulsion was prepared from successive additions of polypropylene glycol (PPG) and methyl ethyl ketoxime (MEKO) which was chosen as a blocking agent to Diphenyl-methane-diisocyanate (MDI). Then, the polyurethane particle emulsion was used to treat cotton fabrics. Under exact treated temperature, blocked polyurethane will generate free isocyanates and further react with the hydroxyl group of cotton fabrics (Fig. 4.15). Polyurethane obtains a crosslinking structure during the finishing process. Hence, the modified cotton improves resistance to creasing and shrinking and/or has improved handle.

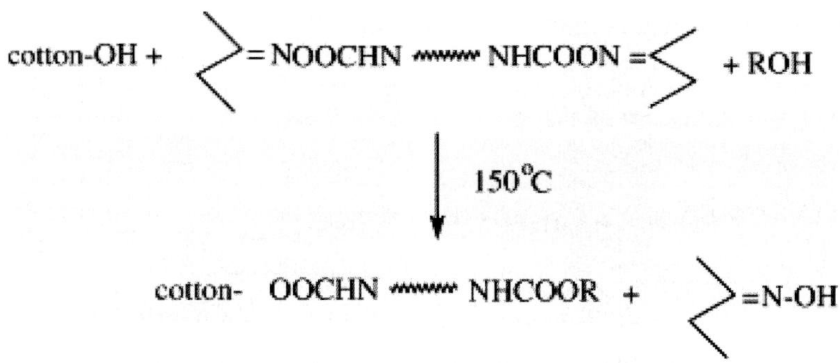

Fig. 4.15 Reaction of blocked polyurethane with cotton.[63]

4.5.1.4 *Micro-encapsulation on textiles*

Microencapsulation of particles can provide textiles with new properties and added value.[17] Rodrigues *et al.* introduced Scentfashion®,[9] which is a technology that uses microencapsulated perfumes in textile applications. In their research, perfume formulation was performed by taking into account the target market and type of textile substrate for man suits. Interfacial polymersation was used to produce polyurethane/urea (PUU) microcapsules. During dry cleaning of lab-scale impregnated fabrics, the loss of limonene was 38% in the first cycle and up to 87% after five dry cleaning cycles.

Monllor *et al.* studied flavour microcapsules in cotton fabrics,[64] incorporating a nanotechnology process that has been adopted by the textile field. They found that the composition of the wall material determines the release of the product into the fabric and is the object of study in their present work. The behaviour of microencapsulated peppermint was analysed on cotton fabrics. The microcapsules are applied on the surface of cotton fabrics by bath exhaustion and impregnation (Fig. 4.16). Their results proven that the standard fabric that is used to rub the fabric containing microcapsules can remove microcapsules from its surface and lower the active effect. Some odor intensity remained after some washing cycles and the rubbing test, which means that these methods do not destroy all of the microcapsules.

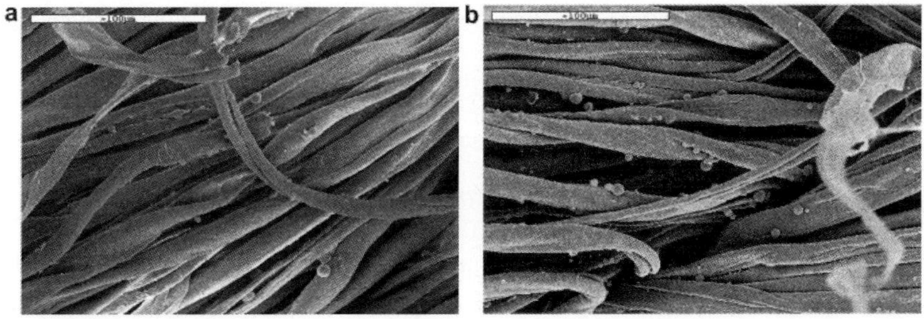

Fig. 4.16 SEM micrographs of cotton fabric with microcapsules. (a) Microcapsules applied by bath exhaustion. (b) Microcapsules applied by impregnation.[64]

4.5.1.5 *Phase change material particles*

The technology for incorporating phase change material (PCM) microcapsules into textile structures to improve their thermal performance was developed in the early 1980s under a NASA research programme by Nelson *et al.*[65] Microcapsules with walls that are less than 2 μm in thickness and 20-40 μm in diameter are useful in fibre applications as shown in Fig. 4.17. The microcapsules can be produced by depositing a thin polymer coating on small solid particles or on dispersions of solids in liquids. The core contents may be released by friction, pressure and diffusion through the polymer wall, dissolution of the polymer wall coating, or even biodegradation. In their application in textiles, the paraffins are either in a solid or liquid state. In order to prevent dissolution of the paraffin while in the liquid state, the paraffin is enclosed into a small plastic sphere with a diameter of only a few micrometres.

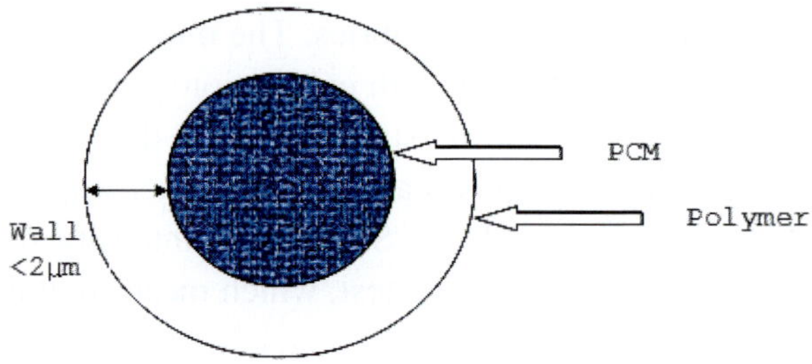

Fig. 4.17 Microencapsulation: paraffinic PCM core material with a hard polymeric shell.[65]

4.5.2 *Applications in medicine*

4.5.2.1 *Biodegradable polymer micro-hydrogels*

Polymer drug carriers have been made from a variety of biodegradable polymers. Common biodegradable polymers include poly(lactic acid) (PLA), poly(glycolic acid) (PGA), their copolymers; poly(lactic acid-co-glycolic acid) (PLGA) and poly(ε-caprolactone). Other biodegradable polymers include poly(orthoesters), poly(anhydrides), poly(amides), poly(phosphazenes) and poly(phosphoesters), as well as their various copolymers. Copolymers of these materials with poly(ethylene glycol) (PEG), poly(ethylene oxide) (PEO) or poly(propylene oxide) (PPO) have also been used to make drug delivery particles. The choice of polymers impacts several aspects of the carrier, including the types of drug encapsulated, mode of degradation and drug release, and biocompatibility and physical properties. A variety of methods, including single and double emulsification, spray drying and drop break-up of a liquid stream, have been used to produce spherical polymer particles. Collectively, these methods offer control over basic parameters, such as particle diameter, encapsulation efficiency and polydispersity, as well as advanced parameters, such as porosity and drug compartmentalisation.

4.5.2.2 *Nanoparticles in gene therapy*

Cai *et al.* used particles as carriers for the delivery of therapeutic materials to target tissues. This has become popular in recent years and demonstrated great potential for treatments in a wide range of diseases.[66] Particular attention is paid to the CK30-PEG compacted DNA particles that have been successfully tested in the eyes, lungs and brain. These particles result in higher transfection efficiency and longer duration of expression than other non-viral vectors without any toxicity or other side effects. They have been safely used clinically and are efficient for a broad range of gene therapy applications.

4.5.2.3 *Particles in drug delivery*

Physically bonded nanoparticle networks can also act as a novel drug delivery system as shown in Fig. 4.18.[67] Monodispersed nanoparticles consisting of interpenetrating polymer networks (IPNs) of polyacrylic acid (PAAc) and PNIPAAm are prepared by a seed-and-feed method. The temperature-dependent viscosity measurements reveal that the IPN nanoparticle dispersions with polymer concentrations above 2.5 wt% undergo an inverse thermoreversible gelation at about 33°C. Dextran markers of various molecular weights as model macromolecular drugs are mixed with IPN nanoparticle dispersion at room temperature. At body temperature, the dispersion becomes a gel. The drug release profiles as a function of particle size and polymer concentration are measured by using a UV–visible spectroscopy. The drug delivery model presented is significant because such a mixture of dispersion and drug without chemical reaction at room temperature forms a drug delivery liquid. This liquid could then be injected into a body to form *in situ* a gelled drug depot that releases the drug slowly.

4.5.3 *Application to cosmetics*

The application of particles in the field of cosmetics goes back to the blending of ultrafine particles of titanium dioxide in sunscreen during the 1980s.[49] Ultrafine titanium dioxide particles that were 30 nm in size provided the first sunscreen with a high UV protection capacity as well as high transparency. While 10-30 nm of titanium dioxide or zinc oxide particles are now being used, any ultrafine particle sized less than 100 nm will be referred as a nanoparticle here. Apart from cosmetics where nanoparticles are dispersed in this manner, compound particles where nanoparticles are arranged in the base particles are realising various optical characteristics, and thus used for makeup. The application of nanoparticles in cosmetics that include emulsified particles and the application of compound particles which can implement form control are described here.

With the advent of nanotechnology, it has been possible to develop polymeric particles, such as dendrimers and fullerenes. More recently developed polymers contain lattice-like matrices that hold the drug or entrap it between the polymer strands.

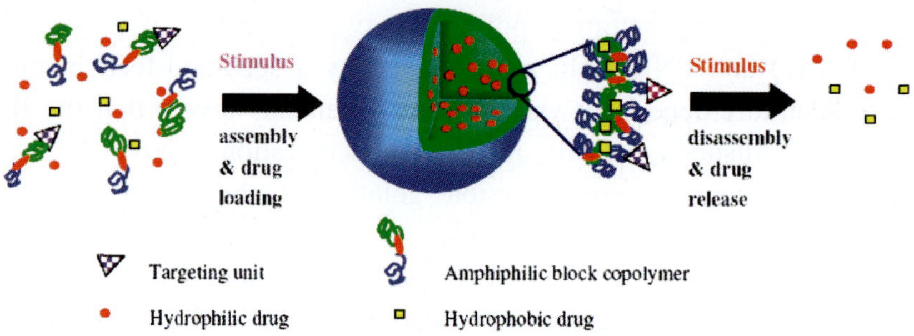

Fig. 4.18 Schematic presentation of the formation and dissociation of stimuli-sensitive polymersomes.[68]

4.5.3.1 *Particles on makeup cosmetics*

Structural colours that give out colour due to interference of light, such as the colours in the wings of a butterfly, are being used actively. It is known that blue or red moles, such as Nevus of Ota or vascular neoplasm, disappear, giving a natural appearance through the use of titanium dioxide coated mica with interference colours. However, since titanium dioxide coated mica has gloss, it gives an unnatural finish when used in foundations. In order to improve its shine, a compound particle was developed with the surface of titanium dioxide coated mica covered with spherical particle of PMMA. While correcting the reflectivity of the red colour by using its red interference colour, it is possible to realise a fine makeup texture on skin that also encompasses transparency since the gloss of titanium dioxide coated mica is controlled by the diffused reflection of spherical PMMA particles. An example of skin colour correction using fluorescence is the reduced form of zinc oxide. A reflectivity of about 500 nm is increased due to the UV light present in sunlight or fluorescent.

4.5.3.2 *Particles on skin care cosmetics*

Emulsification and solubilisation technologies are used in skin care cosmetics, such as skin lotions, milks and creams. For example, by changing the sequence of mixing or the pressure during the process of mixing water and oil, it is possible to control the size of the emulsified particles from the usual 1-10 um to several tens of nanometres. Müller *et al.* studied solid lipid nanoparticles (SLN) on the application of skin care.[69-71] In their study, they determined that the occlusive effect of SLN depends on the particle size by using an in vitro occlusion. Compared to liposomes and emulsions, solid particles possess some advantages, e.g. protection of incorporated active compounds against chemical degradation and more flexibility in modulating the release of the compound. Freely diffusible nanoparticles in low concentrated dispersion can collide and aggregate; in highly concentrated dispersions, the particles are fixed in a network, while dilution with water releases non-aggregated definite nanoparticles (Fig. 4.19). In hexagonal packing, about 76% of the surface are covered, 24% are uncovered, meaning the uncovered surface is identical for both the microparticles and nanoparticles. Water condensates in the pores due to their small size and reduces vapour pressure (Laplace's equation), and thus the pores in the SLN film attract rather than lose water. Recent investigations with an electron microscopy showed that after evaporation of water from an SLN dispersion, a continuous and poreless film is formed.

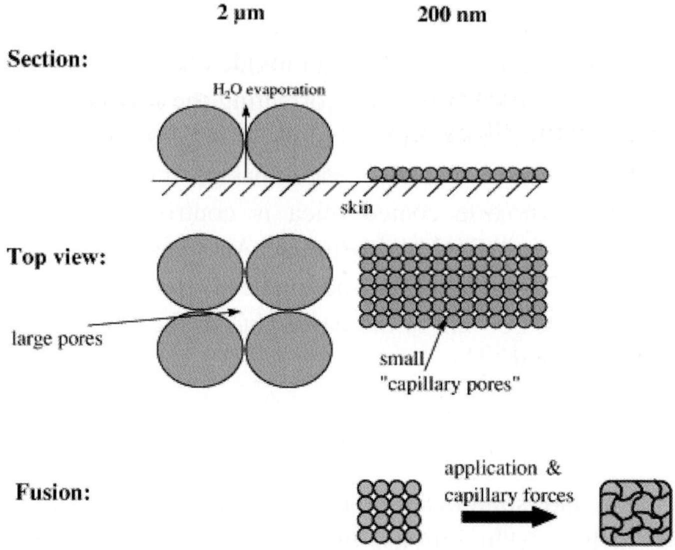

Fig. 4.19 Model of film formation on skin for lipid 2-µm particles and lipid 200 nm particles shown in Section (upper), from a top view (middle), and a new model of fusion of the nanoparticles to a poreless film (lower).[69]

While the major purpose of cosmetics is skin care, the so-called 'skin care particles' have been developed to improve skin problems using dermatology for this purpose. This skin care particle has a much higher skin care effect compared to only using zinc oxide or silica independently. Another type of skin care particle is called a nano-lipid particle (NLP). The main application of NLP is to make particles small enough to penetrate the skin. The secret is lyphazome nanospheres, which are only 1/50th the size of human skin cells, and penetrate the epidermis, delivering ingredients deep into the skin, where they are most effective.

4.6 Summary

We have discussed recent work on the properties, preparation and modification of particles and their applications in medicine, cosmetics, and textiles. All applications discussed so far require different properties in the particles. The properties of adaptive polymeric particles are determined by their physical and chemical structures. Morphology and size also have an influence on the properties of particles. In particular, application to textile fibres has been described for the preparation of core-shell particles, micro-capsules, surface modified particles and phase chapter materials, and shape memory particles on the finishing fabrics. There may be the following research prospects for the future. The first is the relationship between structure and physical/chemical properties of polymeric particles. The second is the fundamental theory research of polymeric particles. Finally, there are the industrial techniques and applications of polymeric particles. If all of these aspects are able to experience substantial development in the near future, then there may be a better understanding of adaptive polymeric particles.

References

1. Nomura, M., Tobita, H., and Suzuki, K., *Emulsion Polymerization: Kinetic and Mechanistic Aspects*, in *Polymer Particles*. 2005. 125: pp. 1-128.
2. Rodríguez-Hernández, J., Chécot, F., Gnanou, Y., and Lecommandoux, S., Toward 'smart' nano-objects by self-assembly of block copolymers in solution. *Progress in Polymer Science*, 2005. 30(7): pp. 691-724.
3. Ballauff, M. and Lu, Y., "Smart" nanoparticles: Preparation, characterization and applications. *Polymer*, 2007. 48(7): pp. 1815-1823.
4. Jakobs, E. and Hanein, Y., Micrometer scale gel patterns. *Colloids and Surfaces A: Physicochemical and Engineering Aspects*, 2006. 290(1-3): pp. 33-40.
5. Kumar, A., Srivastava, A., Galaev, I. Y., and Mattiasson, B., Smart polymers: Physical forms and bioengineering applications. *Progress in Polymer Science*, 2007. 32(10): pp. 1205-1237.
6. Xiang, Y. and Chen, D., Preparation of a novel pH-responsive silver nanoparticle/poly(HEMA-PEGMA-MAA) composite hydrogel. *European Polymer Journal*, 2007. 43(10): pp. 4178-4187.
7. Kawaguchi, S. and Ito, K., *Dispersion Polymerization*, in *Polymer Particles*. 2005. pp. 299-328.

8. Qiu, Y. and Park, K., Environment-sensitive hydrogels for drug delivery. *Advanced Drug Delivery Reviews*, 2001. 53(3): pp. 321-339.
9. Rodrigues, S. N., Martins, I. M., Fernandes, I. P., Gomes, P. B., Mata, V. G., Barreiro, M. F., and Rodrigues, A. E., Scentfashion® Microencapsulated perfumes for textile application. *Chemical Engineering Journal*, 2009. 149(1-3): pp. 463-472.
10. El-Zawahry, M. M., El-Shami, S., and El-Mallah, M. H., Optimizing a wool dyeing process with reactive dye by liposome microencapsulation. *Dyes and Pigments*, 2007. 74(3): pp. 684-691.
11. Oda, M., Chikara, H., Ryozi, U., and Akira, T., *Synthesis of Compound and Individually Separated Ultra-Fine Particles*, in *Ultra-Fine Particles*. 1995, William Andrew Publishing: Westwood, NJ. p. 133-159.
12. Kamath, K. R. and Park, K., Biodegradable hydrogels in drug delivery. *Advanced Drug Delivery Reviews*, 1993. 11(1-2): pp. 59-84.
13. Holmes, N. S., A review of particle formation events and growth in the atmosphere in the various environments and discussion of mechanistic implications. *Atmospheric Environment*, 2007. 41(10): pp. 2183-2201.
14. Neelmeijer, C. and Mader, M., The merits of particle induced X-ray emission in revealing painting techniques. *Nuclear Instruments and Methods in Physics Research Section B: Beam Interactions with Materials and Atoms*, 2002. 189(1-4): pp. 293-302.
15. Hu, X., Zhou, J., Zhang, N., Tan, H., and Gao, C., Preparation and properties of an injectable scaffold of poly(lactic-co-glycolic acid) microparticles/chitosan hydrogel. *Journal of the Mechanical Behavior of Biomedical Materials*, 2008. 1(4): pp. 352-359.
16. Dong, Y., Bai, Z., Liu, R., and Zhu, T., Decomposition of indoor ammonia with TiO_2-loaded cotton woven fabrics prepared by different textile finishing methods. *Atmospheric Environment*, 2007. 41(15): pp. 3182-3192.
17. Mondal, S., Phase change materials for smart textiles - An overview. *Applied Thermal Engineering*, 2008. 28(11-12): pp. 1536-1550.
18. Heller, D. S., Westhoff, C., Gordon, R. E., and Katz, N., The relationship between perineal cosmetic talc usage and ovarian talc

particle burden. *American Journal of Obstetrics and Gynecology*, 1996. 174(5): pp. 1507-1510.

19. Provder, T., Clara, D. C., Charles, E. C., and Jr, *Challenges in particle size distribution measurement past, present and for the 21st century*, in *Applied Polymer Science: 21st Century*. 2000, Oxford: Pergamon. pp. 873-892.
20. Takeuchi, H., Thongborisute, J., Matsui, Y., Sugihara, H., Yamamoto, H., and Kawashima, Y., Novel mucoadhesion tests for polymers and polymer-coated particles to design optimal mucoadhesive drug delivery systems. *Advanced Drug Delivery Reviews*, 2005. 57(11): pp. 1583-1594.
21. Herrero-Vanrell, R., Rincón, A. C., Alonso, M., Reboto, V., Molina-Martinez, I. T., and Rodríguez-Cabello, J. C., Self-assembled particles of an elastin-like polymer as vehicles for controlled drug release. *Journal of Controlled Release*, 2005. 102(1): pp. 113-122.
22. Buzea, C., Pacheco, I. I., and Robbie, K., Nanomaterials and nanoparticles: Sources and toxicity. *Biointerphases*, 2007. 2(4): pp. MR17-MR71.
23. Nahal, A., Miloslavsky, V. K., and Ageev, L. A., Influence of photoinduced gyrotropy on the formation of spontaneous periodic structures in light-sensitive AgCl-Ag thin films. *Optics Communications*, 1998. 154(4): pp. 234-242.
24. Li, Z., Xiong, D. a., Xu, B., Wu, C., An, Y., Ma, R., and Shi, L., Fabrication of an asymmetric hollow particle with a thermo-sensitive PNIPAM inside corona. *Polymer*, 2009. 50(3): pp. 825-831.
25. Gohy, J.-F., Lohmeijer, B. G. G., Varshney, S. K., Decamps, B., Leroy, E., Boileau, S., and Schubert, U. S., Stimuli-Responsive Aqueous Micelles from an ABC Metallo-Supramolecular Triblock Copolymer. *Macromolecules*, 2002. 35(26): pp. 9748-9755.
26. Fowkes, F. M. and Mostafa, M. A., Acid-Base Interactions in Polymer Adsorption. *Industrial & Engineering Chemistry Product Research and Development*, 1978. 17(1): pp. 3-7.
27. Persans, P. D., Hayes, T. M., and Lurio, L. B., Size-dependent composition of semiconductor nanoparticles in glass. *Journal of Non-Crystalline Solids*, 2004. 349: pp. 315-318.
28. Huang, H., Tang, C., Zeng, Y., Yu, X., Liao, B., Xia, X., Yi, P., and Chu, P. K., Label-free optical biosensor based on localized surface

plasmon resonance of immobilized gold nanorods. *Colloids and Surfaces B: Biointerfaces*, 2009. 71(1): pp. 96-101.
29. Atia, A. A., Donia, A. M., and Al-Amrani, W. A., Adsorption/desorption behavior of acid orange 10 on magnetic silica modified with amine groups. *Chemical Engineering Journal*, 2009. 150(1): pp. 55-62.
30. Preining, O., Rajiv, K., and Mittal, K. L., *The Physical Nature of Very, Very Small Particles and its Impact on Their Behavior*, in *Developments in Surface Contamination and Cleaning*. 2008, William Andrew Publishing: Norwich, NY. pp. 3-24.
31. Ford, J. L., Rubinstein, M. H., and Hogan, J. E., Propranolol hydrochloride and aminophylline release from matrix tablets containing hydroxypropylmethylcellulose. *International Journal of Pharmaceutics*, 1985. 24(2-3): pp. 339-350.
32. Dendukuri, D., Tsoi, K., Hatton, T. A., and Doyle, P. S., Controlled Synthesis of Nonspherical Microparticles Using Microfluidics. *Langmuir*, 2005. 21(6): pp. 2113-2116.
33. Sozzani, P., Bracco, S., Comotti, A., Simonutti, R., Valsesia, P., Sakamoto, Y., and Terasaki, O., Complete shape retention in the transformation of silica to polymer micro-objects. *Nat Mater*, 2006. 5(7): pp. 545-551.
34. Champion, J. A., Katare, Y. K., and Mitragotri, S., Particle shape: A new design parameter for micro- and nanoscale drug delivery carriers. *Journal of Controlled Release*, 2007. 121(1-2): pp. 3-9.
35. Shengqing Xu et al, Generation of Monodisperse Particles by Using Microfluidics: Control over Size, Shape, and Composition13. *Angewandte Chemie International Edition*, 2005. 44(5): pp. 724-728. Copyright Wiley-VCH Verlag GmbH & Co.KgA.
36. Dendukuri, D., Pregibon, D. C., Collins, J., Hatton, T. A., and Doyle, P. S., Continuous-flow lithography for high-throughput microparticle synthesis. *Nat Mater*, 2006. 5(5): pp. 365-369.
37. Manoharan, V. N., Elsesser, M. T., and Pine, D. J., Dense Packing and Symmetry in Small Clusters of Microspheres. *Science*, 2003. 301(5632): pp. 483-487.
38. Y. Yin, Y. X., Self-Assembly of Monodispersed Spherical Colloids into Complex Aggregates with Well-Defined Sizes, Shapes, and Structures. *Advanced Materials*, 2001. 13(4): pp. 267-271.

39. Ho, C. C., Keller, A., Odell, J. A., and Ottewill, R. H., Preparation of monodisperse ellipsoidal polystyrene particles. *Colloid & Polymer Science*, 1993. 271(5): pp. 469-479.
40. Feil, H., Bae, Y. H., Feijen, J., and Kim, S. W., Effect of comonomer hydrophilicity and ionization on the lower critical solution temperature of n-isopropylacrylamide copolymers. *Macromolecules*, 1993. 26(10): pp. 2496-2500.
41. Cao, G. and Brinker, C. J., *Annual review of nano research*. 2008: World Scientific Publishing Company.
42. Satish Nayak, L. A. L., Soft Nanotechnology with Soft Nanoparticles. *Angewandte Chemie International Edition*, 2005. 44(47): pp. 7686-7708.
43. Pelton, R., Temperature-sensitive aqueous microgels. *Advances in Colloid and Interface Science*, 2000. 85(1): pp. 1-33.
44. Ngai, T., Behrens, S. H., and Auweter, H., Novel emulsions stabilized by pH and temperature sensitive microgels. *Chemical Communications*, 2005(3): pp. 331-333.
45. Hoare, T. and Pelton, R., Highly pH and Temperature Responsive Microgels Functionalized with Vinylacetic Acid. *Macromolecules*, 2004. 37(7): pp. 2544-2550.
46. Yan Lu, A. W. M. B. M. D., Preparation of Polystyrene-Poly(*N*-isopropylacrylamide) (PS-PNIPA) Core-Shell Particles by Photo-emulsion Polymerization. *Macromolecular Rapid Communications*, 2006. 27(14): pp. 1137-1141.
47. Sol & Gel. http://www.chemat.com/4.tpl.
48. Whitmore, F. C., Pietrusza, E. W., and Sommer, L. H., Hydrogen-Halogen Exchange Reactions of Triethylsilane. A New Rearrangement of Neopentyl Chloride1. *Journal of the American Chemical Society*, 1947. 69(9): pp. 2108-2110.
49. Hosokawa, M., Nogi, K., Naito, M., Yokoyama, T., Masuo, H., Kiyoshi, N., Mario, N., and Toyokaz, U. Y., *Development of new cosmetics based on nanoparticles*, in *Nanoparticle Technology Handbook*. 2008, Elsevier: Amsterdam. pp. 521-527.
50. Pérez-Carrillo, L. A., Puca, M., Rabelero, M., Meza, K. E., Puig, J. E., Mendizábal, E., López-Serrano, F., and López, R. G., Effect of particle size on the mechanical properties of polystyrene and poly(butyl acrylate) core/shell polymers. *Polymer*, 2007. 48(5): pp. 1212-1218.

51. Qi, D., Bai, F., Yang, X., and Huang, W., Synthesis of core-shell polymer microspheres by two-stage distillation-precipitation polymerization. *European Polymer Journal*, 2005. 41(10): pp. 2320-2328.
52. Green, B. K., *Oil-containing microscopic capsules*. 1960, The National Cash Register Company: USA No. 2800458.
53. Yeo, Y. and Park, K., A new microencapsulation method using an ultrasonic atomizer based on interfacial solvent exchange. *Journal of Controlled Release*, 2004. 100(3): pp. 379-388.
54. Sarier, N. and Onder, E., The manufacture of microencapsulated phase change materials suitable for the design of thermally enhanced fabrics. *Thermochimica Acta*, 2007. 452(2): pp. 149-160.
55. Hegemann, D., Hossain, M. M., and Balazs, D. J., Nanostructured plasma coatings to obtain multifunctional textile surfaces. *Progress in Organic Coatings*, 2007. 58(2-3): pp. 237-240.
56. Uddin, M. J., Cesano, F., Scarano, D., Bonino, F., Agostini, G., Spoto, G., Bordiga, S., and Zecchina, A., Cotton textile fibres coated by Au/TiO_2 films: Synthesis, characterization and self cleaning properties. *Journal of Photochemistry and Photobiology A: Chemistry*, 2008. 199(1): pp. 64-72.
57. Yuranova, T., Mosteo, R., Bandara, J., Laub, D., and Kiwi, J., Self cleaning cotton textiles surfaces modified by photoactive SiO_2/TiO_2 coating. *Journal of Molecular Catalysis A: Chemical*, 2006. 244(1-2): pp. 160-167.
58. Bozzi, A., Yuranova, T., Guasaquillo, I., Laub, D., and Kiwi, J., Self-cleaning of modified cotton textiles by TiO_2 at low temperatures under daylight irradiation. *Journal of Photochemistry and Photobiology A: Chemistry*, 2005. 174(2): pp. 156-164.
59. Meilert, K. T., Laub, D., and Kiwi, J., Photocatalytic self-cleaning of modified cotton textiles by TiO_2 clusters attached by chemical spacers. *Journal of Molecular Catalysis A: Chemical*, 2005. 237(1-2): pp. 101-108.
60. Bozzi, A., Yuranova, T., and Kiwi, J., Self-cleaning of wool-polyamide and polyester textiles by TiO_2-rutile modification under daylight irradiation at ambient temperature. *Journal of Photochemistry and Photobiology A: Chemistry*, 2005. 172(1): pp. 27-34.
61. Wang, R. H., Xin, J. H., and Tao, X. M., Multi-functional protective nano-finishing of textile materials 1: Self-cleaning surfaces.

Abstracts of Papers of the American Chemical Society, 2005. 230: pp. 273-INOR.

62. Weijun Ye, Xin, J. H., Li, P., Lee, K. L., Kwong, T. L., Durable antibacterial finish on cotton fabric by using chitosan-based polymeric core-shell particles. *Journal of Applied Polymer Science*, 2006. 102(2): pp. 1787-1793.
63. Liu, Y. Q., Hu, J. L., Zhu, Y., and Yang, Z. H., Surface modification of cotton fabric by grafting of polyurethane. *Carbohydrate Polymers*, 2005. 61(3): pp. 276-280.
64. Monllor, P., Bonet, M. A., and Cases, F., Characterization of the behaviour of flavour microcapsules in cotton fabrics. *European Polymer Journal*, 2007. 43(6): pp. 2481-2490.
65. Nelson, G., Microencapsulation in textile finishing. *Review of Progress in Coloration and Related Topics*, 2001. 32: pp. 57-64.
66. Cai, X., Conley, S., and Naash, M., Nanoparticle applications in ocular gene therapy. *Vision Research*, 2008. 48(3): pp. 319-324.
67. Xia, X., Hu, Z., and Marquez, M., Physically bonded nanoparticle networks: a novel drug delivery system. *Journal of Controlled Release*, 2005. 103(1): pp. 21-30.
68. Meng, F., Zhong, Z., and Feijen, J., Stimuli-Responsive Polymersomes for Programmed Drug Delivery. *Biomacromolecules*, 2009. 10(2): pp. 197-209.
69. Müller, R. H., Radtke, M., and Wissing, S. A., Solid lipid nanoparticles (SLN) and nanostructured lipid carriers (NLC) in cosmetic and dermatological preparations. *Advanced Drug Delivery Reviews*, 2002. 54(Supplement 1): pp. S131-S155.
70. Pardeike, J., Hommoss, A., and Müller, R. H., Lipid nanoparticles (SLN, NLC) in cosmetic and pharmaceutical dermal products. *International Journal of Pharmaceutics*, 2009. 366(1-2): pp. 170-184.
71. Wissing, S. A. and Müller, R. H., Cosmetic applications for solid lipid nanoparticles (SLN). *International Journal of Pharmaceutics*, 2003. 254(1): pp. 65-68.

Chapter 5

Adaptive Textiles Using Adaptive Polymers

Adaptive polymer textiles are defined as textiles that made of adaptive polymer alone or adaptive polymer and other natural, synthesis materials can sense and react to environmental conditions or stimuli from mechanical, thermal, chemical, electrical or magnetic sources. Thermoregulation, shape memory, chameleonic, laminate, conductive, thermo regulated and other functional polymer textiles are introduced in this chapter.

Keywords: thermoregulation adaptive textiles, shape memory textiles, adaptive chameleon textiles, luminescent adaptive textiles, conductive polymer textiles, functional textiles

5.1 Adaptive Textiles for Thermoregulation

The new textile insulation materials can regulate temperature according to the environment and body temperature. These adaptive insulation textiles are called thermo-regulated textiles, and this effect could be achieved by phase change material (PCM).

For textile applications, the melting temperature of PCMs (heat-absorbing) is 15 to 35°C close to body temperature.[2] Paraffin and PEG are the most preferred kind of PCM used in textiles. Textiles containing phase change materials react immediately to changes in and body temperatures. When temperature rises, the PCMs react by absorbing heat and storing energy in the liquified phase change materials. When the temperature falls, the PCMs release this stored heat energy and the phase change materials become solid again.[3]

5.1.1 *Types of phase change materials*

According to the report, the required properties of a PCM for a high efficiency thermal regulation system in textile fields are as follow:[2]

(i) Melting point between 15 and 35°C, (ii) Large fusion heat, (iii) Little temperature difference between the phase change process, (iv) Environmental friendly, (v) Lower toxicity, (vi) Non-flammable, (vii) Stability and repetition, (viii) effective heat transfer; (ix) Ease of availability, (x) Low cost.

PCMs are divided into organic PCMs and inorganic PCMs. Organic PCMs have more uses due to their outstanding properties, as shown in Table 5.1. They include three subgroups, namely linear long chain hydrocarbons, such as parrafins (alkanes); polyethylene glycol; fatty acids and their binary mixtures.

Table 5.1 Organic PCMs.[4,2]

Compound	Melting temperature(oC)	Heat of fusion (kJ/kg)
n-Heneicosane(21)	40.5	-
n-Eicosane(20)	36.8	30.6
n-Octadecan(18)	28.2	244
n-Heptadecane(17)	21.7	171.38
Vinyl stearate	27-29	122

Capric acid	32	153
Methyl-1,2-hydroxy-stearate	42-43	120-126
Lauric acid	42-44	178

Compared with other PCMs, parrafinic hydrocarbons have such outstanding properties as, non-corrosiveness, chemical and thermal stability. They are inexpensive, and have extensive source of raw materials.

Polyethylene glycol (PEG) is an alternative organic compound for use as a PCM and is also one of the most important PCMs. Its melting point and heat depend on its molecular weight when its molecular weight is lower than 20,000.

5.1.2 *Incorporation method of PCMs in textiles*

The PCMs changing phases just above and below human skin temperature would be suitable for application in textiles. Coating, lamination, finishing, melt spinning, bi-component synthetic fibre extrusion, injection molding, foam techniques are some of the convenient processes for PCMs' incorporation into the textile matrix.

5.1.2.1 *Fibre Technology*

Filling hollow fibres by PCM

Vigo and Frost filled hollow rayon fibres with $LiNO_3.3H_2O$, Zn $(NO_3)_2.6H_2O.CaCl_2.6H_2O$ /$SrCl_2.6\ H_2O$ and $Na_2\ SO_4.10\ H_2O$ /$NaB_4O_7.10\ H_2O$.[5] It is obvious that the decrease of heat capacity of the fibre is greater after more heat-cool cycles.

Polyethylene glycols (PEG with average molecular weights of 400, 600, 1,000 and 3,350) incorporated as 57% solutions into hollow rayon and polypropylene fibres, after drying and conditioning, produced heat contents 1.2-2.5 times greater than untreated hollow polypropylene and 2.2-4.4 times greater than untreated hollow rayon fibres.[5]

Vigo and Frost filled hollow rayon and hollow polypropylene fibre with 2, 2-dimethyl-1, 3-propanediol (DMP) and impregnated non-hollow rayon with DMP. The fibre is not suitable for clothing textiles, because of its high phase change temperature.[5] There are still some defects in the water resistance, durability and handle of the heat-storage textiles produced by these processes.

Spinning method

The fibre spinning process has developed quickly since the 1990s. Watanabe[6] used the mixture of two aliphatic polyesters as the core component, PET as the sheath component, and melt span to produce heat-absorbing and heat-releasing synthetic conjugate for heat-insulating garments. PTMG was used as the core component in the composite fibre-spinning process. A composite fibre that uses PTMG as the core and PET as the sheath was designed.[7]

Zhang *et al.* had studied the melt spin ability of PEG alone, and PEG mixed with ethylene-vinyl acetate as the core component, and polypropylene as the sheath.[8] The mixture of paraffin and polyethylene was melt spun into the fibre directly. The surface of the fibre was coated with epoxy resin in order to prevent leakage of the paraffin.

Microencapsulation is the process of involving small amounts of phase change materials in a shell material. The thermo regulated fibres are prepared with melt spinning and wet spinning process.

In the methods mentioned above, the composite spinning is simple to operate, and the fibres have high latent heat, but PCMs in the fibres could easily be lost in use. The thermal regulating fibres made through the microencapsulated PCM composition spinning method also has some disadvantage, such as a relatively low latent heat value and very complicated operation. In addition, particle size and size distribution of microencapsulated PCMs also affect the spin ability and mechanical properties of thermal regulating fibres.[9] This method is restrained in large-scale application. Now *in situ* microencapsulation method is adopted. In this method, PCM and membrane-forming reagents are added into polymer melts or solutions to form lacteal spinning dopes which are spun via conventional spinning techniques to form as-spun fibres. The PCM microcapsules in the fibre are formed by post-treatment.

Mengjin Jiang *et al.* chose polyvinyl alcohol (PVA as fibre matrix, ethyl orthosilicate (TEOS) as membrane-forming reagent and paraffin as PCM to produce thermal regulating PVA fibre through wet spinning and in situ microencapsulation. In this method, it is not necessary to add PCM microcapsules in the spinning dopes, and the advantages of high latent heat, easy operation and low cost are obvious, which will lead to the potential applications of the thermal regulating fibre or other functional fibres.

Recently, with the development of nano-science and technology, ultrafine fibres of PCM/polymer composite,[10] have been developed via electrospinning technique. McCann *et al.* has developed a method based on melt coaxial electrospinning for preparing phase change nanofibres consisting of long-chain hydrocarbon cores and TiO_2–Poly(vinyl pyrrolidone) (PVP) sheaths.[11] Chen *et al.* has reported the ultrafine fibres of the polyethylene glycol/cellulose acetate (PEG/CA) composite from the mixture solution of PEG and CA by conventional electrospinning method.[12] And the composite fibres showed a good thermal stability.

5.1.2.2 *Coating method*

Coating method to make PCM fibre and textile is very easy and useful. Microspheres containing phase change material are dispersed in water solution containing a surfactant, a dispersant, an antifoam agent and a polymer mixture. The coating would be applied to a textile substrate. There are various coating processes available such as knife-over-roll, knife-over-air, pad-dry-cure, gravure, dip coating and transfer coating.

Zuckerman *et al.*[13] suggested one more process and also evaluated the properties of fabrics coated with PCMs and without PCMs. They used a knife-over-roll coating technique to achieve foam. Some techniques are not suitable for coating of microencapsulated PCMs, such as solvent-based gravure printing, thermoplastic gravure printing, thermoplastic spray and thermoplastic extrusion techniques.

Shin[14] studied melamine–formaldehyde microcapsules added to polyester knit fabrics by a conventional pad-dry-cure process to develop thermo-regulating textile materials. The fabric has good thermo-regulating efficiency. But the finishing process including binder types and loading methods needs to be studied further to improve laundering durability.

5.1.2.3 *Laminating method*

PCM would be incorporated into polymer film and applied to the inner side of the fabric system by lamination. Laminating method was introduced by Pause.[15] He directly incorporated PCMs into a polymer film (0.3 mm thickness) that is then laminated to the non-woven fabric system. Pause has claimed that when lamination is compared with the other PCM applications in garments, the method has the following advantages: a high PCM concentration per unit area, low cost of the micro-encapsulation procedure of the PCM and low weight of the garment is minimised.

5.1.3 *Applications in adaptive garments*

Textiles and garments containing PCM are still being developed and investigated. Many possible applications have been reported by various investigators. The possible applications reported are summarised in Table 5.2.

Table 5.2 Applications of PCMs in Textiles[4]

Applications	Brief description
Casual clothing	Underwear ; Jackets ; Sports wear; Skiwear
Professional clothing	Firefighters'uniforms; Bullet-proof vests; Divers' coveralls; Space suits; Sailor suits
Household	Curtains; Quilts; Batting; Sleeping bags
Shoe linings	Ski boots; Golf shoes
Medical uses	Surgical gauze; Bandage;
Building materials	In proof , In concrete
Other uses	Automotive interiors; Battery warmers;

There are some commercial garments that possess microcapsule of PCM for example the registered mark Outlast®. Outlast Technologies microencapsulated PCMs are called Thermocules™. Thermocules™ can be applied as finishing on fabrics or infused into fibres during the spinning process. The microcapsules are located inside the fibre. The fibres are spun to yarns then to socks, underwear or knitwear. Outlast® Thermocules® can be coated on fabric surface. Nonwovens are coated, and used in jackets linings, but it can also be applied to mid-layers between the first layer and the lining. In that case manufacturers are free to choose any design and first layer they want.

Wang *et al.*[16] reported a special clothing system for thermal protection against extreme cold-weather conditions, consisting of four layers. The first layer is cotton fabric and the second is a non-woven polyester fabric treated with PCM enclosed in small polymer spheres with diameters of only a few micrometres. Non woven polyester fabric makes up the third layer, and the outermost layer consists of a waterproof breathable fabric. The second layer keeps the first cotton layer from becoming saturated with perspiration. When the PCM layer's temperature increases above the PCM's melting point, the PCM melts and becomes liquid. Thermal energy is absorbed and stored during this process. When the temperature of the PCM layer falls, the liquid PCM becomes solid and releases heat energy.

The main challenge in developing textile-PCM structures is their application method. Encapsulation of PCMs in a polymeric shell is an obvious choice but it adds weight to the active material. Efficient encapsulation, stability during use and integration of capsules onto fabric structure are some of the technological issues that would be considered.

Treatment of polyester, nylon 66, cotton, and wool fabrics with aqueous solutions of polyethylene glycol phase change materials and of plastic crystal compounds by a conventional pad-dry procedure produced modified fabrics with thermal storage and release properties 2-2.5 times greater than those of untreated fabrics at the same temperature intervals. Application of another plastic crystal compound in aqueous solution to the fabrics under these conditions was not suitable, since it sublimed during the fabric drying process. Fibre type and heating rate appeared to have little effect on the overall heat content or thermal performance of the treated fabrics.[5]

New types of clothing of heat-storage garment came onto the market. The PEG-coated fabrics produced by the Mitsui Corporation were used as ski and sportswear.[17] Other thermo-regulated textile products, such as blankets, sleeping bags, underwear, jackets, sports garments, socks, ski boots, helmets, etc., have been on the market since 1997.[11]

5.2 Shape Memory Polymeric Textiles

Several kinds of shape memory polymers can be employed to prepare shape memory polymer fibres. The fibres made of shape memory polymers are spun by wet spinning, dry spinning, melt spinning, reaction spinning and electric spinning.

Shape memory yarns include the shape memory fibres alone and blended, warped or core yarn of shape memory polymer fibres and natural, regenerated, or synthetic fibres. Shape memory fabrics include woven, knitted, braided and nonwoven fabric. The fabric can be applied to textiles depending on the properties and uses, such as the collars, cuffs of shirts and any other apparel which need shape fixity; elbow, knee of apparels and any other cloths which need the recovery of bagging and shape fixity of denim, velvet, cord, knitting fabric and any other fabrics.

Shape memory finishing fabric can be acquired with coating shape memory emulsion or combining shape memory film. A Japanese patent[23] published a nonwoven fabric combined with shape memory polymer and shape memory adhesive. A US patent[24] published a shape memory fabric which adhesives ordinary fabric by shape memory polymer powder. Hong Kong Polytechnic University has studied shape memory finishing chemicals and technologies for cotton fabrics, wool fabrics and garment finishing. This method is more environmental, simpler finishing technology than other traditional finishing methods and the fabrics have good temperature sensitivity.

5.2.1 *Shape memory polymeric fibres*

As shown in Table 5.3, the SMPU fibres were synthesised by pre-polymersation, spun by wet spinning method and treated with high pressure steaming as described in our reported literature.[16,16,18,22]

Table 5.3 Mechanical properties of SMPU fibres.

Sample code	Hard segment Content (%)	Linear Density (tex)	Maximum Strain (%)	Tenacity (cN/tex)
SMPU64	64.1	7.18	147.0	7.4
SMPU55	54.7	5.28	172.4	7.5
SMPU48	47.7	5.14	215.9	8.7

Dynamic mechanical properties of the samples show that a difference between SMPU fibre and conventional man-made fibres is the variation of E' in normal using temperature range. For SMPU fibres, the variation of E' is very significant. Namely, when the temperature was increased above the T_g, the E' will sharply decrease and the rubbery state plateau will appear and be extended to above 160°C as shown Fig. 5.1 and Fig 5.2. Therefore, this switch point imparts the heating responsive shape memory properties to the SMPU fibre in normal using temperature just as reported T_g used as the switch temperature in the shape memory polyurethane film. Figure 5.3 illustrates the temperature dependence of loss tangent, which was used to pinpoint the glass transition temperature of SMPU fibre. The T_g of soft segment reflected by the peak of tanδ can be controlled with solely changing hard segment content. It will be used in fabricating SMPU fibre with different 'switch' temperature. In Fig.5.4, it can be observed that the fixity ratio to temporary elongation decreases with the decrease of hard segment content. It might be due to the decreasing trend of T_g of soft segment with the decrease of hard segment content.

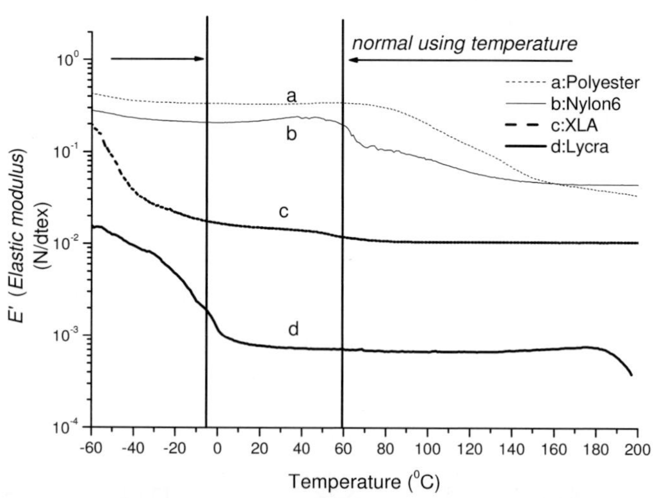

Fig. 5.1 Temperature dependence of storage modulus (E') of existing man-made fibres.

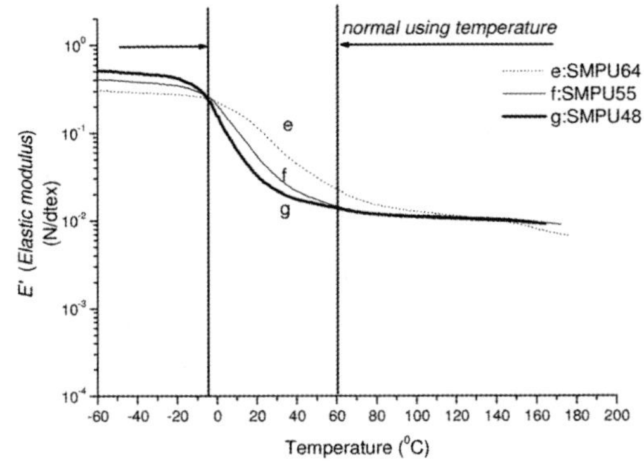

Fig. 5.2 Temperature dependence of storage modulus (E') of SMPU fibres.

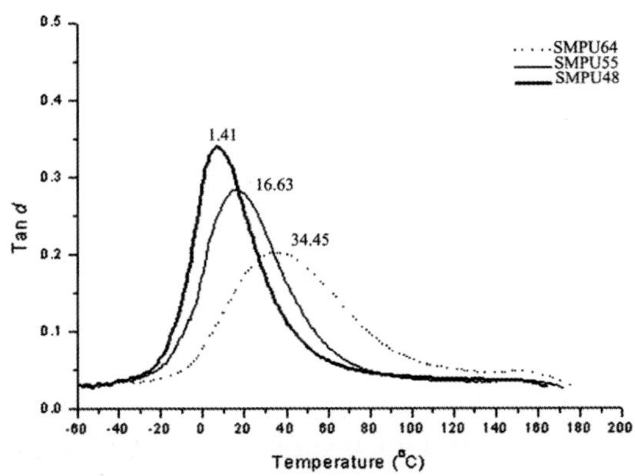

Fig. 5.3 Loss tangents (tanδ) of SMPU fibres.

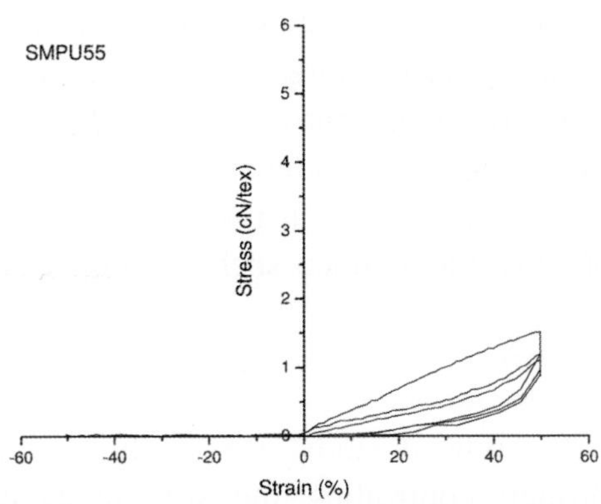

Fig. 5.4 Cyclic tensile tests for evaluating shape memory effect of SMPU fibres.

5.2.2 Shape memory function from fibre to fabrics

5.2.2.1 Self-adaptability of shape memory fabrics

For wet spinning shape memory fibres, the glass transition (T_g) of amorphous soft segment phase was used as switch temperature to control the fixity of temporary elongation and recovery temperature. Whereas, the physical crosslink of hard segment phase was responsible for memorising the original length of fibres. In this way, the fabric composed of SMPU fibres was designed to possess the following advantages. When the garment with given size is enlarged to fit different wearers' figures, the fixity ability to temporary deformation caused in wearing process was expected to diminish most of pressure sensation to wearer. After wearing, heating in washing or drying process will give rise to the recovery of the original size. The mentioned mechanism stems from the shape memory function of SMPU fibres, which was expected to be imparted into fabric, garment in this study.

The pressure in vertical direction to testing cylinder surface of enlarged fabrics was detected. The fabric column specimens made with SMPU fibres and other commercial fibre products were firstly enlarged with various elongation ratios in perimeter and then fixed in the deformed size for ten minutes. After that, specimens were set onto the testing cylinder surface. The Pliance-X system recorded the average pressure. Although the effect of fabric structure of used fabric specimens was tentatively ignored, the resultant data illustrates that the fabric made with SMPU fibres after stretching with various elongation ratio can effectively diminish the pressure in vertical direction in comparison with fabrics made with Lycra or PA/Lycra. Hence, through this study, it is expected to offer a novel smart textile material to automatically fit customised size without uncomfortable tension to wearers. It potentially was used to improve the comfort sensation of textile products such as intimate apparel.

5.2.2.2 Self-repairability of shape memory fabrics

With respect to the related study, we synthesised the shape memory polyurethane through controlling the soft segment length and hard segment content so as to fabricate the SM polyurethane filaments

possessing the soft segment glass transition temperature close to or higher than body temperature.[16] Three kinds of knitted fabrics were prepared by using (a) 100% SM fibre, (b) shape memory core yarn (50% cotton and 50% SM fibre) and (c) two ply yarn (100% cotton). The bagging recovery test was conducted to evaluate the thermal responsive shape memory function of the knitted fabrics.
Basic fabric parameter is illustrated in Table 5.4. The shape memory specimens were then heated inside the curing machine to form the permanent flat shape and fixed at ambient temperature. The strain recovery curves of shape memory fabrics are shown in Figs. 5.5, 5.6 and 5.7.

Table 5.4 Basic Fabric Parameter (Plain Knit) (90×90mm).

Fabric Code	Fibre Content	Yarn Type	Count (tex)	Tightness factor
S1	100%SM Polymer PU56 (Y1)	shape memory fibre	9	13.72
S2	100%SM Polymer PU66 (Y4)	shape memory fibre	20	12.08
S3	100%SM Polymer PU66 (Y4)	shape memory fibre	20	10.4
C/S 1	50% SM Polymer PU66 (Y8) & 50% Cotton	shape memory core yarn	16	13.72
C/S 2	50% SM Polymer PU66 (Y8) & 50% Cotton	shape memory core yarn	16	10.4
C1	100%Cotton (Y13)	Two-ply yarn	63	19.25
C2	100%Cotton (Y13)	Two-ply yarn	63	18.04
C3	100% Cotton (Y13)	Two-ply yarn	63	16.75

Bagging test was designed for evaluating performance of different knitted fabrics by Instron machine, and to find out the influence of shape memory effect by temperature. Bagged fabrics were formed and were heated in the oven at a constant temperature such as 30°C, 75°C respectively for three minutes. Non-recoverable bagging height was measured after heating every minute. Strain fixity (%) and Strain recovery ratio (%) were calculated by the bagging height.

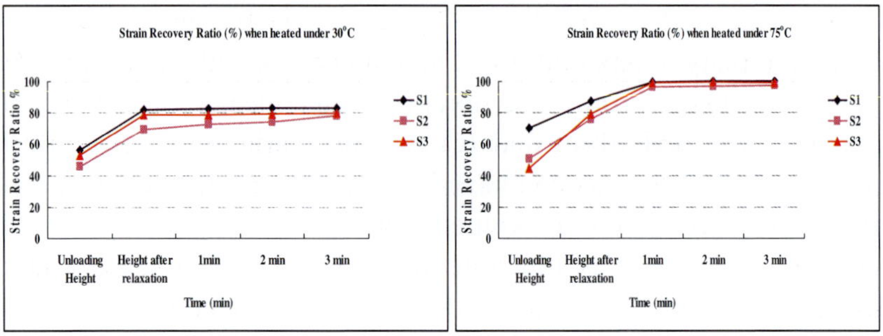

Fig. 5.5 Strain recovery ratio (%) of 100% shape memory knitted fabric solely composed of SM fibres (75°C, 30°C).

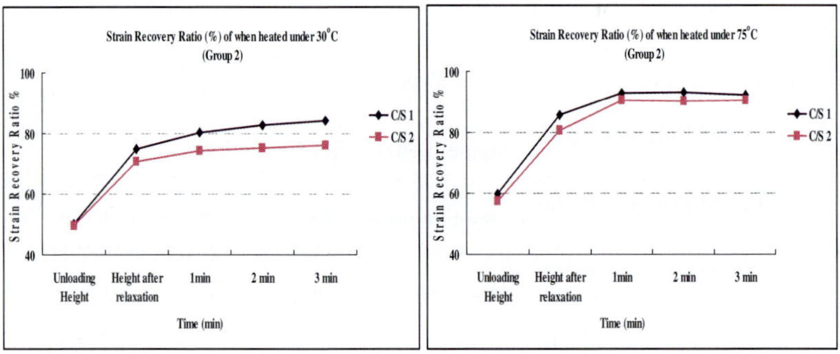

Fig. 5.6 Strain recovery ratio (%) of 50% SM fibre and 50% cotton knitted fabric (75°C, 30°C).

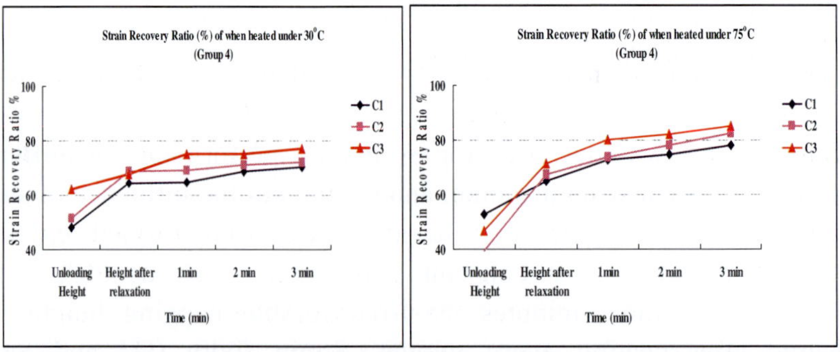

Fig. 5.7 Strain recovery ratio (%) of 100% Cotton Knitted Fabric (75°C, 30°C).

In the investigation of knitted fabric, it can be found SM knitted fabrics containing 100% SM fibre and shape memory core yarn (50% cotton and 50% SM fibre) presents thermal sensitive shape memory performance in comparison with the knitted fabric solely composed of two ply yarn (100% cotton). The recovery ratio for knitted fabrics composed of 100% SM fibres can be approximately 100% after heating above the transition temperature of SM fibres. Even the fabrics containing 50% shape memory fibres can possess above 90% recovery ratio after heating to 75 °C.

5.2.3 *Shape memory garments*

Different shape memory effects can be inspired and achieved according to fabric structure, yarn configuration and shape memory fibre contents. For general fabrics with shape memory fibres, such as plain or twill fabric, an original shape can be set above a hard segment transition temperature, which is higher than the normal use temperature and the switch temperature, and then cooled to a lower temperature so that the original shape is fixed. In the case where the switch temperature is higher than the normal use temperature, even if the fabric deforms, it easily returns to the original shape it remembers when heated to a temperature higher than the switch temperature. In the case where the switch temperature is lower than the normal use temperature, the article can return to its original shape when the deforming force is removed. A series of fabrics retain their original shape such as flat or bagging and can be set an original shape, such as crease, at a temperature higher than a hard segment transition temperature, then, cooled to a lower temperature so that the original shape is fixed. Even if the fabrics are deformed, they easily return to their original shape when heated to a temperature higher than the switch temperature.

As compared the bagging recoveries of knitting fabrics with or without shape memory fibres, Fig. 5.8 explains the results of these comparisons. For all tested fabrics, with the time increasing, the bagging height decreases and bagging recovery rate increases. The fabrics with shape memory fibres have better heat sensitivity and distinct bagging recovery rate when heated than other fabrics. The fabrics with shape memory fibres have lowest begging residual in the same fabric series (PET series; PA series; cotton series). The fabric with shape memory fibres has lower original bagging height than the fabrics with pure PET fibres or PA fibres or cotton. It means the fabric with shape memory fibres has some elasticity under common temperature.

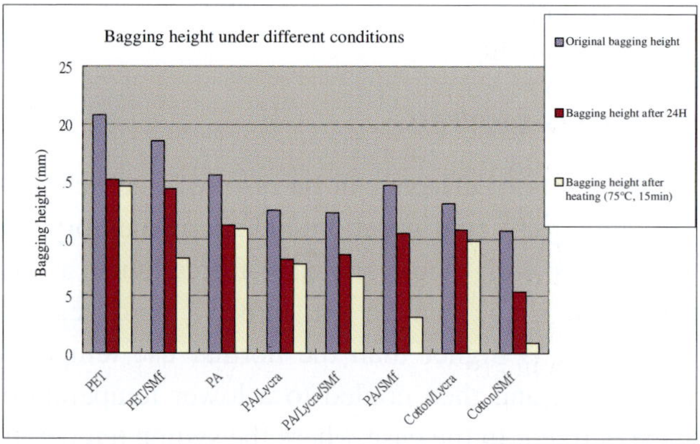

Fig. 5.8 Bagging recoveries of different fabrics.

Shape memory fibres and yarns can be used in knitted garment such as 3D textiles, underwear, sportswear, skirts and gloves which have good shape memory properties and advantages. A corset was knitted from the friction yarn and set at a temperature higher than the switch temperature to get the original dimension. The corset can be stretched to allow ease of dressing. However, the switch temperature of the so called fabric is below body temperature. Therefore, when put on the body, the corset became tighter and tighter and recovered the original shape to achieve a satisfactory tying effect. A shaped sports bra was knitted by shape memory fibres or shape memory yarn. The bra has a switch temperature of about 30°C. The body temperature becomes higher than the switch temperature when people do sports. So the bra became tight and had good protective function to the body, but people didn't feel too tight and uncomfortable.

5.2.4 *Shape memory finishing fabrics and garments*

Hong Kong Polytechnic University has studied shape memory finishing chemicals and technologies for cotton fabrics. The finished cotton fabrics are originally flat. After washing or being stored in closet, the crease and bagging will appear. The switch temperature of shape memory emulsion is about 60°C. The shape memory finished fabric feels soft, the crease and wrinkle will disappear and recover the original shape when washed in hot water or dried at higher than 60°C.

The advantages of shape memory finishing fabric are as follows:

(1) has good temperature sensitivity
(2) has higher breaking strength
(3) has better performance and improved wrinkle resistance
(4) has better wash ability, flat appearance and crease retention increase with increased number of washes times.

Wool fabrics and sweaters have serious felting phenomenon and shape shrinkage due to the oriented friction. The finishing methods to increase the dimensional stability of wool fabrics include resin coating, chlorination shrink proofing, oxidation anti-felting and other methods. Hong Kong Polytechnic University has studied shape memory finishing chemicals and technologies for wool fabrics and garment finishing. This method is more environmentally friendly, simpler finishing technology than other traditional finishing methods and the fabrics have good temperature sensitivity. Now this finishing method can meet the following demands: flat appearance; bagging recovery; dimensional stability and fastness.

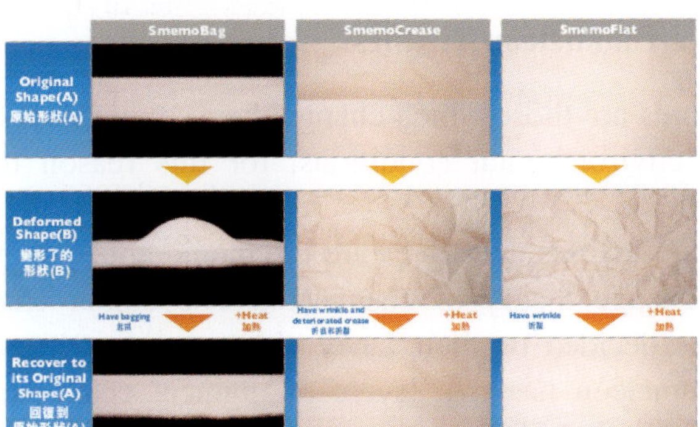

Fig. 5.9 Shape memory recovery of cotton fabric.

Shape memory finishing chemicals and technologies for wool can increase the dimensional stability of wool fabrics and felting.

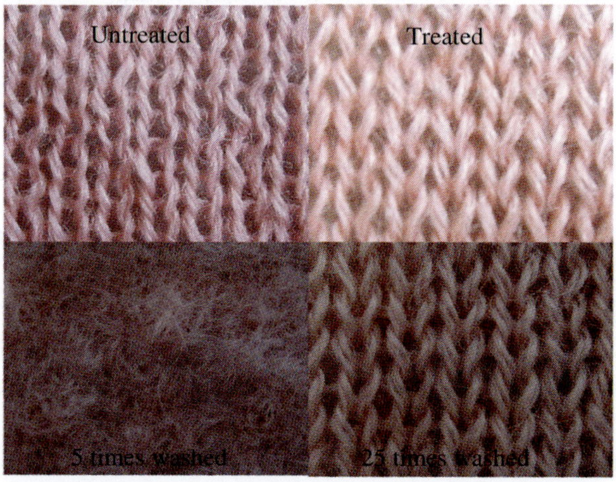

Fig. 5.10 Texture of untreated and treated wool fabric.

Figure 5.10 shows the texture of untreated fabric and treated fabric with shape memory polymers. The texture of treated fabrics is clear, but it is distorted in the untreated fabrics. After washing, the texture of treated fabrics is still clear and it can stand 25 times washing and tumble drying according to AATCC wool washing standard, but untreated fabrics have serious felting after five times washing.

5.3 Adaptive Chameleon Textiles

Adaptive fibres are those which change their colour reversibly according to external environmental conditions, for this reason they are called chameleon fibres.[19]

Chameleon textiles and fibres are the bionic applications according to the colour change theory of chameleon. The colour change ability of chameleon is decided by their fancy pigment cell. Different from other animals, chameleon has two layers of pigment cell under cuticle. The upper pigment cell is like shutter which can expand or contract. When expanding, the shutter is closed and we only see the colour of shutter.

When contracting, the shutter is open and the below colour shows. So a pigment-containing or pigment-producing cell, that by expansion or contraction can show different colour of the skin.

Traditionally, the industry has been concerned with maintaining the colour regardless of environmental influences. Yet there are new markets, which alter their colour on demand. Chameleon fibres will allow for the creation of value added products in traditional industry, as well as entry into new areas. There are many methods for chameleonic textiles. Molecules can change colour in the presence of acids and bases, but these reagents and solvent required to transport them make this method difficult to implement in applications. Oxidation states changes are also effective, but require the migration of ions. The response time can be fast in solvents, but this complicates the device. Gel-type devices might also be possible, although physical robustness, oxygen stability and response times represent serious engineering challenges.

There are a number of systems that undergo reversible bond breaking, bond forming processes that result in colour changes. Most commonly, these are light initiated processes. Certain compounds have been shown to undergo colour changes as a result of applied stress and a chromic system is constructed by surface modification of conducting polymers. Some highly polaraisable systems have been observed to change colour in the presence of electric of magnetic fields.

5.3.1 *Types of chameleon materials for adaptive textiles*

The chromic materials which can radiate the colour erase the colour or change colour. Because its induction is caused by the external stimulus, we can classify chromic materials depending on the stimulus [20]

Photochromic: external stimulus is light.

Thermochromic: external stimulus is heat. Thermochromic materials change colour as a result of reaction to heat.

Solvatechromic: external stimulus is liquid or gas.

Electrochromic: external stimulus is electricity.

Piezorochromic: external stimulus is pressure.

Carsolchromic: external stimulus is electron beam.

Photochromic materials are generally reversible unstable organic materials that change molecular configuration with the influence of a special radiation.

Thermo-chromic materials are those which colour changes when heated, especially thermo-chromic dyes which can change colours at particular temperatures. There are two types of thermo-chromic systems that have been used successfully in textiles: the liquid crystal type and the molecular rearrangement type. In two cases, the dyes are entrapped in microcapsules and applied to garment fabric. The most important types of liquid crystal for thermo-chromic systems are cholesterin types, whose adjacent molecules are arranged to form helices. The selective reflection of light of the liquid crystal leads to the thermo-chromism. The refractive index and the helical arrangement of the liquid crystal govern the wavelength of the light reflected. Because the length of the helical varies with temperature, the wavelength of the reflected light is altered, and colour changes. Disadvantage of liquid crystal thermo-chromic material is that it is expensive and sensitive to chemicals which weaken the colour change effects.

An alternative method of inducing thermo-chromism is by a rearrangement of the molecular structure of a dye. The most common dyes which exhibit thermo-chromism through molecular rearrangement are the spirolactones. Crystal violet lactone and related compounds were used in colour changing printing. It is a reversible colour system with temperature change. Under some temperature, bisphenol A emits proton and crystal violet lactone open rings and combines with the proton to make π system and shows colour. The colour varies with the substituent. When substituent is H, the colour is violet; when R is CH_3 and X is OCH_3, the colour is blue.

The commercial thermo-chromic products include multi-components. It generally consists of three components, namely the colour former, an acidic catalyst and a non-polar co-solvent medium. The thermochromism centres around the interaction of the colour former and developer. The organic thermo-chromic materials are easy to obtain and they are cheaper than thermo-chromic liquid crystal but not sensitive and their colour change range is narrow. These materials can be used on some thermo-chromic printing without high quality demands.

In electro-chromism, a reversible and visible colour change is observed in a material as a result of electrochemical oxidation or reduction. When oxidation or reduction occurs, the electronic structure of the electro-chromic π-conjugated molecule changes, shifting the π–π^* electronic absorption. The synthesis of electro-chromic polymeric materials has been researched in many laboratories over the last years. Reynolds and co-workers have been successfully developing electro-chromic materials with tailored colours. By synthesising derivatives with a range of band gaps, a broad range of colours has been achieved. Clemson University researchers have prepared bis(3,4-ethylenedioxythiophene)-based materials (BEDOT-Q) for electro-chromic applications that provide a broad range of emission wavelengths.

5.3.2 *Application methods of chameleon materials in textiles*

The pioneering and simple methods of integrating chameleon materials into textiles are printing, coating and dying. For printing and coating, the chameleon materials are en-microcapsuled fist then coated or printed on fabric surface by common methods such as pad-dry-cure process.

Fibre method was developed later than printing and dyeing method for chameleon application. Fibre with chameleon has obvious advantages such as good hand feeling, washing durability and chameleon permanence. The main manufacturing method of chameleon fibres includes solution spinning, melt spinning, treatment and graft copolymersation. Chameleon polymer and anti-transfer agent are added in solution and spin directly. A Japanese company uses the solution of acrylonitrile/styrene/vinyl chloride copolymer and chromic to spin the photochromic fibre. These fibres have no colour without sunlight and show dark green under sunlight or UV. It can be used in garments, curtains, carpets and toys.

Melt spinning method is composed of polymersation method, polyblending method, sheath-core compound spinning method. Post-treatment method also make the fibre has chromic property. For example, fibres or fabrics are dipped into solution which includes styrene monomer with pyrone and the fibres have photochromic property due to the monomer polymerising.

Chromic groups are introduced into polymer. These polymer can be spun and produce chromic fibres. This method is a polymersation method. Polyblending method is using chromic polymer and polyester, polyamide and other polymers blended and melt spinning. This method is simple but has high demand for the chromic polymer because of the high spinning method.

Sheath-core compound spinning is the main method for chromic fibre manufacturing. The core component includes chromic agent and common fibre as sheath. The core contains 1-40% chromic agent and the melting point is below 230°C. This chromic fibre has good hand feeling, washability and colour reaction.

Pre-treatment and graft copolymersation method also are used on the chromic fibres production. This method is simple, easy to control and extended. The pioneering and simple methods combing chromic material and fabrics are printing and dying. The chromic materials should be encapsulated first and mixed into resin solution. Fabrics are printed by the mixing solution and get the chromic fabric. The drying temperature is about 80-90°C. Curing temperature is about 140-150°C and time is about 3-10min.[20]

The formation of polyaniline hollow-fibre membranes were studied this year. The potential usefulness of these hollow-fibre membranes is as the transport of ionic charge carriers to electrochromic coatings. The ionic charge carriers act as dopants to turn the colour on or off.

Toray Industries reported in 1987 the development of a temperature sensitive fabric using microcapsules heat sensitive dyes, which are coated homogeneously over fabric surface agent (electron acceptor) and of basic four colours and combined 64 colours. This fabric can reversibly change colour at temperature greater than 5°C and is operable from - 40 to 80°C. The change of colour with temperature in these fabrics is designed to match the application.[21]

Some of these fibres present the phenomenon called solvate chromism,[22] whose colour changes when in contact with liquid, for example water. These materials are used for 'designer' swimsuits. Apart from this, the most important application for chromic materials is to create fantasy designs which change its colour depending on the volume of incident light.

5.4 Luminescent Adaptive Textiles

5.4.1 *Introduction of luminescent materials*

Luminescence is the emission of light by substances that have absorbed energy. The difference between chromic and luminescent materials is that the first one changes colour when the second one emits light.[23] There are several types of luminescent effects.

Photoluminescence: external stimulus is light. There are two types of photoluminescent materials, the organic and the mineral. The organics photoluminescent are compounds which have a good molecular conjugation and the relaxation mode allows the emission of a photon.[24] Photoluminecent materials are generally used in textiles for application in dress for a night club and more interestingly in the marking of labels with UV revelation materials for the detection of imitation goods and security labels. Phosphorescent materials have been applied in inks which are used in work clothes for road works/repairs in bad light situations. The obtained effect is generally known as glow in the dark.[25]

Electroluminescence: external stimulus is electricity. Electroluminescence includes inorganic semi-conductors, organic small molecules and conjugated polymers. Polymers as electroluminescence have processing and structural advantages. Most electroluminescence polymers are soluble in water or common organic solvents. They can be used to fabricate large area films at low cost. In addition, amorphous polymer films degrades the long term stability of small molecule based, thin film, organic electroluminescence devices. Schott have studied the emission spectrum which is nearly identical to that of photoexcited fluorescence material. It is indicated that the emitting state in electroluminescence and ordinary fluorescence are the same. The emission spectrum is generally fairly broad and not satisfactory saturated pure colours in the sense of the trichromic colour system. One way for improvement is material modification; the other is the use of micro-cavities.[26]

Chemiluminescence: external stimulus is a chemical reaction. Generally, chemiluminescent reactions are oxidation reactions. Chemiluminescent compounds are exceited by oxidation with oxygen or hydrogen peroxide. The chemiluminescent processs of organic molecules include the following three steps: (1) formation of intermediate compounds by chemical reactions; (2) excited molecular formation by an intermediate through a transition state; (3) light emission from an excited molecule or energy transfer and subsequent light emission from a fluorescent energy acceptor.[27] The main chemiluminescence materials include luminol derivatives, dioxetanes, lophine, acridine derivatives, tetrakis(dimethylamino)ethylene, indole derivatives, Schiff bases, oxalates, oxamide, and diphenoylperoxide.

Opticoluminescence is the typical effect encountered in optical fibres. The use of these kinds of fibres is now implanted for manufacturing textiles that emit light. There are also applications with optical fibres at the development stage for the creation of screens.[28]

5.4.2 *Applications of luminescent in adaptive textiles*

Electroluminescent compounds and opticoluminescent materials are more used in textiles. The most application results from the use of electroluminescent yarn (constituted by mineral compounds) in the area of fashion garments. The electroluminescent phenomena studied in the area of smart textiles thanks to the emergence of the organic light emitting diodes[29,30] that possess a flexible character and that are envisaged for the manufacture of flexible screens adequate for the wearable computer.

Fibre-based devices such as microring lasers and electrically excited LEDs were produced. High-quality-factor lasers were made using micro-cavities such as cylindrical coatings of 2, 5-dioctyloxy-substituted poly(*p*-phenylene vinylene) (DOO-PPV) deposited on various substrates including optical fibres, metal wire, and polyaniline fibres.

Optico-luminescent fibres, optical fibres are widely used in textiles at present. Various applications of optico-luminescent textiles include structural health monitoring and wearable sensing, and clothes with unique appearance, flexible and wearable displays. Integration of optical fibre-based sensor elements into clothes can accurately monitor bodily and environmental conditions, which is of importance to various hazardous civil occupations and military. Advantages of optical fibre sensors include resistance to corrosion, high flexiblility and suitable visible spectral range. Both silica[31] and polymer-based PBG fibres[32] are now used for textile applications. Total Internal Reflection (TIR) fibres have been used in textiles,[33] as well as backlighting panels for medical and industrial applications.[34,35]

Recently photonic crystal fibres (PCFs) have been introduced. Their cross-section contains either periodically arranged micron-sized air voids[36] or a periodic sequence of micron-sized layers of different materials.[37-39] The fibres appear coloured due to optical interference effects in the microstructured region of a fibre. Application of photonic crystal fibres in textiles was only in the context of distributed detection and emission of mid-infrared radiation for security applications.

B. Gauvreau, N. Guo, K. Schicker, etc. have made solid and hollow core PBG Bragg fibres using layer-by-layer deposition of polymer films, around the core mandrel.[40] Such fibres are easier to manufacture than hollow core PBG fibres, and their cost is much lower.

A RGB yarns based on PBG fibre changing its emissive colour can be used on textiles. The colour of the RGB yarns will be very stable over time and largely independent of the fluctuations in the intensity of a light source. Moreover, fibres having the same reflective colour[41] can have different emissive colours. Such colour-changing textiles could be used in uniforms, signage and machine vision.[42]

5.5 Conductive Polymer Textile

Conductive textiles are a broad range of products with widely differing specific (surface) conductivity. Electrically conductive textiles have become a topic of research.

The apparels with electrically conductive properties also named 'intelligent apparels' by utilising special textiles or electronic devices integrated into the textile structures. The broad principle topics are:

(1) Professionals (the need for 'hands free function'); (2) Health care (monitoring); (3) Everyday life (telephone); (4) Sports (training, performances measurement); (5) Leisure (aesthetic customisation network games).

These new achievements of the textile industry make possible for electronic devices to be integrated into the textile, and modify the function of the apparel. Apart from the main functions of apparel, the clothes become an interface between the individual and the environment. For all the applications of electrically conductive fibres and textiles in sensors, electrostatic discharge, interference shielding etc. it is important to modify the fibre properties (electrical and other) in order to modify the general properties of the textile structure and to add new functions.

5.5.1 *Conducting polymer fibres, fabrics and sensors*

Conducting polymers have attracted great interests in research due to their electrical properties and responsiveness to external stimuli. But the poor mechanical properties of chemically or electrochemically synthesised conducting polymer films hinder their applications. Now this limitation has been overcome by the polymersation on textile surfaces to produce conductive fabrics with good mechanical properties.

The pseudocapacitance of electronically conducting polymers arises due to the fast and reversible oxidation and reduction processes related to the p-conjugated polymer chains.[43,44] In addition, p-dopable polymers are more stable against degradation than n-dopable polymers.[45] Hence investigations with p-dopable polymers are welcome. The conductive polymer used as modified membrane is coated to the well conductive material to reduce resistance, combined as inorganic–organic hybrid electrode material, which fully utilises their respective advantages.[46,47]

A number of groups have reported promising findings on electrolytic capacitors (ECs) based on polymer electrodes. However, the long-term stability during cycling may be a problem. Swelling and shrinking of electro active polymers are well known and may lead to degradation during cycling, and the charge-storage mechanism in polymer electrodes is still not completely understood.[48] Conductive polymers fibres show electrical properties due to their conjugated double bond chain structures. But they are insoluble and infusible because of their strong intermolecular interactions.

A new method is using intrinsically conductive organic polymers to produce highly conductive textile materials. One common route is to apply dispersions or powder of conductive polymers as coatings. Those methods usually lead to low conducting materials. An interesting alternative method is to create the conductive polymers by polymersation of monomers on the textile. The textiles sensors and electronics textiles are the integration of electronics as sensors and microchips, in order to detect and analyse stimuli and provide the adequate response.

Polypyrrole (PPy) is one of the most important conductive polymers, which is ease of preparation, high conductivity, good environmental stability and non-toxicity. PPy filament can be used as chemical sensors for detection of gases and vapors and PPy films can be used as artificial muscles with tactile sensitivity. Because of the good adhesion with different substrates, PPy can form composites with non-conductive fibres or fabrics to prepare the electrical conductive textiles. The composites can be used as flexible smart materials in electromagnetic interference (EMI) shielding, broadband microwave absorbing, static charge dissipation, biomedical and tissue engineering, etc.

Polyaniline (PANI) is another important material because of its excellent environmental, thermal and chemical stability. A doped polyaniline has been mainly used to make conductive fibres. It can be used as textile sensors with all the mechanical properties as elongation, bending, shearing and twisting corresponding to a real textile yarn. These textile properties are very important because the conductive fibres should be woven and knitted to form textiles structures.[49,50]

PPy as well as PANI are common conductive organic polymers, but handling them is cumbersome or hazardous. Unwanted side reactions occur during polymersation, and little is known about long-term stability of the end product. PANI coated PET yarns preserve their original strength and flexibility. PANI molecules have more possibilities to penetrate into the spun yarns due to their inherent bulky surface formation and free volumes.[49] Dierk Knittel studied the electrically conductive layers of poly(ethylenedioxythiophen)(PEDOT) on textiles made of cotton and synthetic fibres by an on-substrate polymersation.

The presented results show the promising application on heating elements, shielding materials for EMI-protection, or signal transducers, and even enhance further galvanic metal deposition on the modified fabrics.[51]

Richard used conductive poly(α,ω-bis(3-pyrrolyl)alkanes) to coat on wool fabrics.[35] The coated wool had the highest conductivity compared to those produced by solution polymersation and vapour polymersation.

Conductive textiles coated with poly (1, 8-bis (pyrrolyl) octane) had the second highest conductivity, only succeeded by poly (1, 5-bis (pyrrolyl) pentane). Rossi *et al.* developed a glove based on the fabrics of polypyrrole-coated Lycra/cotton, but the sensor was aged in air and the conductivity decreased continuously. Kim *et al.* also proposed the PPy-coated PET/Spandex which can be used as a strain sensor for a large deformation up to 50%.[52] Y. Li, X. M. Tao, etc. reported the chemical vapor deposition of thin coatings of polypyrrole on fabric surfaces, which have high strain sensitivity of the sensor and good stability. The PPy-coated fabrics can be used in sensing garment, wearable hardware and rehabilitation, etc.

The methods for producing fibres electrically include wet spinning, melt spinning from conductive polymer polyaniline coating fibres with electrically conductive materials.[53,54] The coating techniques are simple in process and easy to handle. The textiles produced not only have controllable electrical properties, but also maintain their physical properties, such as mechanical strength and flexibility. P. Xue and X. M. Tao, etc have researched conductive fibres by using the carbon nanotubes on different substrates.[54] PVA/CNTs-coated polyester yarn, PP yarn and silk yarn have much better electrical properties than the other conductive yarns and preserve their original strength and flexibility. The tensile property of wool/nylon mixed yarn decreased greatly after acetylisation treatment.

5.5.2 *Applications in textiles*

Georgia Tech has developed a smart shirt for use in combat conditions.[55] The garment uses optical fibres to detect wounds and special sensors to monitor vital signs during combat conditions.

The types of sensors could be used in a large variety of fields such as medical monitoring, disease monitoring, infant monitoring and athletics military uses.

The most famous commercial products are the KENPO jacket that has integrated MP3 lectors and the Ipods jeans by Levis.[56] A few prototypes have been made. The products of gleno-humeral joint, elbow joints and the joints of the hand have been made. In these early prototypes, sensors were posted in correspondence to each joint in a number equal to the degrees of freedom. In the new generation, a large set of sensors is allocated in the garment.[53] The LifeShirt System is the first non-invasive, continuous ambulatory monitoring system. It can collect pulmonary, cardiac and other physiologic data, and correlate them over time.

5.6 Other functional textiles

5.6.1 *Nano-technology applications in textiles*

The application of nano-technology in textiles and apparels is gaining increasing importance. Nanotechnology uses structures that have at least one dimension that is nano-sized. Consequently, nanotechnology improves the properties of materials due to such a small size. Nano particles have high surface area to volume ratio and have high affinity for fabric which leads to improved durability. Some of the finishing based on nano particles is found to maintain the breathability and soft hand. Some of the most successful applications are water repellence, soil resistance, UV protection, antibacterial and antistatic finishes as well as self cleaning which are described below.

5.6.1.1 *Water repellence*

The water repellence property can be achieved by nano-whiskers added to the fabric, nanophere coated and plasma treatment methods.

Nano-tex resists spills is using the Nano-whiskers treatment on the fabric surface. Nano-whiskers are hydrocarbons and 1/1,000 of the size of cotton fibre, which create a peach fuzz effect of the fabric. The spaces between the whiskers on the fabric are smaller than the drop of liquid, but larger than liquid molecules; thus liquid remains on the top of the whiskers and above the surface of the fabric.[33,57]

NanoSphere finishing has a three-dimensional surface structure with gel-forming additives which can repel water and prevent dirt particles from attaching themselves. The mechanism is similar to the lotus surface which is hydrophobic, rough and textured. Once water droplets fall onto them, water droplets bead up and will roll off if the surface slopes slightly. So the surfaces stay dry during a heavy shower and the droplets pick up small particles of dirt as they roll, so the leaves of the lotus plant keep clean.[58]

A hydrophobic property can be used on a cotton fabric by coating a nanoparticulate plasma film. PAN textiles can be made water and oil repellent using a low-pressure plasma process which induces simultaneously the grafting and the polymersation of monomers on the surface of the substrate in one step.[59]

5.6.1.2 *Anti Bacterial Property*

Nano-sized silver, titanium dioxide and zinc oxide are used to impart antibacterial properties on textiles. Metallic ions and metallic compounds display a certain degree of sterilising effect. The oxygen in the air or water is considered to become active oxygen relying on catalysis with the metallic ion, and dissolving the organic substance to create a sterilising effect.[60] The number of nano-sized particles per unit area is increased, and thus antibacterial effects can be maximised.

Nano-silver particles are very reactive with proteins and have an extremely large relative surface area. When contacting bacteria, it will inhibit cell growth and multiplication.[61,62,18] Titanium dioxide is a photocatalyst. When illuminated by light, the electron (e-) and electric hole (h+) pairs will appear on the surface of the photo catalyst. The negative electrons and oxygen will combine into O_2; the positive electric holes and water will bring hydroxyl radicals. These unstable chemical substances will combine with the organic compound on and turn into carbon dioxide (CO_2) and water (H_2O). This cascade reaction is called 'oxidation-reduction'.[63] A fabric treated with nano-TiO_2 could have effective protection against bacteria and the discolouration of stains, because of the photocatalytic activity of nano-TiO_2 Zinc oxide is also a photocatalyst, and has the similar photocatalysis mechanism to that of titanium dioxide.[1,64,65]

5.6.1.3 *Self cleaning*

TiO_2 could generate highly oxidative radical species (HO_2•–HO•) and oxidants (H_2O_2) from oxygen and water vapor under light irradiation. These reactive substances are able to discolour partially or totally organic stains like wine, coffee and make-up.

A team of researchers at Monash University in Australia and the Hong Kong Polytechnic University have developed a process that make keratin fabrics to self clean and protect themselves from UV degradation. Researchers have used a near room temperature sol-gel process to carefully apply nano-crystals of titanium dioxide to the protein fibres. In this way, the fibres will maintain their intrinsic properties. The fabrics could confer self cleaning properties and will remain protects against UV degradation.[33]

Except for the keratin fabrics, cotton and polyester fabric have been treated by TiO_2 to obtain the self cleaning fabrics and textiles. John H. Xin, etc. have studied single-phase anatase-coated polyester fibres pretreated with low-temperature oxygen plasma which showed significant improvement in self cleaning performance under simulated day light irradiation. The adhesion between TiO_2 layers and polyester substrates was improved by plasma treatment. The UV absorption of the titanium-coated polyester was significant enough to promote excellent UV protection to polyester.[66]

5.6.1.4 *UV Protection*

Nano-particles with a larger surface area per unit mass and volume leads to the increase of the effectiveness of blocking UV radiation. According to Rayleigh's scattering theory, the optimum particle size to scatter UV radiation between 200 and 400 nm will be between 20 and 40 nm.[67]

Inorganic UV blockers are better than organic UV blockers because of non-toxic and chemically stable under exposure to high temperatures and UV. TiO_2, ZnO, SiO_2 and AL_2O_2 are usually used for inorganic UV blockers, and titanium dioxide (TiO_2) and zinc oxide (ZnO) are commonly used. Nano-sized titanium dioxide and zinc oxide were more efficient at absorbing and scattering UV radiation than the conventional size and were better able to block UV. Various research works of UV-blocking treatment to fabric using nanotechnology were conducted. The sol-gel method was used on UV-blocking treatment for cotton fabrics. A thin layer of titanium dioxide is formed on the surface of cotton fabric which presents excellent UV-protection; the effect can be maintained after 50 home launderings.[68]

Apart from titanium dioxide, zinc oxide nanorods of ten to 50 nm in length were also applied to cotton fabric to provide UV-blocking property.[69]

5.6.1.5 *Antistatic Properties*

Nano size titanium dioxide, zinc ozide, nano antimony dioxide, tin dioxide are conducting materials causing the effective decay of static charge and have good antistatic effect.

As reported, electrically conductive materials such as nano-sized titanium dioxide, zinc oxide whiskers,[70] nanoantimony-doped tin oxide (ATO),[71] and silane nanosol[72] could improve the antistatic properties of synthetic fibres. Such material could effectively eliminate the static charge which accumulated on the fabric. On the other hand, silane gel has antistatic properties, because it absorbs water and moisture in the air by amino and hydroxyl groups and bound water. Nanotechnology has been applied in manufacturing an antistatic garment. W.L. Gore and Associates GmbH used nanotechnology and polytetrafluroethylene (PTEE-Dupont's Teflon®) to produce an antistatic membrane for protective clothing. Electrically conductive nano-particles are durably combined on the fabrics and cannot be washed off after a few laundry cycles.[73]

5.6.2 *Intelligent impact protection textiles*

Traditional impact protections in active sports were bulky, rigid protectors. But these are not ideal because the compromises between the amount of protection you can accept and the freedom of movement you can enjoy.

Developed and manufactured by UK based d3o Lab, d3o (dee-three-oh) has a unique composition containing 'Intelligent Molecules' which flow when you move but upon shock lock together to absorb the impact energy. The ability of d3o is lightweight and flexible whilst under shock. Although currently mainly used in the sports arena, d3o could be used in a multitude of situations.[33]

The Dow Corning® Active Protection System is another premium 'intelligent' textile that delivers unprecedented levels of impact safety and comfort for the wearer. It is comprised of innovative materials that instantly become rigid on impact, but flow with body movements when protection is not required. The Active Protection system can be sewn directly into garments and accessories and integrate easily and quickly into standard manufacturing processes. The patented System consists of a spacer textile treated with a responsive silicone coating. This 'intelligent' textile is soft and flexible under normal conditions, but hardens instantly on impact. If removing the impact force, the material immediately returns to a flexible state.[74]

5.7 Perspective

For PCM textiles, the main challenge in developing is the method of their application. Weight adding, efficient encapsulations, stability during use are some of the technological issues that would be considered. For shape memory polymer textiles, the properties of shape memory fibres and solution are the key points for the applications and final products development. For chameleon and luminescence polymer fibres and textiles, the efficient, stability and applications should be considered. Conductive polymer textiles are interesting research areas which attract more attention. These textiles endowed with the functions like sensors should be controlled accurately and ingeniously.

A few years ago, adaptive textiles were presented as imaginary products and as a non-competitive market. Nowadays they are an implanted customer interest and as the future of the textile industry. A lot of commercial products are available and, as it was presented this chapter, a lot of scientists are developing new ideas and concrete products.

References

1. Brahim, S., Narinesingh, D., and Guiseppi-Elie, A., Bio-smart hydrogels: co-joined molecular recognition and signal transduction in biosensor fabrication and drug delivery. *Biosensors and Bioelectronics*, 2002. 17(11-12): pp. 973-981.
2. Mondal, S., Phase change materials for smart textiles - An overview. *Applied Thermal Engineering*, 2008. 28(11-12): pp. 1536-1550.

3. Rupp, J., Interactive textiles regulate body temperature. *InternationalTextile Bulletin*, 1999. 45(1): pp. 58-59.
4. Erkan, G., Enhancing The Thermal Properties of Textiles With Phase Change Materials. *Research Journal of Textile and Apparel* 2004. 8(2): pp. 57-64.
5. Vigo, T. L., and Frost, C. M., Temperature adaptable hollow fibers containing polyethylene glycols. *Journal of Coated Fabrics*, 1983. 12(4): pp. 243-254.
6. Magill, C, M., Hartmann, H, M., Haggard, and S, J., *Multi-component fibers having enhanced reversible thermal properties and methods of manufacturing thereof*. 2005. United States Patent 6855422
7. Goda, H., *Machine crimped synthetic fiber having latent three-dimensional crimpability and method for production thereof*. 2004. United States Patent Application 20040234757
8. Zhang, X. X., Zhang, H., and Niu, J. J., *Thermoregulated fiber and its products*. China Pat ZL96105229.5, 2004. (This is a patent)
9. Zhang, X.Z., Jo Lewis, P., and Chu, C. C., Fabrication and characterization of a smart drug delivery system: microsphere in hydrogel. *Biomaterials*, 2005. 26(16): pp. 3299-3309.
10. Chen, C., Wang, L., and Huang, Y., Morphology and thermal properties of electrospun fatty acids/polyethylene terephthalate composite fibers as novel form-stable phase change materials. *Solar Energy Materials & Solar Cells*, 2008. 92: pp. 1382-1387.
11. Anon, A new class of hydrogels. *Chemical Week*, 2002. 164(22): p. 26.
12. Chen, Z., Mo, X., He, C., and Wang, H., Intermolecular interactions in electrospun collagen-chitosan complex nanofibers. *Carbohydrate Polymers*, 2008. 72(3): pp. 410-418.
13. Zuckerman, J. L., Pushaw, R. J., Perry, B. T., and Wyner, D. M., *Fabric coating containing energy absorbing phase change material and method of manufacturing same*. 2003. United States Patent 6514362
14. Shin, Y., Yoo, D. I., and Son, K., Development of thermo-regulating textile materials with micro encapsulated phase change materials (PCM). IV. Performance properties and hand of fabrics treated with PCM microcapsules. *Journal of Applied Polymer Science*, 2005. 97(3): pp. 910-915.
15. Pause, B., Nonwoven protective garments with thermo-regulating properties. *Journal of Industrial Textiles*, 2003. 33(2): pp. 93-99.

16. Urayama, K., Ogasawara, S., and Takigawa, T., Pure shear deformation of physical and chemical gels of poly(vinyl alcohol). *Polymer*, 2006. 47(19): pp. 6868-6873.
17. Lavin, M. and Preston, J., *Fibers with multifunctional properties: a historic approach Handbook of fiber Science and Technology, Volume III, High Technology Fibers Part C*. 1993, CRC press.
18. Liu, M., Bian, F., and Sheng, F., FTIR study on molecular structure of poly(N-isopropylacrylamide) in mixed solvent of methanol and water. *European Polymer Journal*, 2005. 41(2): pp. 283-291.
19. Gregory, R. V., Samuel, R. J., Hanks, T., and National Textile Center Annual Report (USA). 2001.
20. Bamfield, P., *Chromic Phenomena: Technological Application of Colour Chemistry* 2002: Royal Society of Chemistry Ed.
21. Jiyuuji, K. and Nobuyasu, N., Oakes, J., Batchelor, S. N., and Dixon, S., Azo dye location in textiles: a novel UV–vis approach. *Coloration Technology*, 2005. 121(5): pp. 237-244.
22. Krasovitskii, B. M. and Bolotin, B. M., *Organic Luminescent Materials*. 2002: Weinheim NY, VCH.
23. Lakowicz, J. R., *Principle of Fluorescence Spectroscopy*. 1999: Springer , second.
24. Gravisse, P. and Schoote, J. F., *Photoluminescent textile materials*.US patent 4211813
25. Schott, M., Introduction to the physics of organic electroluminescence. *J. Applied Physics*, 2000: pp. 381-402.
26. Wu, Z., Wang, X., Tao, T., Zhang, Y., Liu, Y., and Fong, H., Synthesis and Luminescence Studies of Poly(Vinyl Acetate)–Eu (III) Coordination Compound *Journal of Applied Polymer Science*, 2007. 103: pp. 3617-3622.
27. Koncar, V., Optical Fiber Fabric Displays. *Optics and Photonics news*, 2005. 16(4): pp. 40-45.
28. Kato, K., Uchida, E., Kang, E.-T., Uyama, Y., and Ikada, Y., Polymer surface with graft chains. *Progress in Polymer Science*, 2003. 28(2): pp. 209-259.
29. Ernevi, A. and Jacobs, M., *(Forthcoming 2005) IT+Textiles*. 2005, Helsinki, Finland.: IT Press/Edita Publishing.
30. Vienne, G., Xu, Y., Jakobsen, C., Deyerl, H. J., Jensen, J., Sorensen, T., Hansen, T., Huang, Y., Terrel, M., Lee, R., Mortensen, N., Broeng, J., Simonsen, H., Bjarklev, A., and Yariv, A., Ultra-large

bandwidth hollow-core guiding in all-silica Bragg fibers with nano-supports. *pt. Express*, 2004. 12: pp. 3500-3508.

31. Argyros, A., Eijkelenborg, M., Large, M., Zagari, J., Nicorovici, N. A., R.McPhedran, and Sterke, C. M. d., Ring structures in micro-structured polymer optical fibres. *Optics Express*, 2001. 9: pp. 813-820.
32. Available from: http://www.lumigram.com/.
33. Selem, B., Rothmaier, M., Camenzind, M., Khan, T., and Walt, H., Novel flexible light diffuser and irradiation properties for photodynamic therapy, *Journal of Biomedical Optics*, 2007. 12.
34. Kaynaka, A., and Pfeffer, F. M., Conductive poly(α,ω-bis(3-pyrrolyl)alkanes)-coated wool fabrics. *Synthetic Metals*, 2007. 157: pp. 534-539.
35. Knight, J. C., Birks, T. A., Russell, R. S. J., and Rarity, J. G., Bragg scattering from an obliquely illuminated photonic crystal fiber. *Applied Optics.*, 1998. 37: pp. 447-452.
36. Schwartzberg, H. G. and Hartel, R. W., *Physical chemistry of foods*. 1992: Marcel Dekker. p. 269.
37. Benoit, G., Hart, S. D., Temelkuran, B., Joannopoulos, J. D., and Fink, Y., Static and dynamic properties of optical micro-cavities in photonic bandgap yarns. *Advanced Materials.*, 2003. 15: pp. 2053-2056.
38. Dupuis, A., Guo, N., Gauvreau, B., Hassani, A., Pone, E., Boismenu, F., and Skorobogatiy, M., Guiding in the visible with "colorful"solid-core Bragg fibers. *Optics Letters.*, 2007. 32: pp. 2882-2884.
39. Gao, Y., Guo, N., Gauvreau, B., Rajabian, M., Skorobogata, O., Pone, E., Zabeida, O., Martinu, L., Dubois, C., and M.Skorobogatiy, Consecutive solvent evaporation and co-rolling techniques for polymer multilayer hollow fiber preform fabrication. *Journal of Materials Research*, 2006. 21: pp. 2246-2254.
40. S.Hardaker, S. and Gregory, R. V., Progress toward Dynamic Color-Responsive "Chameleon" Fiber Systems. *MRS Bulletin*, 2003. 8.
41. Gauvreau, B., Guo, N., Schicker, K., Stoeffler, K., Boismenu, F., A.Ajji, Wingfield, R., Dubois, C., and Skorobogatiy, M., Color-changing and color-tunable photonic bandgap fiber textiles. *Optics Express*, 2008. 16: pp. 15677-15693.

42. S-W, W., K, D., and K, K., Composite electrode composed of bimodal porous carbon and polypyrrole for electrochemical capacitors. *Journal of Power Sources*, 2008. 185: pp. 1589-1593.
43. BC, K., JM, K., and GG, W., A novel capacitor material based on Nafion-doped polypyrrole. *Journal of Power Sources*, 2008. 177: pp. 665-668.
44. Nowak, A. P., Breedveld, V., Pakstis, L., Ozbas, B., Pine, D. J., Pochan, D., and Deming, T. J., Rapidly recovering hydrogel scaffolds from self-assembling diblock copolypeptide amphiphiles. *Nature*, 2002. 417(6887): pp. 424-428.
45. A, L., M, Z., E, F., and F, B. g., Supercapacitor based on activated carbon and polyethylene oxide–KOH–H_2O polymer electrolyte. *Electrochim Acta*, 2001. 46: pp. 2777-2780.
46. AD, F. and A, G., Mastragostino M, Soavi F. Carbon-poly(3-methylthiophene) hybrid supercapacitors. *Joural of The Electrochemical Society*, 2001. 148A: pp. 845-850.
47. Li, H. Q., Zou, Y., and Xia, Y. Y., A study of nitroxide polyradical/activated carbon composite as the positive electrode material for electrochemical hybrid capacitor. *Electrochim Acta*, 2007. 52: pp. 2153-2157.
48. Kima, B., Koncara, V., Devauxa, E., Dufourb, C., and Viallierc, P., Electrical and morphological properties of PP and PET conductive polymer fibers. *Synthetic Metals*, 2004. 146: pp. 167-174.
49. G.G.Wallace, Spinks, G. M., Kane-Maguire, and Teasdale, P. R., *Conductive Electroactive Polymers, Intelligent Materials Systems,2nd ed.* 2003, Boca Raton, USA: CRC Press.
50. Knittel, D. and Schollmeyer, E., Electrically high-conductive textiles. *Synthetic Metals*, 2009. 159: pp. 1433-1437.
51. Li, Y., Cheng, X. Y., Leung, M. Y., Tsang, J., Tao, X. M., and Yuen, M. C. W., A flexible strain sensor from polypyrrole-coated fabrics. *Synthetic Metals,* 2005. 155: pp. 89-94.
52. Mazzoldi, Rossi, D. D., Lorussi, F., Scilingo, E. P., and Paradiso, R., Smart Textiles for Werable Motion Capture Systemsa. *AUTEX Research Journal*, 2002. 2(4): pp. 199-203.
53. Xue, P., Park, K. H., Tao, X. M., Chen, W., and Cheng, X. Y., Electrically conductive yarns based on PVA/carbon nanotubes *Composite Structures,* 2007. 78: pp. 271-277.
54. Jayaraman, *Advancements and Applications of Mechatronics Design in Textile Engineering*. 1995: Kluwer Academic Publishers.

55. Available from: http://www.the-gadgeteer.com/review/kenpo_jacket _ for_ipod.
56. Moad, G., Rizzardo, E., and Thang, S. H., Living radical polymerization by the RAFT process. *Australian Journal of Chemistry*, 2005. 58(6): pp. 379-410.
57. Russell, E., Nanotechnologies and the shrinking world of textiles. *Textile Horizons*, 2002(9): pp. 7-9.
58. Hochart, F., Jaeger, R. D., and Graft-polymerization of a hydrophobic monomer onto PAN textile by low-pressure plasma treatments. *Surface and Coatings Technology*, 2003. 165: pp. 201-210.
59. Saito, M., Antibacterial, Deodorizing, and UV Absorbing Materials Obtained with Zinc Oxide (ZnO)Coated Fabrics. *Journal of Coated Fabrics*, 1993. 23: pp. 150-164.
60. Hu, J. L., Ji, F. L., and Wong, Y. W., Dependency of the shape memory properties of polyurethane upon thermomechanical cyclic conditions. *Polymer International*, 2005. 54: pp. 600-605.
61. Hu, J. L. and Mondal, S., structural characterization and mass transfer properties of segmented polyurethane: influence of block length of hydrophilic segments. *Polymer International*, 2005. 54(764-771).
62. Yang, J. H., Chun, B. C., Chung, Y. C., and Cho, J. H., Comparison of thermal/mechanical properties and shape memory effect of polyurethane block-copolymers with planar or bent shape of hard segment. *Polymer*, 2003. 44(11): pp. 3251-3258.
63. Chen, K. S., Tsai, J. C., Chou, C. W., Yang, M. R., and Yang, J. M., Effects of additives on the photo-induced grafting polymerization of *N*-isopropylacrylamide gel onto PET film and PP nonwoven fabric surface. *Materials Science and Engineering: C*, 2002. 20(1-2): pp. 203-208.
64. Wang, R. H., Xin, J. H., Yang, Y., Liu, H. F., Xu, L. M., and Hu, J. H., The characteristics and photocatalytic activities of silver doped ZnO nanocrystallites. *Applied Surface Science*, 2004. 227: pp. 312-317.
65. Xiang, Y. and Chen, D., Preparation of a novel pH-responsive silver nanoparticle/poly(HEMA-PEGMA-MAA) composite hydrogel. *European Polymer Journal*, 2007. 43(10): pp. 4178-4187.

66. Burniston, N., Bygott, C., and Stratton, J., Nano Technology Meets Titanium Dioxide. *Surface Coatings International Part A*, 2004: pp. 179-814.
67. Daoud, W. A. and Xin, J. H., Low temperature sol-gel processed photocatalytic titania coating. *Journal of Sol-Gel Science and Technology*, 2004. 29: pp. 25-29.
68. Wang, R. H., Xin, J. H., Tao, X. M., and Daoud, W. A., ZnO nanorods grown on cotton fabrics at low temperature. *Chemical Physics Letters*, 2004. 398: pp. 250-255.
69. Hu, X., Zhou, J., Zhang, N., Tan, H., and Gao, C., Preparation and properties of an injectable scaffold of poly(lactic-co-glycolic acid) microparticles/chitosan hydrogel. *Journal of the Mechanical Behavior of Biomedical Materials*, 2008. 1(4): pp. 352-359.
70. Wu, Y., Chi, Y. B., Nie, J. X., Yu, A. P., Chen, X. H., and Gu, H. C., Preparation and application of novel fabric finishing agent containing nano ATO. *Journal of Functional Polymers*, 2002. 15: pp. 43-47.
71. Xu, P., Wang, W., and Chen, S. L., Application of nanosol on the antistatic property of polyester. *Melliand International*, 2005. 11: pp. 56-59.
72. Lampert, C. M., Chromogenic smart materials. *Materials today*, 2004. 7(3): pp. 28-35.
73. Corning, D., *Superior Defense and Comfort for High-Performance Apparel and AccessoriesNEW Dow Corning® Active Protection System*. 2007.

Chapter 6

Adaptive Polymeric Composites and Applications

Adaptive composites are a very important type of smart material. Their shape and mechanical, mass transfer and electrical properties can adapt to different stimulations. This chapter comprehensively reviews the different kinds of adaptive composites with heat, electricity, light, magnetic field or moisture as stimulus. Due to the versatility of composites, adaptive composites have been broadly used in artificial muscles, electrochemical devices, robotics, biomimetics, miniature manipulators, actuators and sensors, active sound absorbing materials, flexible speakers, dust-wipers, various intelligent medical instruments and auxiliaries, smart textiles and apparels, heat shrinkable materials for electronics packaging, micro-systems and self-deployable solar sails in spacecrafts. The applications of adaptive composites in smart textiles are highlighted in this chapter.

Keywords: adaptive polymeric composite; thermal adaptive; electro adaptive; light adaptive; magnetic adaptive; moisture/water adaptive; textile

6.1 Introduction

Adaptive polymeric composites are a very important type of smart material. They can rapidly change their properties (shape, mechanical properties, mass transfer properties or electrical properties) in a predefined way under appropriate stimuli such as heat,[1] electricity,[2] light,[3] magnetic field[4] and moisture.[5] Adaptive polymers have been used in artificial muscles,[6] electrochemical devices,[7] robotics,[8] biomimetics,[9] miniature manipulators,[10] actuators and sensors,[11,12] active sound absorbing materials, flexible speakers,[13] dust-wipers and various intelligent medical instruments and auxiliaries,[14-17] smart textiles and apparels,[18,19] heat shrinkable materials for electronics packaging,[20] micro-systems[21], and self-deployable solar sails in spacecrafts.[22] Depending on the stimulus that is employed to trigger property changes and their different application environment, they are also known as thermal, electrical, light, magnetic and moisture/water adaptive polymeric composites. This chapter gives a review of thermal, electrical, light, magnetic field and moisture adaptive polymeric composites and their present applications in smart textiles. Their potential applications in smart textiles are also discussed.

6.2 Thermal Adaptive Polymeric Composites

6.2.1 *Introduction of thermal adaptive polymeric composites*

Shape memory polymers (SMPs) are the most widely studied thermal adaptive function polymers. They have found broad applications in areas such as smart textiles and apparels, intelligent medical devices, heat shrinkable packages for electronics, sensors and actuators, high performance water vapor permeability materials, self-deployable structures in spacecrafts, and micro-systems in the forms of solution, emulsion, film, foam or bulk.[18,19,23,24] SMPs have many advantages over shape memory alloys and ceramics, such as lightweightness, relatively low cost, good processing ability, high shape deformability and high shape recoverability.[25] Another prominent advantage of SMPs is their tailorable switch temperatures that are related to the glass or melting transition temperature of the polymers, which allow usage in environments with different temperatures.

In order to increase the versatility of adaptive SMPs, blending, prepreging, *in situ* polymerisation, grafting, segmenting, and interpenetrating network-based functional polymers have been investigated. The compounding of SMP is mainly to meet five aims: (1) reinforcing to improve shape recovery stress and mechanical properties; (2) increasing thermal conductivity to improve shape memory effect; (3) decreasing thermal expansion to improve shape memory effect; (4) blending to prepare novel polymer/polymer blends with shape memory effect, tailorable switch temperature and mechanical properties; and (5) incorporating functional particles to prepare other adaptive polymers such as electricity, magnetic, light or moisture adaptive SMPs.

6.2.2 *Thermal adaptive SMP reinforcement*

Although SMPs have found broad applications, their applications are restricted due to their fatal weaknesses compared with shape memory metallic alloys (SMAs). The shape memory behaviour of SMAs arises from the phase transition of a martensite to its parent phase. This phase transition is an instant variation of a crystallographic structure. However, the shape memory of SMPs comes from glass or melting transitions, and the phase change occurs in a wide range. As a result, the shape recovery of SMAs can be very sharp, while the shape recovery of SMPs is not so sensitive. Intrinsically, SMAs have a much higher mechanical strength and elastic modulus than that of SMPs. In addition, the driving force for shape recovery of SMPs comes from the stress stored in the hard phase or network during deformation. This stored internal stress partially relaxes due to the molecule slippages during the shape fixity and recovery process. The recovery stress of SMAs can reach as high as 800 megapascals.[26,27] However, the recovery stress of SMPs is usually very low in the range of a few tenths of a megapascal to several tens of megapascals.[28-31] Furthermore, unlike SMAs, the elastic modulus of SMPs decreases significantly with increasing temperature, which also confines the applications of SMPs in high temperature.

One of the immediate methods to reinforce pristine SMPs for improving mechanical properties and recovery stress is by using high modulus inorganic or organic fillers. The functions of microfibres, micro-size organic/inorganic particles, nanofibres, nanoparticles, and nanolayered inorganic materials on mechanical property improvements of SMPs have been studied. In general, microscale fillers can improve the mechanical strength of SMPs. However, they severely reduce the shape memory effect. Nanoscale fillers can more effectively improve the mechanical properties of SMPs than microscale fillers. The results of the influences of nanoscale fillers on the shape recovery effect of SMPs are not consistent. Some researchers find that certain fillers are unable to improve the recovery stress and simultaneously improve the shape recovery ratios of SMPs, while other researchers show that nanofibres do not improve the shape recovery ratio although shape recovery stress is increased. The different results may be related to the complexity of preparing the composite materials. The properties of the final composite products are affected by many factors, such as the processing techniques, filler distribution, interface, filler size and aspect ratio.

6.2.3 *Polymer/polymer blending with thermal adaptive effect*

Polymer/polymer blending is a widely used method to prepare new materials with tunable properties. Generally, there are two aims of polymer/polymer blending in the SMP area: to tune or improve the properties of available SMPs, such as switch temperature, mechanical strength and shape memory effects and to prepare novel polymers with shape memory effects with one polymer forming the fixing phase while the second polymer forms the reversible phase.

By blending SMP with another polymer, the shape memory effect of the SMP may be maintained and desirable properties of the second polymer may be obtained. SMP composites with adjustable switch transition temperatures, mechanical properties and degradability can be obtained by blending different polymers with a shape memory effect.

Thermoplastic shape memory polyurethane (SMPU) can be blended with phenoxy resin with good miscibility if the SMPU is synthesised using poly(caprolactone) (PCL) as the soft segment.[32] The glass transition temperature (Tg) of the blends varies in smoothness with composition change, which indicates that the switch temperature of the blends can be tuned smoothly by varying the relative fraction of the PCL, pheonoxy and hard segment. Poly(vinyl chloride) (PVC) is also miscible with PCL and can be used to improve the mechanical strength of SMPU and tune the switch temperature.

Blending amorphous and semi-crystalline polymers is an effective strategy to prepare novel SMPs. The crystalline polymers which have been used include poly(vinylidene fluoride) (PVD), poly-(hydroxxybutyrate), polyethylene-co-vinyl acetate, poly(ethylene glycol) polyethylene, polylactide, poly(vinyl chloride), poly(vinylidene chloride) and copolymers of poly(vinylidene chloride) and poly(vinyle chloride). The amorphous polymers include poly(vinyl acetate), poly (methyl acrylate), poly(ethyl acrylate), atactic poly(methyl methacrylate), isotactic poly(methyl methacrylate), syndiotactic poly(methyl methacrylate) and other poly(alkyl methacrylates). These blending polymers with shape memory effects are of great industrial interest compared to covalently crosslinked SMPs because they can be obtained immediately by using conventional polymer processing methods.

Through reactive blending, SMPs with interpenetrating polymer network (IPN) structures which have much higher mechanical strength can be prepared. An IPN SMP with the potential for usage in clinical or medical applications is prepared through this strategy.[33] In this composite, poly(ethylene glycol) dimethacrylate (PEGDMA) is introduced to poly(lactideco-glycolide) PLGA/isophorone diisocyanate (IPDI) systems to form IPNs with good shape memory and hydrophilic properties. The hydrophilicity, transition temperatures and mechanical properties of IPNs can be conveniently adjusted through variation of compositions.

6.2.4 *SMP/thermoexpanded graphite (TEG)*

In some polymeric composite systems with weak interphase interactions, the shape memory effect is accompanied by an obvious increase in volume. This phenomenon is associated with the relaxation of micro-stresses at interfaces which results in material loosening. The phenomenon is first observed in thermoexpanded graphite (TEG)/epoxy polymer composites.[34] Carbon materials with less bulk density mean a more pronounced phenomenon. This material makes it possible to prepare products with an unusual behaviour under heating: the rods increasing in length with diameter are kept constant; bushes increasing in internal diameter with length and external diameter are kept constant and disks increasing in diameter with height are kept unchanged. Similarly, the unusual shape memory effect is found in other carbon materials of different bulk density filled polymers. The carbon material fillers include TEG, cast graphite and carbon, and the polymer matrixes include epoxy polymer, ultrahigh-molecular polyethylene and polypropylene.[35]

6.2.5 *Thermal expansion of SMP composites*

During thermomechanical deformation and recovery processes, SMPs are subjected to heat. Like other polymers, SMPs have a high coefficient of thermal expansion which is, in fact, much higher than that of SMAs. The study of thermal expansion during a shape recovery process is necessary in order to obtain accurate shapes during applications. Fillers, which are mostly intended for increasing the stiffness of SMPs, also influence the coefficient value of the thermal expansion of SMPs. During the shape recovery process, thermal expansion contradicts shape recovery shrinkage because of heating. After the incorporation of fillers, especially high aspect ratio fillers, the thermal expansion of SMPs decreases due to the decreased coefficient of the thermal expansion.

The investigation on the influences of thermal expansion on the shape memory effect of organoclay, such as carbon nanofibre (CNF), silicon carbide (SiC), and carbon black (CB) reinforced SMPUs, indicates that the thermal expansion coefficient below and above the melting region of pure SMPUs is a constant designated as a coefficient of linear thermal expansion (CLTE). The shape memory composites with nano-sized fillers have a more moderating CLTE than that of microsized filler filled SMPs. High aspect ratio fillers, such as organoclay and CNF, are more effective in decreasing the CLTE than spherical fillers.[36]

6.2.6 *Thermal conductivity of SMP composites*

Compared with SMAs, SMPs have another disadvantage-low thermal sensitivity. Being of organic nature, polymers are thermal insulators with thermal conductivity usually below 0.30 W/m.K.[37] Rapid heating to release and rapid cooling to freeze the morphology of an oriented polymer sample is in fact, a challenge in SMPs. An approach to increase the thermal conductivity of polymers is to incorporate fillers with high heat transfer properties, such as alumina, fused silica, SiC, boron nitride and glass fibre, into SMPs. A similar problem happens because at an overly high filler content, the fillers deteriorate the shape memory and mechanical properties of SMPs.

6.2.7 *Micro-mechanics of SMP composites*

During the shape fixity and recovery cycle, an SMP is first subjected to deformation at a temperature above the switch transition temperature. If the material is cooled to a temperature below the switch temperature, the modulus of the material increases because of nitrification or crystallisation of the fixity phase. Consequently, internal stress is stored in the material among the physically or chemically crosslinked structures. If the material is reheated to a temperature above the switch temperature, the fixity phase becomes soft because of the glass or melting transition. Therefore, internal stress is released with the result where the polymer returns to its original shape.

Internal stress storage and release in shape memory epoxy nanocomposites reinforced with nanometer-scale SiC ceramic have been studied.[38] During the shape recovery process, the elastic-plastic stress-strain properties of the SMP composite are measured and calculated through a micromechanics technique. The calculated stress is in reasonable agreement with the measured change during deformation and recovery. The mechanical behaviour of laminate composites has also been investigated. The applicability and limitations of using a commercial general-purpose finite-element code are explored through the analysis of a component-level shape memory composite structure. The results are very helpful for structural design and improvement of key performance of SMP composites.

6.2.8 *SMP composites with special functions*

Many researchers have studied the novel properties of CNT reinforced SMPs and explored the potential applications. Shape memory composites filled with CNT can have a good electromagnetic interference shielding effect(EMISE). Studies on multi-wall-carbon-nanofibre (VGCF) reinforced SMPUs show that the composites demonstrate a low electrical percolation threshold and high level of conductivity because of the high aspect ratio of VGCFs.[39] According to the standards, VGCF/SMPU nanocomposites are considered to be excellent shielding materials for electromagnetic interference.

Due to the ultraviolet (UV) absorption property of nanocarbon tubes, CNT reinforced SMPs can also have good UV protection properties. Mondal and Hu[40] treated cotton fabrics with a hydrophilic polyurethane solution, which contained a certain amount of multiwalled carbon nanotube (MWNT), hoping to fabricate fabrics with not only smart water vapor permeability (WVP), but also good UV protection. A dense CNT/polyurethane layer was formed continuously over the surface of the fabrics. The coated fabrics with 1 wt% of MWNT had excellent protection against UV radiation.

The influence of CNTs on the WVP of SMPU is a complex phenomenon. Below the soft segment melting transition temperature, a small amount of CNT decreases the WVP because the nano-sized CNTs act as a nucleating agent, and thus, enhance the soft segment ordered crystal structure. In contrast, with an increasing CNT content, the WVP increases. This is because in one aspect, MWNT constrains the formation of ordered soft segment phase structures; in another aspect, the straight CNTs with large aspect ratios offer a relatively straight 'free' pathway for water molecule diffusion on the surface of or inside the CNT to pass through, the WVP increase. At temperatures above the soft segment phase melting transition temperature, the WVP increases markedly with increasing CNT content which is also attributed to the pathway effect of MWNT by forming a channel through which water molecules can diffuse rapidly.

6.2.9 *Development trends of thermal adaptive SMP composites*

In general, the switches to fix or recover the shapes of SMPs are either glass or melting transition temperature for all SMPs reported. Researchers continue efforts to find other switches to replace traditional glass or melting transition as the shape switches for SMPs.

6.2.9.1 *SMP composites based on non-thermal phase transitions*

The first structure which may be used as a switch for SMPs is the thermal reversible covalent reaction such as the thermo-reversible Diels–Alder reaction. Figure 6.1 shows a thermally reversible Diels–Alder cycloaddition of a multi-diene (multi-furan) and a multidienophile (multi-maleimide).[41] Monomer 1 has four furan moieties. Monomer 2 has three maleimide moieties. The thermal reversibility is archived by the retro- Diels–Alder reaction.

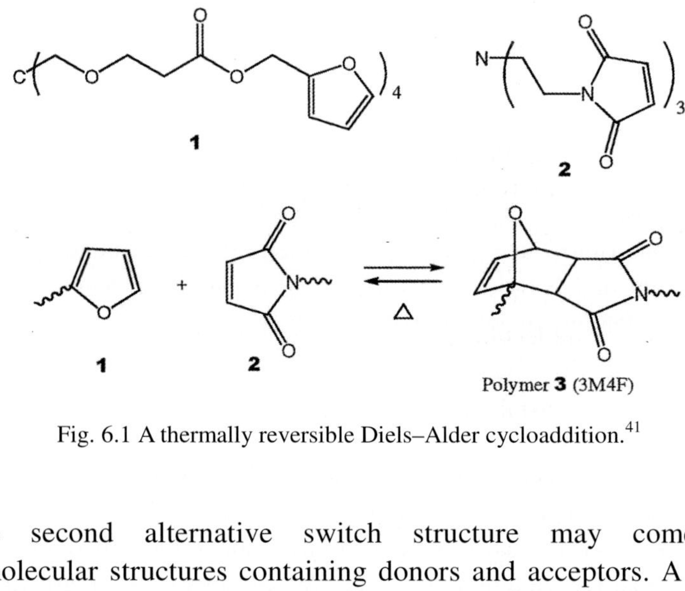

Fig. 6.1 A thermally reversible Diels–Alder cycloaddition.[41]

The second alternative switch structure may come from supermolecular structures containing donors and acceptors. A thermal reversibly associating structure; quadruple H-bonding, has the potential to be used as a switch for SMPs. Quadruple H-bonding can be incorporated into a rubbery network polymer and thus create an SMP with thermal-reversible hydrogen bonding as the switch. An example is a crosslinked poly(n-butylacrylate)s containing quadruple H-bonding ureidopyrimidinone (UPy) side-groups synthesised by free radical copolymersation.[42] The hydrogen bonding is capable of stabilising mechanically strained states. Unlike conventional shape memory polymers, these dynamic networks lack a well-defined shape recovery temperature.

6.2.9.2 *Two-way SMP composites*

The shape memory effect of the above thermal adaptive polymers is a one-way shape memory effect. Researchers are making great efforts to prepare SMPs with a two-way shape memory effect. The cooling-induced crystallisation of crosslinked poly(cyclooctene) films under a tensile load can result in elongation and subsequent heating to melt the network yields contracting (shape recovery).[43] Stress has to be applied during the shape extension. Furthermore, this phenomenon can only be observed in poly(cyclooctene) with high molecular orientation.

Lately, a two-way shape memory effect was observed on a carbon nanotube–elastomer nanocomposite.[44,45] The light induced schematic two-way shape memory effect of the carbon nanotube–elastomer nanocomposite is shown in Fig. 6.2. The direction of the shape change is determined by the magnitude of the pre-strain. If the composite is pulled at a low strain, it expands if exposed to infrared light. If the composite is deformed with a large strain, it contracts if exposed to infrared light. This process is reversible. The mechanism of the bimodal response has not been completely identified. For both of the above two-way shape memory effects, external force for molecular orientation of pre-deformation is essential.

An automatic two-way shape memory effect is achieved in a self-assembled liquid crystal fibre by using a triblock liquid-crystal polymer. The large central block is made of a main-chain nematic polymer. The polymer has large spontaneous elongation along the nematic director.[44] Effective crosslinking is obtained by small terminal blocks formed of terphenyl moieties. The terphenyl moieties phase is separated into semicrystalline micelles which can act as junctions of the network. The resulting transient network retains the director alignment and shows a significant two-way shape memory effect.

6.3 Electro Adaptive Polymeric Composites

6.3.1 *Introduction of electro adaptive polymeric composites*

Since the 1990s, researchers have found a series of electro active polymers which can significantly change shape or size as a response to electrical stimulation. The electro adaptive effect is like the natural motion of human muscles. Using these materials, scientists can mimic the movements of human limbs.[46,47] Electro adaptive polymers can induce strains that are as high as two orders of magnitude greater than the striction-limited, rigid and fragile electro active ceramics. Furthermore, electro active polymers are also superior to metallic alloys with higher response speed, lower density, and greater resilience.[13]

Electro adaptive polymers fall into two categories based on the different activation mechanisms: the electronic electro adaptive polymers whose shape change is triggered by electric fields or coulomb forces; and the ionic electro adaptive polymers, whose shape deformation is triggered by the mobility or diffusion of ions. Electronic electro adaptive polymers include electrostrictive graft elastomers, electrostrictive cellulose papers, electro-viscoelastic elastomers, ferroelectric electro adaptive polymers, dielectric electro adaptive polymers, all-organic composites, liquid crystal elastomers, electro adaptive conductive polymers and conductive thermal adaptive polymeric composites. Ionic electro adaptive polymers consist of ionomeric polymer-metal composites, ionic polymer gels and carbon nanotubes.

6.3.2 *Electrostrictive graft elastomers*

Electrostrictive graft elastomers have a flexible backbone macromolecule and a grafted polymer which can form crystalline structures. The grafted crystalline polar phase provides moieties, which can respond to an applied electric field and crosslinking sites for the elastomer system backbones. The molecular structure of the electrostrictive graft elastomer is shown in Fig. 6.2. Both the backbone and graft unit have polarised monomers containing atoms with electric partial charges. The electric charges generate dipole moments. When an electric field is applied to the polymer, force moments are induced into the polymer. As a result, rotation of the dipole moment is caused. However, because the force moments of backbones cancel each other out, the whole backbone does not rotate. At the same time, the monomer dipole moments of the crystal graft unit accumulate. Therefore, the electrical field rotates the whole unit. Consequently, the local backbone reorientation and crystal unit rotation cause electrostrictive deformation. The significant advantage of electrostrictive graft elastomers is high stiffness compared to other electrostrictive polymers, such as silicon rubber and polyurethane. It is reported that the modulus of the electrostrictive graft elastomers is about 30 times that of polyurethane.[49,50]

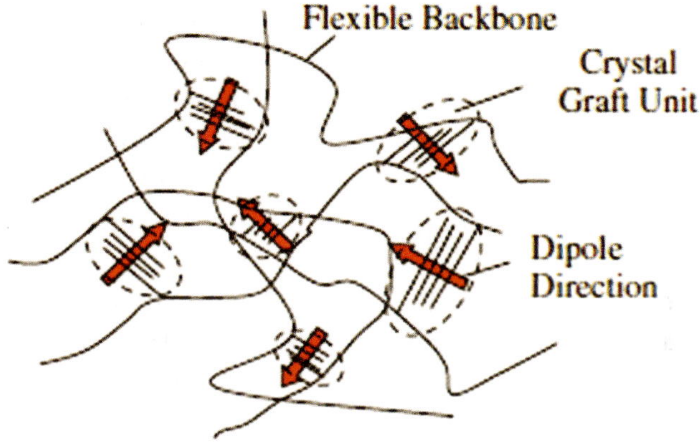

Fig. 6.2 Structure of the grafted elastomer.[50]

6.3.3 *Electrostrictive cellulose paper*

Electrostrictive cellulose paper actuators are made by depositing very thin electrodes on a cellulose paper strip. If an electric field is applied through the thickness of an electrostrictive cellulose paper strip, a large bending displacement with low force can be produced. The tip displacement increases linearly with the voltage and reaches a saturated value at the maximum displacement.[51] The working principle of electro adaptive cellulose paper is that the piezoelectric and ionic migration effects are, at the same time, accompanied by the dipole moment of the cellulose paper ingredients. Cellulose is composed of linked glucose residues arranged in linear chains. Every other glucose residue is rotated about 180°. Cellulose exists as a crystalline array of many parallel orientated chains-microfibrils. The microfibril size ranges from 36~1200 chains.[52] They are highly organized and can diffract as a single crystal. The amorphous chains on the microfibrils form surface disorder regions. Cellulose papers made with regenerated cellulose have a large fraction of disordered chains. In the disordered chains, water molecules can attach to the hydroxyl groups to form hydrogen-bonding. Ions, such as sodium and sulfur, can be injected into the paper fibre during the paper making process. If an electric field is applied to the paper, these ions migrate to the anode. Free water molecules in disordered regions move along the electric field due to the strong polarity of water molecules. The ionic and water movement across the cellulose paper under the electric field leads to bending. Furthermore, the cellulose molecules have a dipole nature. The crystal structure of cellulose II is monoclinic and noncentro-symmetric. It shows piezoelectric and pyroelectric properties.

6.3.4 *Electro-viscoelastic elastomers*

Electro-viscoelastic elastomers are composites of polymer elastomers and a polar phase. An electric field is applied to orient and fix the polar phase in the elastomeric matrix during curing. During applications, if an electrical field is applied to the material, the polar particles electrically bind together and cannot slip past each other.[53,54] Thus, an interaction between dipoles occurs and become strong as the dipoles come close to each other, which apparently increases the elastic modulus of the material. To have good electro activity, the modulus of the elastomer has to be very low. If the elastic modulus of the elastomer is larger than the total interactions of the particles, the electro-rheological effect cannot be observed macroscopically.

6.3.5 *Ferroelectric electro adaptive polymers*

Ferroelectric electro adaptive polymers are mainly PVD and its derivatives, such as poly(vinylidene fluoride–trifluoroethylene) (P(VDF-TrFE)). Their crystals show polymorphism which can be controlled by preparation methods. In P(VDF-TrFE), a large strain (~10%) can occur due to the phase transformation between the ferroelectric and paraelectric phases arising from molecular conformation change. This confirmation change is triggered by electric fields at a temperature above the ferroelectric-paraelectric phase transition due to the ferroelastic coupling in the P(VDF-TrFE) copolymer. In addition, at a temperature just above the first order transition of the ferroelectric and paraelectric phase, a polar phase can be electrically induced. A very large electromechanical coupling factor can be obtained for P(VDF-TrFE) in the electric field induced phase. All of these are responsible for a large electrostriction observed in ferroelectric electro adaptive polymers.[55,56]

6.3.6 *Dielectric electro adaptive polymers*

A dielectric electro adaptive actuator is made of dielectric elastomers with a compliant electrode coated on both sides. The modulus of the dielectric elastomer, such as silicone, acrylic and polyurethane elastomers, is very low, while the dielectric constant is very high. The compliant electrodes can be fabricated by lift-off stenciling techniques for powdered graphite, selective wetting of ionically conductive polymers, and spray coating of CBs and fibrils in polymeric binders. The operation principle of the dielectric electro adaptive actuator is shown in Fig. 6.3. When voltage is applied between the electrodes, the sandwiched elastomer is squeezed in the thickness direction, which causes the expansion of the composite in the plane direction. Prototype actuators are fabricated, such as stretched films, stacks, rolls and tubes.

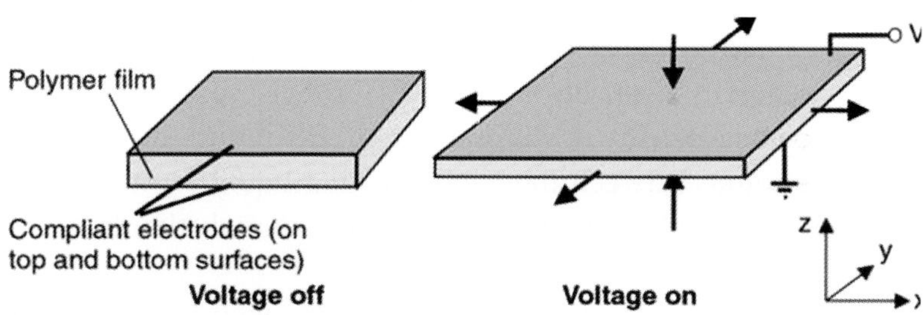

Fig. 6.3 The dielectric elastomers actuate by means of electrostatic forces applied via compliant electrodes on the elastomer film.[57,58]

6.3.7 *All-organic electro adaptive polymers*

The all-organic electro adaptive polymer is a composite of an electrostrictive polymer matrix filled with high-dielectric-constant organic particulates. The dielectric constant of polymeric-like material can reach higher than 400.[59,60] It significantly reduces the applied field to generate high strain with high elastic energy density. The typical all-polymer material with a high-dielectric-constant (K>1,000 at1 kHz) percolative is prepared by the combination of conductive polyaniline particles (K>105) within a fluoroterpolymer matrix (poly(vinylidene fluoride-trifluoroethylene-chlorotrifluoroethylene) terpolymer matrix) (62/26/12 mol%). Researchers[61] have also combined CuPc and conductive polyanline within a polyurethane matrix to fabricate all-organic dielectric-percolative composites with three components. The high-dielectric-constant CuPc particulates in this three-component composite increase the dielectric constant of the polyurethane and the combined two-component dielectric matrix in turn serves as the high-dielectric-constant host for the polyanline to further enhance the dielectric response.

6.3.8 *Electro adaptive liquid crystal elastomers*

Liquid crystal elastomers can form an electro adaptive material that has piezoelectric characteristics and are electrically adaptive by Joule heating. Electro adaptive liquid crystal elastomers are composed of monodomain nematic liquid crystal elastomers and conductive polymers that are dispersed in the network structure. Properties of liquid crystal elastomer materials can be improved through selecting the liquid crystalline phase, density of crosslinking, flexibility of the polymer backbone, coupling between the backbone and liquid crystal group, and coupling between the liquid crystal group and external stimuli.[62]

6.3.9 *Electro adaptive conductive polymers*

The electro activity of conductive polymers is achieved by a reversible counter-ion insertion and expulsion during redox cycling. Oxidation and reduction occurring at the electrodes cause considerable volume changes due to the exchange of ions with an electrolyte. Some additives, such as dopant ClO_4^-, are necessary for actuation. A considerable volume change is induced by oxidation and reduction due to the exchange of ions with an electrolyte.[63,64] If voltage is applied onto the electrodes, oxidation occurs at the anode and reduction occurs at the cathode. The inclusion of the ions causes volume increasing of the polymer and the converse process causes polymer shrinkage. The electro adaptive properties of electro adaptive conductive polymers are determined by the speed of ion insertion and expulsion from the polymer, which is significantly influenced by the size of ions, solvent, voltage magnitude and actuator thickness.

6.3.10 *Electro adaptive composites from conductive thermal adaptive SMPs*

Some degree of conductivity can be achieved after conductive ingredients are incorporated into thermal adaptive SMPs. The current caused Joule heating can raise the polymer temperature to above the switch transition temperature and trigger shape recovery. The added electrical adaptive effect brings more applications to thermal adaptive SMPs, such as biomedical devices, self-deployable aerospace structures, actuators and sensors. The indirect induction heating of SMPs using electricity can also alleviate some heat transfer problems of SMPs. Hu's group members[23] prepared an MWNT/SMPU composite fibre by melt spinning. They found that MWNTs are axially aligned in the polyurethane fibre which is helpful for improving the electrical conductivity of the fibre. The prepared fibres show the electro adaptive shape memory behaviour.[11]

6.3.11 *Electro adaptive ionomeric polymer-metal composites*

Ionomeric polymer-metal composites have an electro adaptive effect because of the mobility of cations in the ionomeric polymer network. Usually an ionomeric polymer-metal actuator has a thin perfluorinated ion exchanged based polymer membrane with noble metallic electrodes on both surfaces. It is neutralised with a certain amount of cations to balance the electrical charge of the anions covalently fixed to the backbone membrane. These composites have high water absorption capabilities and are usually kept in a hydrated state for proper dynamical function. As shown in Fig. 6.4, if DC voltage bias is applied to an ionomeric polymer-metal strip, the strip bends quickly towards the anode, with slow relaxation in the opposite direction. The electro-mechanical activity of an ionomeric polymer-metal composite is affected by the ionomer, counter ion, the nature of the solvent, and amount of its solvent uptake. To improve surface conductivity, metal electrodes are usually platinum with a layer of gold, or maybe pure gold. The cations may include alkali-metal cations, Li+, Na+, K+, Rb+, and Cs+, as well as alkyl-ammonium cations, tetramethyl ammonium and tetrabutyl ammonium.

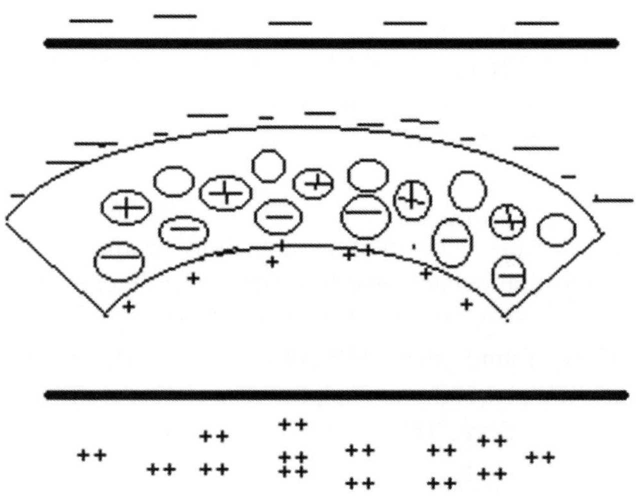

Fig. 6.4 General redistribution of charges in an ionic polymer due to an imposed electric field.[65]

6.3.12 *Electro adaptive ionic polymer gels*

Electro adaptive ionic polymer gels are different from hard solid electro adaptive polymers in that they are swollen matters with solvents. The properties of gels can change from a nearly solid polymer to a solution with low polymer content which still maintains their shape. The typical architecture of electro adaptive ionic polymer gels consists of a network of crosslinked polymers to which electric charges are attached. They can be prepared by crosslinked copolymersation of weakly ionisable comonomers in a polymer network.[66] Employment of an external electric field induces the changes in the degree of ionisation of the gel and the distribution of ions, which causes the swelling or shrinkage of the gel.

6.3.13 *Electro adaptive carbon nanotubes*

Carbon nanotube is a very promising actuator because the voltage required is as low as several voltages, working temperature can be as high as 1,000°C, has good mechanical and electrical properties, and strain induced above 1.0%. The actuation principle of a single-walled-carbon-nanotube (SWNT) is illustrated in Fig. 6.5, where charges of the opposite sign are injected into the nanotube electrodes by applied potential. The nanotube electrodes are in a liquid or solid electrolyte (blue background). The charges in the nanotube electrodes are balanced by ions of electrolyte as the charged spheres on the nanotube cylinders. The single nanotube electrodes can be an arbitrary number of nanotubes in each electrode which mechanically and electrically act in parallel. The opposite electrode deformation depends on the potential and the relative number of nanotubes.

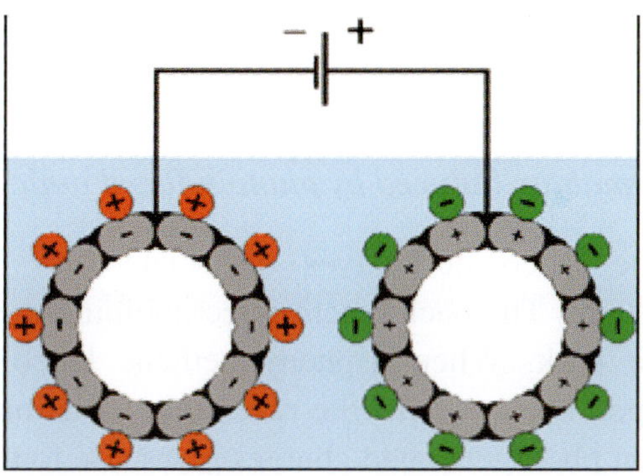

Fig. 6.5 Schematic illustration of charge injection in a nanotube-based electromechanical actuator.[67,68]

6.4 Light Adaptive Polymeric Composites

6.4.1 *Introduction of light adaptive polymeric composites*

The extensive study of light adaptive polymers started in the 1960s. Since then, several types of photodeformable polymers have been developed with different photodeformation mechanisms: a) photoisomerizable molecules, such as azobenzenes, b) triphenylmethane leuco derivatives that undergo photoinduced ionic dissociation, x may be –OH or -C$\equiv$N and c) photoreactive molecules, such as cinnamates.[69]

6.4.2 *Photodeformability induced by photoisomerisation*

Azobenzenes exist in two different states: cis and trans conformations. Upon exposure to light of appropriate wavelengths, azobenzenes show reversible trans-cis isomerisation and exhibit a large change in molecular length. The distance between 4- and 4'-carbons changes between 9.0 Å (trans) to 6.5 Å (cis). When azobenzene groups are linked to macromolecules, the interconversion between the two photoisomers can induce macroscopic changes in the polymeric material. A photo-deformable polymer film synthesised by a liquid-crystal monomer and a diacrylate crosslinker, has been reported.[70] Both of the liquid crystal monomer and crosslinker possess azobenzene moieties. The film of the liquid-crystal network containing the azobenzene chromophore can be repeatedly and precisely bent along any chosen direction by using linearly polarised light.

6.4.3 *Photodeformablity induced by photoinduced ionic dissociation*

Triphenylmethane leuco derivatives dissociate into ion pairs upon exposure to UV light. The back reaction recombining the ion pair occurs thermally in the dark. When triphenylmethane leuco derivatives are incorporated in polymers or gels, the reversible variation of electrostatic repulsion between photogenerated charges gives rise to the photoinduced expansion and shrinkage of the polymers or gels. An example is an N-isopropylacrylamide-based gel incorporated with triphenylmethyl moieties.[71] For fixed appropriate temperatures, the gels discontinuously swell in response to the irradiation of UV light and shrink when the light is removed.

6.4.4 *Photodeformability induced by photoreactive molecules*

Photochemically reactive molecules are able to form photoreversible covalent crosslinks in polymers.[69] Polymers containing cinnamic groups could be deformed by UV light, and they can recover their original shape at ambient temperatures when exposed to a UV light of a different wavelength.[72]

6.4.5 *Light adaptive composites from thermal adaptive SMPs*

Another class of light adaptive polymers is thermal adaptive SMPs that use light to increase the material temperature to above the switch transition temperature. The heat absorbance can be enhanced by the incorporation of CB, and carbon nanotubes. Heat absorbed by the light adaptive polymers raises the internal temperature and melts polymer crystallites or changes the material from a glassy to rubber state to trigger the release of the stored strain energy.

6.5 Magnetic Adaptive Polymeric Composites

6.5.1 *Introduction of magnetic adaptive polymeric composites*

Generally, there are three types of magnetic adaptive polymeric composites: magnetic adaptive elastomers, gels and SMP composites. A magnetic adaptive elastomer is a composite of a highly elastic polymeric matrix filled with small (mainly nano- and micron-sized) magnetic particles. A magnetic adaptive gel, also called ferrogels, can be considered as a chemically crosslinked polymer swollen by a ferrofluid. A magnetic adaptive SMP composite is actually a magnetic particle filled thermal adaptive SMP.

6.5.2 *Magnetic adaptive elastomers*

A magnetic adaptive elastomer is a composite of a highly elastic polymeric matrix filled with small (mainly nano- and micron-sized) magnetic particles. When a magnetic field is applied to the composite, all of the forces acting on the magnetic particles are transmitted to the polymer, which leads to the instantaneous deformation of the composites. If the field is non-uniform, magnetic particles experience a magnetophoretic force. Consequently, the particles are attracted to regions of stronger field intensities. This will cause macroscopic shape changes of the magnetic adaptive elastomer. The shape deformation disappears abruptly when the external magnetic field is removed. In addition to shape deformation, the magnetic effect can also cause a considerable rise in Young's modulus, pseudo-plasticity or shape memory in soft elastomers.[73-75] To obtain high magnetic sensitivity, the polymer needs to have low elastic modulus and high initial susceptibility as well as high saturation magnetisation. Elastic materials with tailor-made anisotropy particles may be prepared under external fields. The magnetic active elastomers with anisotropy particles have anisotropic mechanical, electrical transport, and magnetic properties due to the chain-like particle structures resulting from the magnetic dipole interactions between particles.

6.5.3 *Magnetic adaptive gels*

Magnetic field adaptive polymer ferrogels, also called ferrogels, can display an abrupt deformation under a non-uniform magnetic field. A ferrogel can be considered as a superparamagnetic particle filled, swollen network. It can also be regarded as a chemically crosslinked polymer swollen by a ferrofluid. In the magnetic adaptive polymer gel, the finely distributed ferromagnetic particles are attached to the flexible network by adhesive forces, which causes direct coupling between the magnetic and mechanical properties. Significant elongation, contraction and curvature can be induced within less than one second.[76-78] The shape distortion can disappear abruptly when the external magnetic field is removed. Compared with other stimulus adaptive polymers, magnetic adaptive gels are gentle and flexible.[79-81] A typical magnetic field active polymer gel is chemically crosslinked poly(viny1 alcohol) hydrogel incorporated with monodomain magnetite particles of colloidal size. Preparation of a magnetic poly(vinyl-alcohol) (PVA) gel is similar to that of other filled elastomeric networks. One can precipitate well-dispersed particles in the polymeric material before, during and after a crosslinking reaction. If the magnetic PVA hydrogel is subjected to a non-uniform magnetic field, where the average field gradient is perpendicular to the axis of the gel, the cylinder bends toward higher field strengths.

6.5.4 *Magnetic adaptive SMPs*

For thermally adaptive shape memory, energy from direct heating is usually used to trigger the shape change. A second approach where Joule heating after the SMP is incorporated with ferromagnetic fillers, can cause indirect heating and recovery of SMPs. Compared with direct and electro heating to actuate shape change, the magnetic filled adaptive shape memory effect has more advantages, including: 1) the selection of a ferromagnetic particle material with ferromagnetic material, the Curie temperature is within safe medical limits, and Curie thermoregulation eliminates the danger of overheating; 2) power transmission lines leading to an SMP device are eliminated; 3) more complex device shapes are possible because consistent heating is expected to be achievable for any type of device geometry, 4) selective heating of specific device areas is possible and 5) remote actuation allows for the possibility of embedded devices that can be later actuated by an externally applied magnetic field.

6.6 Moisture/Water Adaptive Polymeric Composites

6.6.1 *Introduction of moisture/water adaptive polymeric composites*

Both moisture/water adaptive polymeric composites are actually thermal adaptive SMP composites. In one strategy, the shape recovery is achieved by decreasing the switch temperature of the thermal adaptive SMPs through water that is absorbed by the materials. In the second strategy, a high hydrophilic segment is employed in a segmented polymer so that the shape recovery can be triggered by water/moisture.

6.6.2 *Moisture/water adaptive polymer with decreased switch temperature*

If T_g type SMPUs are exposed to a high humidity environment at room temperature for several days, their glass transition temperature drops prominently because they absorb moisture. A water driven shape memory effect of the SMPU in water due to the decrease glass transition temperature was observed. A straight SMPU wire is bent into a circular shape at 40°C and the circular shape is fixed after the wire is cooled to an ambient temperature. In the following week, at an ambient temperature and in a dry cabinet humidity of 30 RH, the shape memory wire does not show significant shape recovery. However, after being immersed in water at room temperature for about 30 minutes, the SMPU circular wire starts to recover gradually.[82]

6.6.3 *Moisture/water adaptive polymer with hydrophilic segments*

Another strategy for moisture adaptive shape memory polymer preparation is to use water soluble segments as a soft segment to prepare SMPUs. A water adaptive SMPU is prepared by using polyhedral oligomeric silsesquioxane (POSS) molecules and poly (ethylene glycol) that act as the hydrophobic and hydrophilic groups. The POSS groups aggregate to a certain extent, and this separated phase of inorganic moieties act as a hard phase because the physical interactions between inorganic moieties hinder the irreversible sliding of the organic main chain.[83] Similarly, a water sensitive biodegradable stent using chitosan films crosslinked with an epoxy compound is fabricated following this approach. The raw materials (chitosan and PEO400) for developing the stent are relatively hydrophilic, and the prepared polymer is responsive to hydration. It is reported that the ether bonds (-O-) in the epoxy compound serve as flexible joints in the crosslinking bridges. The biodegradable stent can be used as an alternative to metallic stents and for local drug delivery.

6.7 Applications to Textiles

Medical applications of stimulus adaptive polymeric composites have been studied extensively and intensively. Recent advances in environmentally adaptive polymers have inspired people to create smart textiles with self-regulating structures and performances in response to environmental variation by employing these kinds of materials onto textiles. Processes for making such textiles include finishing, coating, grafting, laminating, blending, spinning and other innovative methods.

6.7.1 *Thermal adaptive polymer composites for textile applications*

Thermal adaptive polymers have broad applications in textiles, automotive parts, building and construction products, intelligent packing, implantable medical devices, sensors and actuators in the form of solution, emulsion, film, fibre, foam and bulk.[17,28,40,84-90] Textile applied adaptive polymers include breathable, thermal insulating, and shrink-resistant apparel, nanofibre coated fabrics and nanofibres.

6.7.2 *Electro adaptive polymeric composites for textile applications*

Electro adaptive polymeric composites are low in cost and can be tailor-made for particular applications. The applications include artificial muscles, electrochemical devices, robotics, biomimetics, miniature manipulators, active sound absorbing materials, flexible speakers, dust-wipers and various medical instruments and auxiliaries. Polypyrrole has good electrical conductivity, chemical and thermal stability, good environmental stability, ease of preparation and fewer toxicological problems. It is suitable for wearing products which may have contact with human skin directly. A wearable system, which is able to read and record posture and movements of a subject wearing them using electro adaptive polymers, has been proposed.[69] In contrast with the smart textile which incorporates bulky sensors, the smart textiles prepared through this strategy are truly wearable and an unintrusive system since they do not cause discomfort to the body.

6.7.3 *Light adaptive polymeric composites for textile applications*

For several decades, light adaptive polymers have been proposed as high-speed actuators for microscale or nanoscale applications, such as in microrobots in medicine or optical microtweezers.[3,70,91-94] As deformations driven by light require neither batteries nor controlling devices on the materials themselves, it may be simple to employ them on textiles. By using photosensitive materials, the textile will have on and off states in response to sunlight or other artificial light. More specifically, since the light adaptive polymer responds to special wavelength light, the textile employing light adaptive polymers may be used to prepare protective clothing against harmful light.

6.7.4 *Magnetic adaptive composites for textile applications*

Magnetic adaptive polymers are mainly used in elastomer bearings and vibration absorbers because their physical properties, such as elastic modulus, stress-strain behaviour along and perpendicular to the chain-like structure of polymers, are significantly different in response to the magnetic field. Recently, a magnetic adaptive polymeric composite fibre with a diameter of 500 μm, tensile strengths ranging from 8 to 20 MPa, and strains ranging from 6 to 20%, was fabricated. The main constituents of this composite fibre are a thermoplastic polymer and embedded carbonyl iron powders.[95] Then the fibre was woven into flexible plain weave woven fabrics. It is suggested that the new magnetic adaptive fibre and fabrics can be used in covers for magnetic shielding, sensors, actuators, functional clothing, drapes, wall papers, sieves, covers and other rigid or flexible objects.

6.7.5 *Water/moisture adaptive polymeric composites for textile applications*

Moisture and temperature management are an important research topic in textiles. When it is difficult to maintain a stable body temperature, which leads to excess perspiration, wearers will feel uncomfortable. It is very promising to use water/moisture adaptive shape memory textiles to offer wearers high comfort by changing textile microshape or microstructures. A humidity activated shape memory polymer, which contains elastomeric and non-elastomeric polymers, such as a moisture absorbing polymer, has been submitted as US Patent 6 627 673. The elastomeric polymer can be from thermoplastic polyurethane, poly(ether-amide) block copolymer, styrene-butadiene copolymer or silicon rubber. The moisture absorbing polymer can be from polyethylene oxide, polyethylene glycol, polyvinyl alcohol, etc. The humidity adaptive shape memory deformable polymer is in the form of films, fibres, filaments, strands, nonwovens and pre-molded elements. This material is proposed for usage in disposable and reusable products such as disposable diapers, training pants, incontinence products, and feminine care products.

6.8 Summary

Many papers regarding stimulus-adaptive polymeric composites have been published, many inventions patented and many potential applications proposed. Yet there are currently only several commercialised products that are available. In fact, some of the proposed applications are too difficult to be accomplished because textile application situations pose strict requirements on the properties of adaptive polymeric composites. The study of stimulus adaptive polymeric composite for textile applications has begun only in the last few years and remains largely unexplored.

Although there are only few textile products of stimulus-adaptive polymeric composites available now, the market for environmental adaptive textiles using these adaptive polymers is quite large. During the past several decades, most researchers have been trying to improve stimulus adaptive polymer properties to their efficiency limit. Research to realise real environmental adaptive textiles by incorporating traditional fabrics with stimulus adaptive polymeric composites is just beginning. Extensive and intensive work needs to be done to study in detail, the property parameters of environmentally adaptive polymers and combine their special properties with textile applications so that real environmentally smart textiles can be achieved.

References

1. Gutowska, A., Bae, Y. H., Jacobs, H., Feijen, J., and Kim, S. W., Thermosensitive Interpenetrating Polymer Network: Synthesis, Characterisation, and Macromolecular Release. *Macromolecules*, 1994. 27: pp. 4167-4175.
2. Asaka, K. and Oguro, K., Bending of Polyelectrolyte Membrane Platinum Composites by Electric Stimuli. *Journal of Electro - analytical Chemistry* 2000. 480: pp. 186-198.
3. Jiang, H. Y., Kelch, S., and Lendlein, A., Polymers Move in Response to Light. *Advanced Materials*, 2006. 18: pp. 1471-1475.

4. Xulu, P. M., Filipcsei, G., and Zrinyi, M., Preparation and responsive properties of magnetically soft poly(*N*-isopropylacrylamide) gels. *Macromolecules*, 2000. 33(5): pp. 1716-1719.
5. Yang, B., Huang, W. M., Li, C., and Li, L., Effects of Moisture on the Thermomechanical Properties of a Polyurethane Shape Memory Polymer. *Polymer*, 2006. 47: pp. 1348-1356.
6. Bar-Cohen, Y. *Electroactive Polymer (EAP) Actuators as Artificial Muscles-reality, Potential and Challenges, The international society for optical engineering, Bellingham, UK.* 2001.
7. Scrosati, B., *Application of Electroactive Polymers.* 1993, London: Chapman and Hall. pp. 29-30.
8. Bar-Cohen, Y., Xue, T., Shahinpoor, M., Simpson, J. O., and Smith, J. *Flexible, Low Mass Robotic Arm Actuated by Electroactive Polymers* in *SPIE's 5th Ann. Int. Symp. Smart Structures and Materials.* 1998. San Diego, USA.
9. Calvert, P. and Ricke, P., Biomimetic Mineralization in and on polymers. *Chemistry of Materials*, 1996. 8: pp. 1715 -1727.
10. Smela, E., Inganäs, O., and Lundström, I., Controlled Folding of Micrometer-size Structures, *Science*, 1995. 268: pp. 1735-1738.
11. Leng, J., Lu, H., Liu, Y., and Du, S. *Conductive Nanoparticles in Electro Activated Shape Memory Polymer Sensor and Actuator.* in *Proceedings of SPIE - The International Society for Optical Engineering, v 6931, Nanosensors and Microsensors for Bio-Systems 2008, 2008, p 693109, San Diego, CA, United States.* 2008.
12. Bar-Cohen, Y. and Zhang, Q., Electroactive Polymer Actuators and Sensors. *MRS bulletin*, 2008. 33: pp. 173-181.
13. Bar-Cohen, Y. *Electroactive Polymers as Artificial Muscles - Reality and Challenges.* in *Proceedings of the 42nd AIAA Structures, Structural Dynamics, and Materials Conference (SDM), Gossamer Spacecraft Forum (GSF)* April 16-19 2001. Seattle WA, USA.
14. Mazzoldi, A. and DeRossi, D. *Conductive Polymer Based Structures for a Steerable Catheter.* in *Proceedings of SPIE.* 1999.
15. Lendlein, A. and Kelch, S., Shape-memory polymers. *Angewandte Chemie International Edition*, 2002. 41(12): pp. 2034-2057.
16. Metcalfe, A., Desfaits, A.-C., Salazkin, I., L. Yahiab, Sokolowskic, W. M., and Raymonda, J., Cold Hibernated Elastic Memory Foams for Endovacular Interventions. *Biomaterials*, 2003. 24: pp. 491-497.

17. Wache, H. M., Tartakowska, D. J., Hentrich, A., and Wagner, M. H., Development of a Polymer Stent with Shape Memory Effect as a Drug Delivery System *Journal of Materials Science: Materials in Medicine*, 2004. 14: pp. 109-112.
18. Hu, J. L., Meng, Q. H., Zhu, Y., Lu, J., and Zhuo, H. T., *Shape Memory Fibers Prepared by Wet, Reaction, Dry, Melt, and Electro Spinning*. October 6, 2007: US Patent 11/907,012.
19. Meng, Q. H., Hu, J. L., Zhu, Y., Lu, J., and Liu, Y., Morphology, Phase Separation, Thermal and Mechanical Property Differences of Shape Memory Fibres Prepared by Different Spinning Methods. *Smart Materials and Structures*, 2007. 16 pp. 1192-1197.
20. Charlesby, A., *Atomic Radiation and Polymers*. 1960, 198, New York: Pergamon Press.
21. Gall, K., Dunn, M. L., Liu, Y., Finch, D., Lake, M., and Munshi, N. A., Shape Memory Polymer Nanocomposites *Acta Materialia*, 2002. 50: pp. 5115-5126.
22. Sokolowski, W. M., Chmielewski, A. B., Hayashi, S., and Yamada, T. *Cold Hibernated Elastic Memory (CHEM) Self-deployable Structures*. in *SPIE International Symposium on Smart Structures and Materials*. 1999. Newport Beach, CA.
23. Meng, Q. H., Hu, J. L., and Yeung, L. Y., An Electro-active Shape Memory Fibre by Incorporating Multi-walled Carbon Nanotubes. *Smart Materials and Structures*, 2007. 16: pp. 830-836.
24. Hu, J. L., Meng, Q. H., Zhu, Y., Lu, J., and Zhuo, H. T., *Shape Memory Fibers Prepared by Wet, Reaction, Dry, Melt, and Electro Spinning*. 2007: US Patent 11907012.
25. Hu, J. L., Ji, F. L., and Wong, Y. W., Dependency of the Shape memory Properties of a Polyurethane upon Thermomechanical Cyclic Conditions. *Polymer International*, 2005. 54: pp. 600-605.
26. Humbeeck, J. V., Shape Memory Alloys: A Material and a Technology *Advanced Engineering Materials*, 2001. 3: pp. 837 - 850.
27. Kainuma, R., Imano, Y., Ito, W., Sutou, Y., Morito, H., Okamoto, S., Kitakami, O., Oikawa, K., Fujita, A., Kanomata, T., and Ishida, K., Magnetic-field-induced Shape Recovery by Reverse Phase Transformation *Nature*, 2006. 439: pp. 957-960.

28. Liu, M., Bian, F., and Sheng, F., FTIR study on molecular structure of poly(N-isopropylacrylamide) in mixed solvent of methanol and water. *European Polymer Journal*, 2005. 41(2): pp. 283-291.
29. Meng, Q., Hu, J., and Zhu, Y., Properties of Shape Memory Polyurethane Used as a Low Temperature Thermoplastic Orthotic Material: Influence of Hard Segment Content. *Journal of Biomaterials Science, Polymer Edition*, 2008. 19: pp. 1437-1454.
30. Miaudet, P., Derré, A., Maugey, M., Zakri, C., Piccione, P. M., Inoubli, R., and Poulin, P., Shape and Temperature Memory of Nanocomposites with Broadened Glass Transition. *Science*, 2007. 318: pp. 1294-1296.
31. Gunes, I. S. and Jana, S. C., Shape Memory Polymers and Their Nanocomposites: A Review of Science and Technology of New Multifunctional Materials. *Journal of Nanoscience and Nanotechnology*, 2008. 8: pp. 1616-1637.
32. Jeong, H. M., Ahn, B. K., and Kim, B. K., Miscibility and Shape Memory Effect of Thermoplastic Polyurethane Blends with Phenoxy Resin. *European Polymer Journal*, 2001. 37: pp. 2245-2252.
33. Zhang, S., Feng, Y., Zhang, L., Sun, J., Xu, X., and Xu, Y., Novel Interpenetrating Networks with Shape-memory Properties. *Journal of Polymer Science, Part A: Polymer Chemistry*, 2007. 45: pp. 768-775.
34. Beloshenko, V. A., Beygelzimer, Y. E., Borzenko, A. P., and Varyukhin, V. N., Shape Memory Effect in the Epoxy Polymer-Thermoexpanded Graphite System *Composites Part A*, 2002. 33: pp. 1001-1006.
35. Beloshenko, V. A., Beigelzimer, Y. E., Borzenko, A. P., and Varyukhin, V. N., Shape-memory Effect in Polymer Composites with a Compatible Filler. *Mechanics of Composite Materials*, 2003. 39: pp. 255-264.
36. Gunes, I. S., Cao, F., and Jana, S. C., Effect of thermal expansion on shape memory behavior of polyurethane and its nanocomposites. *Journal of Polymer Science, Part B: Polymer Physics*, 2008. 46: pp. 1437-1449.
37. Razzaq, M. Y. and Frormann, L., Thermomechanical Studies of Aluminum Nitride Filled Shape Memory Polymer Composites. *Polymer Composites*, 2007. 28: pp. 287-293.

38. Gall, K., Dunn, M. L., Liu, Y., Stefanic, G., and Balzar, D., Internal Stress Storage in Shape Memory Polymer Nanocomposites. *Applied Physics Letters*, 2004. 85: pp. 290-292.
39. Zhang, C. S., Ni, Q. Q., Fu, S. Y., and Kurashiki, K., Electromagnetic Interference Shielding Effect of Nanocomposites with Carbon Nanotube and Shape Memory Polymer. *Composites Science and Technology*, 2007. 67: pp. 2973-2980.
40. Mondal, S., Phase change materials for smart textiles - An overview. *Applied Thermal Engineering*, 2008. 28(11-12): pp. 1536-1550.
41. Chen, X., Wudl, F., Mal, A. K., Shen, H., and Nutt, S. R., New Thermally Remendable Highly Cross-Linked Polymeric Materials. *Macromolecules*, 2003. 36: pp. 1802-1807.
42. Li, J., Viveros, J. A., Wrue, M. H., and Anthamatten, M., Shape-Memory Effects in Polymer Networks Containing Reversibly Associating Side-groups. *Advanced Materials*, 2007. 19: pp. 2851-2855.
43. Chung, T., Romo-Uribe, A., and Mather, P. T., Two-Way Reversible Shape Memory in a Semicrystalline Network. *Macromolecules*, 2008. 41: pp. 184-192.
44. Annaka, M., Tokita, M., Tanaka, T., Tanaka, S., and Nakahira, T., The gel that memorizes phases. *Journal of Chemical Physics*, 2000. 112(1): pp. 471-477.
45. Vaia, R., Nanocomposites: Remote-controlled Actuators. *Nature Materials*, 2005. 4: pp. 429-430.
46. Bar-Cohen, Y., Kim, K. J., Choi, H. R., and Madden, J. D. W., Editorial: Electroactive Polymer Materials. *Smart Materials and Structures*, 2007. 16: pp. E01-02.
47. Bar-Cohen, Y., Editorail: Focus issue on Biomimetics Using Electroactive Polymers as Artificial Muscles. *Bioinspiration and biomimetics*, 2007.
48. Kaetsu, I., Uchida, K., Shindo, H., Gomi, S., and Sutani, K., Intelligent type controlled release systems by radiation techniques. *Radiation Physics and Chemistry*, 1999. 55(2): pp. 193-201.
49. Feil, H., Bae, Y. H., Feijen, J., and Kim, S. W., Effect of comonomer hydrophilicity and ionization on the lower critical solution temperature of n-isopropylacrylamide copolymers. *Macromolecules*, 1993. 26(10): pp. 2496-2500.

50. Wang, Y. Q., Sun, C. J., Zhou, E., and Su, J., Deformation Mechanisms of Electrostrictive Graft Elastomer. *Smart Materials and Structures*, 2004. 13: pp. 1407-1413.
51. Peppas, N. A. and Kim, B., Stimuli-sensitive protein delivery systems. *Journal of Drug Delivery Science and Technology*, 2006. 16(1): pp. 11-18.
52. Kim, J., Ampofo, J., Craft, W., and Kim, H. S., Modeling Elastic, Viscous and Creep Characteristics of Cellulose Electro-Active Paper. *Mechanics of Materials*, 2008. 40: pp. 1001-1011.
53. Zrinyi, M., Szabo, D., and Feher, J. *Comparative Studies of Electro- and Magnetic Field Sensitive Polymer Gels, Proceedings of the SPIE's 6th Annual International Symposium on Smart Structures and Materials, EAPAD Conf. V 3669, P 406-413*. 1999.
54. Kosmatka, J. B. and Biggerstaff, J. M. *Development of Electro-viscoelastic Materials for Active Damping*. in *Collection of Technical Papers - AIAA/ASME/ASCE/AHS/ASC Structures, Structural Dynamics and Materials Conference, v 4, Collection of Technical Papers - 46th* 2005.
55. Huang, C., Klein, R., Xia, F., Li, H. F., and Zhang, Q. M., Poly(vinylidene fluoride-trifluoroethylene) Based High Performance Electroactive Polymers. *IEEE Transactions on Dielectrics and Electrical Insulation*, 2004. 11: pp. 299-311.
56. Xia, F., Cheng, Z. Y., Xu, H., Li, H., Zhang, Q. M., Kavarnos, G., Ting, R., Sedat, G. A., and Belfield, K. D., High Electromechanical Responses in a Poly(vinylidene fluoride-trifluoroethyl-ene-chloro-fluoroethylene) Terpolymer. *Advanced Materials*, 2002. 14: pp. 1574-1577.
57. Pelrine, R., Kornbluh, R., and Joseph, J. P., Electrostriction of Polymer Dielectrics with Compliant Electrodes as a Means of Actuation. *Sensors and Actuators A: Physical*, 1998. 64: pp. 77-85.
58. Pelrine, R., Kornbluh, R., Pei, Q., and J. Joseph, High Speed Electrically Actuated Elastomers with Strain Greater Than 100%. *Science*, 2000. 287: pp. 836-839.
59. Li, J. Y., Huang, C., and Zhang, Q., Enhanced Electromechanical Properties in all-polymer Percolative Composites *Applied Physics Letters*, 2004. 84: pp. 3124-3126.

60. Huang, C. and Zhang, Q. M., All-organic Dielectric-percolative Three-component Composite Materials with High Electro-mechanical Response. *Applied Physics Letters*, 2004. 84: pp. 4391-4393.
61. Huang, C., Zhang, Q. M., and Su, J., High-dielectric-constant All-polymer Percolative Composites. *Applied Physics Letters*, 2003. 82: pp. 3502-3504.
62. Ren, H. W. and Wu, S. T., Anisotropic Liquid Crystal Gels for Switchable Polarizers and Displays. *Applied Physics Letters*, 2002. 81: pp. 1432-1434.
63. Spinks, G. M., Mottaghitalab, V., Bahrami-Samani, M., Whitten, P. G., and Wallace, G. G., Carbon-Nanotube-Reinforced Polyaniline Fibers for High-Strength Artificial Muscles. *Advanced Materials*, 2006. 18: pp. 637-640.
64. Vidal, F., Plesse, C., Palaprat, G., Kheddar, A., Citerin, J., Teyssie, D., and Chevrot, C., Conducting IPN Actuators: From Polymer Chemistry to Actuator with Linear Actuation. *Synthetic Metals*, 2006. 156: pp. 1299-1304.
65. Shahinpoor, M., Bar-Cohen, Y., Xue, T., Simpson, J. O., and Smith, J. *Ionic Polymer-Metal Composites (IPMC) As Biomimetic Sensors and Actuators, Proceedings of SPIE's 5th Annual International Symposium on Smart Structures and Materials, 1-5 March, 1998, San Diego, CA. Paper No. 3324-27*. 1998.
66. Grohmann, B. A., Wallmersperger, T., and Kroeplin, B., Adaptive Strukturen und gekoppelte Mehrfeldprobleme(Adaptive structures and coupled multi-field problems). *Stahlbau*, 2000. 69(6): pp. 446-454.
67. Qu, L., Peng, Q., Dai, L., Spinks, G. M., Wallace, G. G., and Baughman, R. H., Carbon Nanotube Electroactive Polymer Materials: Opportunities and Challenges. *MRS Bulletin*, 2008. 33: pp. 215-224.
68. Eichhorn, S. J., Baillie, C. A., Zafeiropoulos, N., Mwaikambo, L. Y., Ansell, M. P., Dufresne, A., Entwistle, K. M., Herrera-Franco, P. J., Escamilla, G. C., Groom, L., Hughes, M., Hill, C., Rials, T. G., and Wild, P. M., Current international research into cellulosic fibres and composites. *Journal of Materials Science: Materials in Medicine*, 2001. 36(9): pp. 2107-2131.
69. Yu, Y. and Ikeda, T., Photodeformable Polymers: A New Kind of Promising Smart Material for Micro- and Nano-Applications. *Macromolecular Chemistry and Physics*, 2005. 206: pp. 1705-1708.

70. Yu, Y., Nakano, M., and Ikeda, T., Directed Bending of a Polymer Film by Light. *Nature*, 2003. 425: pp. 145-145.
71. Mamada, A., Tanaka, T., Kungwatchakun, D., and Irie, M., Photoinduced Phase Transition of Gels *Macromolecules*, 1990. 23: pp. 1517-1519.
72. Lendlein, A., Jiang, H., Jünger, O., and Langer, R., Light-induced shape-memory polymers. *Nature*, 2005. 434: pp. 879-882.
73. Nikitin, L. V., Stepanov, G. V., Mironova, L. S., and Gorbunov, A. I., Magnetodeformational Effect and Effect of Shape Memory in Magnetoelastics. *Journal of Magnetism and Magnetic Materials*, 2004. 272-276: pp. 2072-2073.
74. Ginder, J. M., Clark, S. M., Schlotter, W. F., and Nichols, M. E., Magnetostrictive Phenomena in Magnetorheological Elastomers. *International Journal of Modern Physics B, Vol. 16*, 2002. 16: pp. 2412-2418.
75. Hu, X., Zhou, J., Zhang, N., Tan, H., and Gao, C., Preparation and properties of an injectable scaffold of poly(lactic-co-glycolic acid) microparticles/chitosan hydrogel. *Journal of the Mechanical Behavior of Biomedical Materials*, 2008. 1(4): pp. 352-359.
76. Szabo, D., Szeghy, G., and Zrinyi, M., Shape Transition of Magnetic Field Sensitive Polymer Gels. *Macromolecules*, 1998. 31: pp. 6541-6548.
77. T Mitsumata, K. F., E Juliac, K Iwakura, K., Compressive Modulus of Ferrite Containning Polymer Gels. *International Journal of Modern Physics B*, 2002. 16: pp. 2419-2425.
78. Zrinyi, M., SzaboH, D., and Kilian, H.-G., Kinetics of the Shape Change of Magnetic field Sensitive Polymer Gels. *Polymer Gels and Networks*, 1998. 6: pp. 441-454.
79. Buckley, P. R., McKinley, G. H., Wilson, T. S., IV, W. S., Benett, W. J., Bearinger, J. P., McElfresh, M. W., and Maitland, D. J. *Inductively Heated Shape Memory Polymer for the Magnetic Actuation of Medical Devices*. in *IEEE Transactions on Biomedical Engineering* 2006: 53.
80. Razzaq, M. Y., Anhalt, M., Frormann, L., and Weidenfeller, B., Mechanical Spectroscopy of Magnetite Filled Polyurethane Shape Memory Polymers *Materials Science and Engineering: A*, 2007. 471: pp. 57-62.

81. Schmidt, A. M., Electromagnetic Activation of Shape Memory Polymer Networks Containing Magnetic Nanoparticles. *Macromolecular Rapid Communications*, 2006. 27: pp. 1168-1172.
82. Jung, Y. C., So, H. H., and Cho, J. W., Water-Responsive Shape Memory Polyurethane Block Copolymer Modified with Polyhedral Oligomeric Silsesquioxane. *Journal of Macromolecular Science, Part B: Physics*, 2006. 45: pp. 453-461.
83. Chen, M. C., Tsai, H. W., Chang, Y., Lai, W. Y., Mi, F. L., Liu, C. T., Wong, H. S., and Sung, H. W., Rapidly Self-Expandable Polymeric Stents with a Shape-Memory Property. *Biomacromolecules 2007, 8, 2774-2780*, 2007. 8: pp. 2774-2780.
84. Hu, J. L., Ding, X. M., and Tao, X. M., Shape Memory Polymers and Their Applications to Smart Textile Products. *Journal of China Textile University*, 2002. 19: pp. 89-93.
85. Russel, D. A., Hayashi, S., and Yamada, T. *The potential use of memory film in clothing*. in *Techtextil Symposium-New Protective textiles (through Textile Technology Index Database)*. April. 1999.
86. Li, Y., Chung, S., Chan, L., and Hu, J. L., Characterization of Shape Memory Fabrics. *Textile Asia*, 2004. 35: pp. 32-37.
87. Chen, H. and Hsieh, Y.-L., Ultrafine Hydrogel Fibers With Dual Temperature- and pH-responsive Swelling Behaviors. *Journal of Polymer Science Part A: Polymer Chemistry*, 2004. 42: pp. 6331-6339.
88. Chen, H. and Hsieh, Y.-L., Enzyme Immobilization on Ultrafine Cellulose Fibers via Poly(acrylic acid) Electrolyte Grafts. *Biotechnology and Bioengineering, Volume 90, Issue 4 (p 405-413)* 2005. 90: pp. 405-413.
89. Tobushi, H., Hayashi, S., Hoshio, K., and Miwa, N., Influence of Strain-holding Conditions on Shape Recovery and Secondary-shape Forming in Polyurethane-shape Memory Polymer. *Smart Materials and Structures*, 2006. 15: pp. 1033-1038.
90. Tyagi, P., Wu, P. C., Chancellor, M., Yoshimura, N., and Huang, L., Recent advances in intravesical drug/gene delivery. *Molecular Pharmaceutics*, 2006. 3(4): pp. 369-379.
91. Ikeda, T., Nakano, M., Lei, Y. Y., Tsutsumi, O., and Kanazawa, A., Anisotropic Bending and Unbending Behavior of Azobenzene Liquid-crystalline Gels by Light Exposure. *Advanced Materials*, 2003. 15: pp. 201-205.

92. Varga, Z., Filipcsei, G., and Zrínyi, M., Magnetic Field Sensitive Functional Elastomers With Tuneable Elastic Modulus. *Polymer*, 2006. 47: pp. 227-233.
93. Qiu, Y. and Park, K., Environment-sensitive hydrogels for drug delivery. *Advanced Drug Delivery Reviews*, 2001. 53(3): pp. 321-339.
94. Mohr, R., Kratz, K., Weigel, T., Lucka-Gabor, M., Moneke, M., and Lendlein, A. *Initiation of Shape-memory Effect by Inductive Heating of Magnetic Nanoparticles in Thermoplastic Polymers.* in *Proceedings of the National Academy of Sciences, v103, 3540-3545, United States*. 2006.
95. Topolkaraev, V. A. and Soerens, D. A., *Methods of Making Humidity Activated Marerials Having Shape-memory* September 30, 2003: US Patent 6627673.

Chapter 7

Adaptive Polymeric Nanofibre and Nanofilm

Adaptive and functional polymers have drawn much attention in many fields due to their unique properties. Through nanotechnology, there have been reports of fabricating various adaptive polymer nanostructures, such as nanofibres and nanofilms, in recent years. In this chapter, fabrication of nanofibres and nanofilms by concentrating on the electrospinning and self-assembly methods, are first introduced. At the same time, the principles and parameters of electrospinning as well as self-assembly are discussed. Secondly, the morphologies and structures of nanofibres and nanofibrous nonwovens are described briefly. Thirdly, we will present a brief review on the development of some typical adaptive polymer nanofibres and nanofilms, and in particularly, adaptive shape memory polymeric nanofibres and nanofibrous nonwovens; nonlinear optical nanofibres; responsive and adaptive porphyrin based nanostructures; adaptive nanocomposites and functional polymer nanofilms. Finally, the applications of nanofibres and nanofilms are introduced. Thus, a comprehensive understanding of adaptive nanofibres and nanofilms can be obtained in this chapter.

Keywords: nanofibres; nanofilms; electrospinning; nanotechnology; adaptive polymer; functional polymer

7.1 Introduction

In the past decades, adaptive polymers have developed extensively in the nanoscale field by using nanotechnology, which is one of the most rapidly growing scientific disciplines in studying and developing objects with characteristic dimensions below approximately 100 nm. With suitable fabrication techniques, various nanostructures, including nanofibres, nanofilms and nanoparticles can be achieved with adaptive polymers. For example, nanofibres with a shape memory effect can be prepared from shape memory polymers (SMPs) by the electrospinning (ELSP) method. Nanofilms, such as those that are pH responsive, can be fabricated through the self-assembly technique. After the adaptive polymers are fabricated into nanostructures, many unique properties may be obtained due to their surface properties and orientation structure. Even novel functionalities which cannot be found in fibres of larger sizes can be achieved in nanofibres or nanofilms. In this way, nanofibres and nanofilms of adaptive polymers may find many applications in many fields.

This chapter is therefore intended to introduce interested scientists or engineers to adaptive polymer nanofibres and nanofilms. In this chapter, we will first introduce the typical fabrication technologies of nanofibres and nanofilms, discuss the fabrication process and their basic principles and then introduce the typical structure and morphology of nanofibres and nonwovens. Thereafter, the development of some typical adaptive and functional polymer-based nanofibres or nanofilms is reviewed briefly. Finally the applications or potential applications of nanofibres and nanofilms are suggested.

7.2 Fabrication Technologies

So far, various nanostructures, such as nanofibres, nanotubes, nano-particles, nanofilms and nanowires, have been developed for use in practical applications. Amongst them, the polymeric nanofibres, which are fibres with oriented polymer chains and diameters in the range of 100 nm or less, and nanofilms with thickness or dimensions in the nanometre scales, have received increased attention because they can be fabricated easily. They are also capable of controlling their compositional, structural and functional properties.

Providing adaptive polymers are selected, adaptive nanofibres with ultra-fine diameters and nanofilms with ultra-thin thickness can be fabricated by the nanostructure fabrication techniques. Until now, a number of different processing techniques have been used to fabricate nanofibres, such as drawing,[1] template synthesis,[2,3] phase separation,[4] self-assembly,[5] electrospinning.[6] Comparatively, the drawing, which is a process similar to dry spinning of fibre industry, can be used to make one-by-one single nanofibres. However, only a viscoelastic material can be made into nanofibres with the drawing method if they can undergo strong deformations while being cohesive enough to support the stresses developed during pulling. For the template synthesis, on the basis of a nanoporous membrane as a template, polymer materials can be made into nanofibres of either solid (a fibril) or hollow (a tubule) shape. Thus, the most important application of template synthesis is that they are generally used to fabricate nanometer tubules and fibrils of various raw materials, such as electronically conducting polymers, metals, semiconductors and carbons. However, this method cannot make one-by-one continuous nanofibres. The phase separation consisting of dissolution, gelation and extraction using a different solvent appears during temperate synthesis, while nanoscale porous foam is resulted from the freezing and drying. So that, it usually takes a relatively long period of time to transfer the solid polymer into the nano-porous foam. The self-assembly refers to the process in which individual, pre-existing components organise themselves into desired patterns and functions. However, similarly to the phase separation, long time is consumed in the self-assembly processing continuous polymer nanofibres. Finally, it seems that the electrospinning is the only method which can be further developed to produce large quantity of one-by-one continuous nanofibres from synthetic polymers or natural polymeric materials. In addition, it was found that the electrospinning is one of the most efficient, simple, and versatile methods because of its relatively simple and cost-effective setup.

However, it should be pointed out that the development of functional nanofibres via self-assembly is an increasing interested area. Until now, many kinds of polymer nanostructures have been developed successfully by self-assembly. Not only the nanofibres, but also the nanofilms including of monolayer and multilayer, even the nano-wires, nanotubes can be achieved through the self-assembly. In addition, the fabrication of hierarchical architectures from multiple functional components and the development of synergistic functions are emerging as challenging issues. Therefore, these two kinds of fabrication technologies, electrospinning and self-assembly, are mainly introduced in this chapter as following:

7.2.1 *Electrospinning*

7.2.1.1 *Definition of electrospinning*

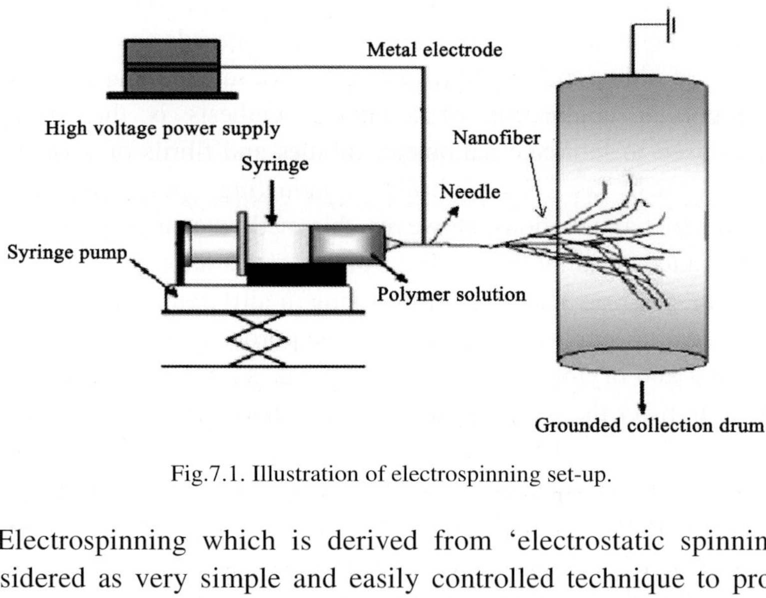

Fig.7.1. Illustration of electrospinning set-up.

Electrospinning which is derived from 'electrostatic spinning' is considered as very simple and easily controlled technique to produce continuous polymer fibres with ultrafine diameters through the action of an external electric field imposed on a polymer solution or melt. In electrospinning, polymer nanofibres ranging from ten to 1,000 nm are collected as a non-woven mesh or membrane on a collector plate acting as the counter electrode. The diameters of fibres with various lengths depend on the solution and the process conditions. The main advantage of electrospun fibres with ultrathin diameters is their large surface-to-mass ratio and high porosity. Moreover, the functionalities of nanofibre may be dominated by the molecules located at the surface of the fibre. Thus, the engineering of fibre properties was allowed by tailoring compositions and morphologies on the surface. Moreover, the properties of nanofibres can be further modified by electro-spinning polymer blends into composite nanofibres. This nanofibre composite is expected to better fulfil specific industrial requirements in terms of their material properties and increases array of potential applications.

7.2.1.2 *Principle of Electrospinning*

The formation of nanofibres through electrospinning is based on the uniaxial stretching of a viscoelastic solution. Similar to conventional fibre spinning methods, the drawing of the solution to form the fibre will continue as long as there is enough solution to feed the electrospinning jet. Thus without any disruption to the electrospinning jet the formation of the fibre will be continuous. However, unlike the conventional fibre spinning methods like melt-spinning, wet-spinning and dry-spinning, electrospinning makes use of electrostatic forces to stretch the solution as it solidifies.

A conventional electrospinning setup consists of a spinneret with a metallic needle, a syringe pump, a high-voltage power supply and a grounded collector (see Fig.7.1). In a typical electrospinning process, the polymer solution is pumped through a thin nozzle which simultaneously serves as an electrode. The nozzle has an inner diameter on the order of 100 mm; and the distance to the counter electrode is usually 10-25 cm in this systems. With the application of a high electric field of 100-500 kVm^{-1}, the current ranging from a few hundred nanoamperes to microamperes appears during electrospinning. At the same time, electrospun fibres are collected on the counter electrode as the substrate. With respect to the electrospinning process, it was noted that the vertical alignment of the electrodes 'from top to bottom' is not insignificant, in principal, the process can also be performed 'from bottom to top' or horizontally. However, usually a cone-shaped deformation of the drop of polymer solution is caused by the applied voltage in the direction of the counter electrode.[7] In the electrospinning, the general value of cone angle is about 30°. If the applied voltages are too high, the deformed drop will form a jet which moves towards the substrate (counter electrode) and becomes narrower in the process.[8] Finally, the solid nanofibres are precipitated with high velocities (of 40 $m.s^{-1}$ or more) on the substrate as the solvent evaporates on the way to the substrate.

For most of polymer electrospinning, solvents are required, while no solvents are used in the melt electrospinning. Therefore, it is attractive, considering the environmental and productivity. However, melt electrospinning is limited to fabricate nanofibres with diameters of less than 400 nm and the resulted nanofibres show a wide diameter distribution. Until now, there are only several polymers, such as polypropylene (PP), polyethylene (PE), polyamide 12 (PA12), polycaprolactone(PCL), polyethylene terephthalate (PET) and polyurethane (PU), have been processed by melt Electrospinning,[9-12] and only the low-melting PCL can be used to fabricate nanofibre with diameters of less than 300 nm which was electrospun directly onto the tissue of living cells.

7.2.1.3. *Parameters of electrospinning*

During the electrospinning process, the production of nanofibres from solution is influenced greatly by the parameters which include:

(I) Solution properties such as viscosity, conductivity, elasticity, and surface tension.

(II) Process parameters such as electric potential at the capillary tip, hydrostatic pressure in the capillary tube and the distance between the tip and the collecting screen.

(III) Ambient condition parameters such as solution temperature, humidity and air velocity.

7.2.1.3.1 Solution concentrations/ viscosity

The solution viscosity will be one of the most significant parameters which influence the nanofibres diameter. Generally, a higher viscosity results in a larger fibre diameter.[13,14] However, in the polymer solutions, the viscosity is proportional to the concentration of polymer. Thus, nanofibres with larger diameters are formed in the higher the polymer concentration, as shown in Fig. 7.2. In fact, it was pointed out that the diameter of nanofibres increased as polymer concentration increased according to a power law relationship.[16] Furthermore, it was reported that the diameter was proportional to the cube of the concentration,[17] because as the nanofibres are formed from evaporation or solidification of polymer fluid jets, the fibre diameters will depend primarily on the polymer contents in the jets as well as on the jet sizes. It has been recognised that nanofibres with different diameters are formed during travelling of a solution jet from the pipette onto the metal collector as the primary jet may[17] or may not[18,19] be split into multiple jets.

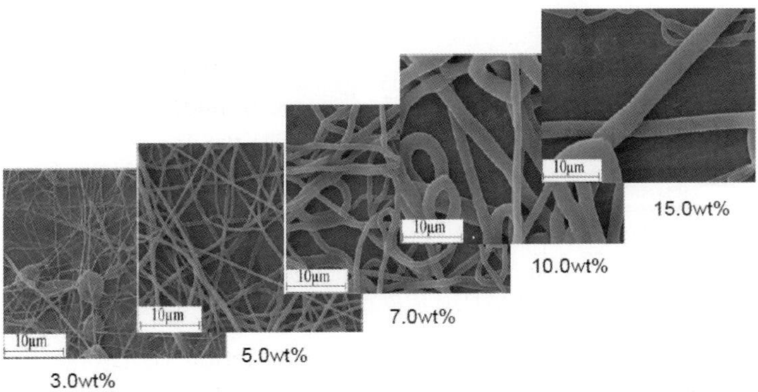

Fig. 7.2. SEM photographs of electrospun SMPU nanofibres from different SMPU/DMF concentration solutions.[15]

7.2.1.3.2 Electrical voltage/applied voltage

In addition, applied electrical voltage is another parameter to influence the fibre diameters significantly. Generally, a higher applied voltage results in a larger fibre diameter as more fluid is ejected in a jet (Fig.7.3). However, the fibre diameters obtained with electrospinning are seldom uniform. This fact is one challenge in the current research of electrospinning. Few reports have been given towards resolving this problem. Until recently, Demir MM *et al.* proposed a useful attempt during the electrospinning polyurethane nanofibres.[16] They found that the diameters of nanofibres obtained at a high temperature (e.g. 70°C) were much more uniform than those obtained at room temperature. However, they did not explain the mechanisms clearly. From the viscosity aspects, at the higher temperature (70°C), the viscosity of the same concentration polyurethane solution is much lower than that at lower temperature (20°C). For example, the upper limited value of concentration for electrospinning into nanofibres is 12.8 wt% at room temperature, whereas it is 21.2 wt% at the high temperature.

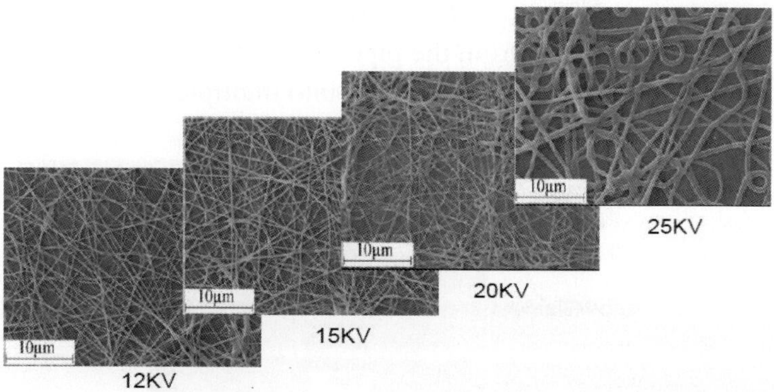

Fig.7.3 SEM images of electrospun SMPU nanofibres at different applied voltage (a) 12KV,(b)15KV,(c)20KV,(d)25KV.[15]

7.2.1.3.3 Feeding rate

It was reported that feeding rate affects volume charge density and electrical current of polymer solution. The electrical current increases with increasing the feeding rate in certain polymer solutions, while it decreases with the feeding rate increasing in other polymer solutions. For instance, an increase feeding flow rate would increase the electrical current in poly (ethylene oxide) (PEO), PVA, PAA, and PU solutions, while it decreases the electrical current in PCL solutions.[20] Thus, increasing the feeding flow rate would decrease the fibre diameter; and the surface charge density decreases as the flow rate increases. It was also pointed out that different morphology could be achieved by controlling the feeding rate at a given electric field, [21] because a certain minimum value of the solution volume suspended at the end of the spinneret should be maintained in order to form an equilibrium Taylor cone. Besides, Um *et al.*[22] had demonstrated that with increasing the feeding rate of polymer solution, the electrospinning performance for creating nanofibres improved, however, the improvement was not sufficient to achieve a consistent production of high quality nanofibres membranes.

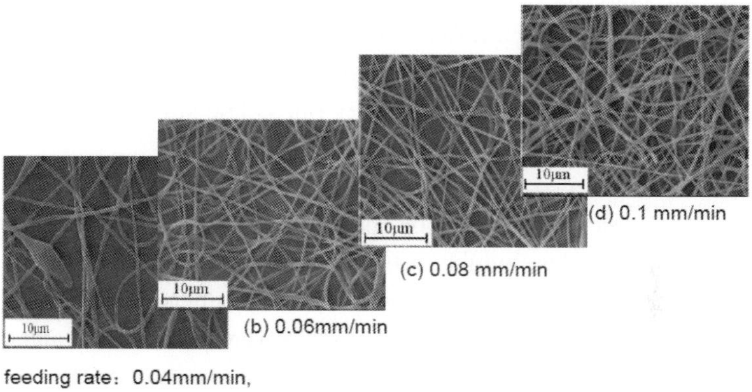

Fig.7.4 SEM images of electrospun SMPU nanofibres at different feeding rate.[15]

7.2.2 *Self-assembly*

7.2.2.1 *Definition of self-assembly*

Self-assembly is the autonomous organisation of components into patterns or structures without human intervention.[23] In the past decades, the concept of self-assembly is used increasingly in many disciplines. Self-assembly is common to many dynamic, multi-component systems, from smart materials and self-healing structures to netted sensors and computer networks. Figure 7.5 presents the examples of typical self-assembly. They involve components from the molecular (crystals) to the planetary (weather systems) scale and many different kinds of interactions. In this chapter, the self-assembly is limit the term to processes that involve pre-existing components (separate or distinct parts of a disordered structure), are reversible and can be controlled by proper design of the components.

7.2.2.2 *Principle of self-assembly*

There are two main kinds of self-assembly: static and dynamic. In static self-assembly, formation of the ordered structure may require energy, e.g. by stirring, but once it is formed, it is stable. Most research in self-assembly has focused on the static self-assembly. For the dynamic self-assembly, the interactions which is responsive to the formation of structures or patterns between components only occur if the system is dissipating energy. The patterns formed by competition between reaction and diffusion in oscillating chemical reactions are simple examples; biological cells are much more complex ones. The study of dynamic self-assembly is in its infancy.

Self-assembly in soft materials relies on the fact that the energy dissipated by fluctuations in the position and the orientation of the molecules or particles, which are the result of Brownian motion, is comparable to the thermal energy. Thermal energy has a dramatic influence on soft materials at the nanoscale as weak non-covalent bonds are broken and sometimes re-formed. This process enables the system to reach thermodynamic equilibrium, which is often a non-uniform state. Because of the relatively weak interactions between molecules, transitions between different structures can readily be driven by changes in conditions, for example, temperature or pH value. Such external triggers that induce phase transitions could lead to a host of responsive materials, or coupled with an appropriate source of energy to nano-mechanical systems.

The driving forces which were used in self-assembly with soft materials are mainly resulted from four classifications: non-covalent interactions; intermolecular packing, biological self-assembly and nanomotors.[23]

7.2.2.2.1 Non-covalent interactions

The non-covalent intermolecular forces are usually much weaker than the covalent bonds between the atoms of a molecule. The weak intermolecular interactions which are responsible to the molecular ordering in soft materials include ionic interactions, hydrogen bonds, dipolar interactions, van der waals forces, coordination bonds in ligands and complexes and hydrophobic interactions. Many types of this noncovalent interaction can be exploited in supramolecular chemistry, such as hydrogen bonding, donor–acceptor binding metal complexation. In this way, diverse methods have been employed to create receptors for ionic and molecular guests. For example, cyclodextrins are used as hosts for the delivery of drugs or pesticides.

7.2.2.2.2 Intermolecular packing

It was observed that at high concentration, the packing of block copolymer or low-molar-mass amphiphilic molecules in solution leads to the formation of lyotropic liquid-crystal phases, such as cubic packed spherical micelles, hexagonal-packed cylindrical micelles, lamellae or bi-continuous cubic phases. Therefore, the intermolecular packing was also used to fabricate nanostructures. In this kind of interaction, the phase formed depends on the curvature of the surfactant–water interface. In one approach to the understanding of lyotropic phase behaviours, the free energy associated with curved interfaces is computed. The curvature is analysed using differential geometry, neglecting details of molecular organisation. In the second main model the interfacial curvature is described by a molecular packing parameter.

7.2.2.2.3 Biological self-assembly

Biological self-assembly is generally used to understand the folding of proteins in nano-biotechnology. For example, in the application of DNA computing, single DNA strands are attached to a silicon chip. Computational operations can then be performed in which certain DNA strands couple to added DNA molecules. In addition, biological self-assembly is also used in the application of DNA-directed assembly of proteins.

7.2.2.2.4 Nanomotor

Nanomotor is a key element of nanomachine. Oscillating chemical reactions is a particularly attractive artificial motor system which is used to drive volume changes in polymer gels. For example, the Belousov–Zhabotinsky (BZ) reaction was used to create an oscillating redox potential. This reaction was coupled to the most familiar polymer–gel system exhibiting a volume phase transition-poly(N-isopropyl-acrylamide) (PNIPAM) in water.[24] The PNIPAM was modified by attachment through covalent bonds of ruthenium tris(2,2'-bipyridine) units which act as catalysts for the BZ reaction. Thus, the oscillations in the BZ reaction were translated into periodic swelling and deswelling of the gel as a result of changes in the charge on the ruthenium complex.

The four interactions were all reported to self-assembly nanostructure in polymer system. However, comparatively, the non-covalent interactions, in particularly, the hydrogen bonding are widely used to fabricate nanofibres, nanofilms and nanotubes etc.

7.2.2.3 *Advantages of self-assembly technology*

According to the principle of self-assembly, lots of nanostructure can be achieved, such as nanofibres, nanofilms, nanoparticles and nanowires. In contrast to conventional nanofibre structures formed from identical molecular components, molecular organisation of such heterogeneous molecular components would allow the design of synergistic functions. The components of supramolecular nanofibres can be selected widely from polymeric, organic, inorganic or bio-molecules. All categories, organic, inorganic and biomolecules, are replete with materials classified according to chemical structures and properties, and molecular components can be selected from each category. There are uninvestigated combinations of molecular components especially in the interdisciplinary area (I–IV), in which a wealth of new phenomena and functions are expected. Ordered self-assembly of hetero-components inevitably produces molecular or nano-interfaces. This would exert decisive roles in determining the property and function of molecular systems. Second, imparting responsive and adaptive natures are important to develop smart molecular systems whose functions are expressed in harmony with the environmental or physicochemical stimuli. They would provide a basis to design neuromorphic properties, as exemplified by self-wiring and self-restoration. They may also lead to molecular systems with self-learning functionalities. Therefore, self-assembly provides a route to achieve several of nanostructures with various components to exhibit many kinds of adaptive or functional properties.

7.3 Morphologies and Structure

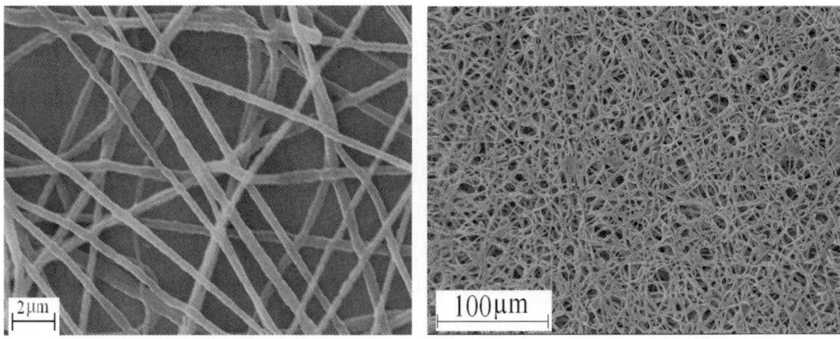

(a) (b)

Fig.7.5 SEM image of SMPU electrospun nanofibres (a) and nanofibrous nonwoven (b).

Nanofibres and nanofilms have drawn wide attention, because they have an ultrafine diameters or ultrathin thickness compared with the traditional fibres or films. Furthermore, these special morphology and structure results in many unique properties, which cannot be found in the other polymer forms. The mechanism in adaptive nanostructure relies on the molecular chain, structure and morphology of selected polymer. Actually, morphology and structure of nanofibres may influence their functionalities greatly. The following will introduce their typical structure observed in nanofibres and its nonwoven films.

7.3.1 *Structure of nanofibres*

7.3.1.1 *Huge surface area to volume ratio*

As mentioned above, the diameter of nanofibres is in nanoscale size (see Fig. 7.5). Consequently, nanofibres or nanowires as well as nanofilms both have a huge surface area to volume ratio, about a thousand times higher than that of a human hair. A typically value of surface area for the nanofibres with diameters range of 10-500 nm is in the range of 500-10 $m^2\ g^{-1}$. Thus, they have the great potential to improve current technology significantly and find application in new areas.

7.3.1.2 *Orientation structure*

During electrospinning, structure is formed within the nanofibres and controlled by the simultaneous processes of the evaporation of the solvent from polymer solutions or the solidification from melts and the extreme elongation of the ultra-fine fibres. A volume element of the jet travels from the spinning nozzle to the substrate in a time frame of 0.1 s in which the processes of evaporation and elongation are completed and the fibre is solid. Through the investigation of electrospinning process, it was found that the elongation magnitude in nanofibres is very high, up to a factor of 10^5, having a elongation rate of up to $10^5 s^{-1}$.[25] Thus, it is estimated that a high degree of chain orientation can be obtained when the product value of the elongation rate with the chain relaxation time exceeds 0.5. By considering the concentration and molecular weight of polymer solutions, the relaxation times were in the range of 0.1 to 0.01 s.[25,26] Thus, chain orientation always occurs in nanofibres during electrospinning.

7.3.1.3 *Surface structures of nanofibres*

The electrospun nanofibres are usually smooth and have a circular cross section. Only in exceptional cases, a different cross section can be produced. However, in some applications, e.g. catalysis, tissue engineering, filtration, drug delivery and nanofibre reinforcement, the porous nanofibre without smooth surface could be advantageous. For example, in tissue engineering, pores of porous nanofibres act as anchoring points for cells. In catalysis or filtration, it also increases the surface area, further modifying the wetting properties, altering the kinetics of drug release and improving the matrix–fibre coupling in fibre strengthening. In addition, pores can also affect the kinetics of biodegradation in the bio-erodible nanofibres. Actually, a variety of fibre topologies can be fabricated with electrospinning by using polymer mixtures, special solvents or solvent mixtures or by varying the humidity. Nanofibres with entirely different topologies could also be obtained if a mixture of immiscible polymers are used to electrospin nanofibres from the same solvent.

7.3.1.4 *Core-Shell nanofibres*

A modified electro-spinning process, as called coaxial Electrospinning, is increasing used to fabricate nanofibres having core-shell structure. In coaxial electrospinning two concentrically aligned nozzles are used for spinning. The same voltage is applied to both nozzles and it deforms the compound droplet. A jet is generated on the tip of the deformed droplet, and in an ideal case, a core-shell nanofibre is created. In the core-shell nanofibre, the functionalisation of nanofibres through the integration of functional molecules or objects can be achieved, such as in the fields of biosensor technology, tissue engineering, drug delivery and nano-electronics. Thus, in a fluid environment, the functionalising agents including of biomolecules, such as enzymes, proteins, drugs, viruses and bacteria, are expected to be maintained. At the same time, the low molecular weight material which is too low to be spun into fibres also can be included in the core fibre if it is fabricated to core-shell structure nanofibres.

7.3.2 *Structure in nanofibrous nonwoven*

7.3.2.1 *Fibre arrangement*

During the electrospinning process, nanofibres are usually deposited onto the counter electrode in a statistical orientation. Spinning for a certain times, the nanofibres form into nonwoven with many open meshes by the layer-by-layer deposition. However, it is not limited to the produce nonwoven with a random planar fibre orientation by electrospinning. Nonwoven with orientated nanofibres along a preferred direction, which is of interest for structural reinforcement and for tissue engineering, can also be fabricated. For example, nonwoven with parallel fibres can be obtained on the counter electrode or combining electrode by using a rapidly rotating cylindrical collectors. The collectors are not limited to a cylinder shape, but also a narrow or wheel shape. In addition, parallel fibres with a very high degree of orientation can also be produced with special electrode arrangements, for example, consisting of two parallel flat plates or with frame-shaped electrodes. It is high possibility to achieve parallel nanofibrous nonwoven in a cross-shaped deposition of nanofibres resulting from the quadratic arrangement of four electrodes. Furthermore, nanofibres with high degrees of orientation can also be produced when the distance of two electrodes is reduced to several centimeters or a few millimeters and if either the spinning electrode or the counter electrode has the shape of a fine tip. Finally, it was also proposed that the common techniques used for the production of conventional nonwoven composed of macroscopic fibres, e.g. needling and water-jet treatment can also be used to produce nanofibrous nonwoven oriented in 3D.

7.3.2.2 *Variation in diameter*

Generally, to assemble nonwoven having a uniform thickness from fibres of the same material is a challenge. However, nonwoven with non-uniform thickness can also find a variety of interesting applications, for example, this kind of nonwoven was used to control the pore-size gradients or to modify the transport properties. Both theoretical and experimental studies showed that the diameter of nanofibre can be varied by a factor greater than ten by selecting different solvent or concentration of the spinning solution. Moreover, using several nozzles, nonwoven composed of significant different diameters nanofibres can be fabricated heterogeneously. Similarly, with multijet arrangements, it can also be produced that the chemically inhomogeneous nonwoven composed of different nanofibers, e.g. polyamide 6 (PA6) and polylactic acid (PLA).

7.3.2.3 *Pore structure*

In the nonwoven (see Figure 7.5), the pore diameters usually show a broad distribution over the range of 10^2-10^4 nm. However, it can be significantly modified by controlling the overall porosity. For instance, the porosity is typically about 90% for planar orientations, whereas a significant compaction of the nonwoven to a porosity of less than 60% can be achieved in elastomers.

Many simulations have been used to investigate the dependency of pore density and pore shape on the number of fibres per layer. It was reported that the average pore diameter was determined by both the average fibre radius and the overall porosity in nonwoven, that is, the volume fraction of the pores depends on the overall volume of the nonwoven. Thus, in the nonwoven with equal radius nanofibre, the higher the porosity, the bigger the average pore size, varied between d =r at a porosity of 20% and d =10r at a porosity of 80%. At lower porosities, d becomes even smaller than r. Moreover, it has been proposed from Monte Carlo simulations that the porosity of nonwoven decreases significantly as the flexibility of the fibres increases; and it also decrease significantly in chemical vapor deposition (CVD), which leads to the construction of a shell around the fibres.

7.4 Development of Adaptive Nanofibres and Nanofilms

In principal, nearly all soluble or fusible polymers can be fabricated into fibres by electrospinning if the molecular parameters, such as glass-transition temperature, solubility, crystallisation velocity, melting point, solvent vapor pressure, molecular weight, pH value, molecular-weight distribution and entanglement density, and the process parameters, such as electrical conductivity, concentration, feed rate, surface tension, geometry temperature, electrode separation and relative humidity are correctly controlled. So far, lots of polymer system including of water-soluble polymers, biopolymers and their derivatives, solvent-based polymer systems and multiphase polymer systems like polymer blends, are reported to be electrospun into nanofibres or nanofibreous nonwoven membranes. Examples of biopolymer include the collagen, cellulase, gelatin, casein, silk fibroin, PLA and their blends with PCL, PEO, PVA(Polyvinyl alcohol), PLA-co-PCL. Examples of water-soluble polymers include PVA, PEO, PVP(Poly (vinyl pyrrolidone)), PAA (Poly(acrylate acid) and PVA/PAA polymer blends; Examples of bioerodible polymers like PLA, PCL etc. Examples of organic soluble polymers or polymers based on organic-soluble precursors include PET, PTT(polytrimethylene terephthalate), PS (Polystyrene), PAN (poly-acrlonitrile), PC(Polycarbonate), PA, PU, PVC (polyvinyl chloride) and PMMA(polymethyl methacrylate) etc. Moreover, it was also reported that nanofibres were electrospun from its melt polymer without any solvent, such as polyethylene (PE), polypropylene (PP), polyamide 12 (PA12), PET, PCL and PU.[27]

In addition to the electrospinning method, nanofibres can also be fabricated through the self-assembly method. For example, nanofibres are self-assembled from properly designed synthetic amphiphile in aqueous or in organic environments. In the aqueous media nanofibres can be formed by the approach of molecular recognition, by molecular pairing or by biological molecular components. In the organic media, organic/inorganic nanofibres self-assembled based on the pseudo one-dimensional halogen-bridged mixed valence complexes, nanofibres are self-assembled at interfaces or self-assembled from lipophilic bridging ligands and metal ions.

In addition to nanofibres, a variety of mesoscopic-scale structures such as vesicles, tubes, disks, lamella, and helices are formed from suitably designed amphiphiles–bolaamphiphiles, gemini surfactants in addition to the classical amphiphiles. Moreover, 1D conjugated electronic structures and nanowires can also be self-assembled by this method.[28]

As mentioned in previous chapter, adaptable polymers are those polymers that can change geometry, flow characteristics, and adsorption properties upon the stimulation of an environmental change, such as temperature, PH, light and electric. The molecular structure of polymer is the fundamental factor for its adaptive functionality in adaptive polymers. Therefore, by selecting suitable adaptive polymers, modern fibre spinning technologies like electrospinning or self-assembly can be used to fabricate the desirable nanostructure like nanofibres, nanofilms. In the following, we would like to introduce some kinds of typical adaptive polymer nanofibres and nanofilms.

7.4.1 *Shape memory polymer nanofibres/non-woven*

As mentioned in previous chapter, shape memory polymers are those polymers that have the capability of fixing a temporary shape and recovering to its original shape. Shape memory polymers (SMP) are very promising smart materials. They can change shapes, hydro-absorbency, water vapor permeability (WVP), optical and other properties when the external stimulus such as heat changes. So far, lots of shape memory polymers are widely studied. Especially, the shape memory polyurethanes (SMPU) have drawn wide attentions of scientists. Recently, using the electrospinning technology, shape memory polyurethane nanofibres are fabricated from the SMPU/DMF solution by Hu *et al.*[15] Furthermore, the nanofibre nonwoven films were prepared after the resulted nanofibres were deposited on the receivers for a long time. For example, electrospun nonwoven of SMPU block copolymers with hard-segment concentrations of 40 and 50 wt % were successfully prepared by electrospinning processing. It was reported that the SMPU nonwoven spun at a lower solution viscosity (ca. 130-180 cPs) had an average fibre diameter of 800 nm and a beaded-on-fibre structure.[29] In this way, the researches of SMPU had been brought into the nanometer scale field through the electrospun nanofbers.

In addition, nanofibres and thin film of main-chain liquid crystal elastomers, which show high response to change of temperature, were electrospun by the electrospinning method.[30] In the resulted electrospun nanofibres, the nematic phase becomes oriented along the fibre axis during electrospinning and the nematic director is found to be oriented macroscopically, i.e. a monodomain fibre is formed. As shown in Fig. 7.6, the birefringent fibre appears bright if the crossed polarisers are at 45° with respect to the long axis of the fibre. By rotating the object carrier by about 45° the whole fibre becomes dark, which means that the fibre has a uniform director alignment along its long axis. This investigation shows that nanofibres with mono-domain structure and high orientation liquid crystalline phase can be achieved by electrospinning. The results show that two-way shape memory effects were expected in this kind of electrospun liquid cyrstalline elastomers (LCEs) nanofibres.

In 2005, Yang *et al.*[31] had also fabricated one kind of main-chain liquid crystalline polymer (MCLCP) nanoparticles. They demonstrate that within a certain range of particle sizes their shape becomes naturally ellipsoidal, with a significant aspect ratio. This shape is determined by the quasi-equilibrium shape memory of entangled MCLCP and changes reversibly back to spherical on heating the polymers into the isotropic phase. The ability to fix the anisotropic shape of the nanoparticles and reversibly change it by stimuli such as temperature or light (in photochromic materials) makes this effect important for nanotech-nological applications such as nano-actuators.

The trigger for mechanical property changes is usually temperature. However, by combining the nanoparticles, radiation can also be used as a trigger for adaptive response. For example, Koerner and co-workers used carbon nanotubes dispersed in shape memory polymers to trigger shape recovery both electrically and optically. In addition, Buckley and co-workers have presented a scheme for shape-changing medical devices activated by radio-frequency radiation.[32]

7.4.2 *PH-responsive nanofibre/nanofilms*

In 2005, Tatsiana *et al.* reported that the Tannic acid can be LbL assembled with weak and strong positive polyelectrolytes into the nanofilms, the stability of which varies with environmental conditions. Capsules and films designed with tannic acid show pH-dependent permeability with strong minimum between pH5 and 7, which is different from that of commonly used PSS/PAH microcapsules.[33,34]

In 2007, Liu *et al.* also reported one kind of pH-responsive hydrogel nanofibres with a diameter of ca.120 nm produced by co-electrospinning from P(AM-MA) and crosslinker DEG aqueous solution. Results show that the hydrogel nanofibre is dually ionic strength and pH responsive. The swelling ratio changes as the ionic strength changes. A two-step water swelling increase of the c-P(AM-MA) hydrogel nanofibre was found at pH 2.5 and 8.1. The maximum water swelling of the c-P(AM-MA) hydrogel nanofibres took place in an aqueous solution of pH 11.0.[34]

In 2008, Zhao *et al.* reported that Peptide RATEA16 can spontaneously self-assemble into higher-order nanofibre hydrogels with extremely high water content (>99.5% (wt/vol)) under physiological condition. The hydrogels could undergo pH-reversible transitions. They change from a viscous solution to elastic hydrogel and to a precipitate. The supramolecular self-assembly and the three phase transitions are driven by hydrophobic interactions, intermolecular hydrogen bonds, and a combination of attractive or repulsive electrostatic interactions.[35]

Most recently, Kim *et al.* also reported one kind of pH-responsive multilayer thin films which was self-assembled through the electrostatic interaction between their polyelectrolyte arms. It was found that the solution pH of the star polymers is the main parameter to tune the thickness, porosity and refractive index of the resulting star/star multilayers. Moreover, as assembled films undergo extensive structural reorganisation upon post-treatment with different pH conditions due to the highly pH-sensitive nature of star polymers.[36]

7.4.3 *Adaptive phase change nanofibres*

Phase changing materials (PCMs) are being used in fabric and foam to store body heat and then release it when needed. In 1987, Triangle R & D Corp. demonstrated the feasibility of incorporating PCMs within textile fibres. Materials used in this work included paraffinic hydrocarbons, such as eicosane and plastic crystals; 2,2-dimethyl-1,3-propanediol (DMP) and 2-hydroxymethyl-2-methyl-1,3-propanediol (DMP). Recently, Meng *et al.* also demonstrated that shape memory thermoplastic polyurethane (PU) synthesised by employing poly(ethylene glycol) (PEG) can also be used as a phase change material (PCM). Furthermore, a temperature-regulating fibre made of PEG-based smart copolymer is further reported by them.[37]

Most recently, a series of ultrafine fibres as a novel shape-stabilised polymer-matrix phase change material (PCM) have been developed via electrospinning technique from PCM/polymer composites based on fatty acids, lauric acid (LA), myristic acid (MA), palmitic acid (PA) and stearic acid (SA). Results show that the electrospun nanofibres showed smooth surfaces and cylindrical shape with diameters ranging from several tens to several hundreds nanometer, and the latent heat of fusion of the fibres is about 70.76 J/g. It was proposed that the resulted PCM nanofibres show suitable and competent tensile strength for the potential applications in solar energy storage and thermo-regulating textile.[21]

7.4.4 *Nonlinear optical nanofibres*

In the last few decades, many researchers have tried to find a variety of nonlinear optical materials for laser applications. Especially, by the nanotechnology, some soft materials are recently developed to nanofibres. For example, in 2004, Quochi *et al.* report the observation of amplified spontaneous emission and random lasing in self-organised crystalline para-sexiphenyl nanofibres. Using subpicosecond excitation, a lasing threshold is observed on the 0-1 emission band near 425 nm at excitation fluences as low as 0.5 J/cm^2 (6×10^{16} cm^{-3} equivalent density), near the onset of density-dependent recombination processes. The dependence of the nonlinear emission spectrum on both the pump intensity and position of the excitation area are attributed to the interplay between random lasing and amplified spontaneous emission occurring along the nanofibres.[38]

In 2005, Schiek *et al.* reported that extraordinarily long and mutually well-aligned semi-conducting nanofibres can be generated on single crystalline mica surfaces from methoxylized quaterphenylenes. Compared to aggregates made from the unsubstituted quaterphenylenes, the nanofibres show superior morphology and orientational properties. Compared to nanofibres from hexaphenylenes, they exhibit blue shifted optical emission as well as more flat top planes.[39]

In addition, Poly(phenylene vinylene) (PPV) has become one of the most famous semiconducting polymer since 1990. The results of research have showed that PPV possesses the excellent photoluminescent (PL) and electroluminescent (EL) properties as well as photovoltaic (PV) and non-linear optical properties. In 2007, Wen Zhang *et al.*[40] had PPV successfully fabricated nanofibres by electrospinning PPV precursor solution in solvent mixtures of ethanol/water. It is interesting to observe the fluorescence photo of PPV fibres that retain their photoluminescent properties.

7.4.5 *Conducting nanofibres/nanofilms*

Nanofibres have been increasingly used in a wide range of applications ranging from filtration to biomaterials. The functionalisation of polymer nanofibres by depositing well-controlled structures on nanofibres would expand the applications of nanofibres, based on changes to both the physical and chemical properties of the nanofibres. Wei *et al.* reported that the coating of functional silver structures on polymer nanofibre surface can modify the surfaces of polymer nanofibres by direct current (DC) sputter coating technique. Thus, the silver coated nanofibres show great potential in the applications as biomaterials, electronics, and medical devices.[41]

In 2008, Tao *et al.* also reported that PA6 electrospun nanofibres can be functionalised with copper (Cu) using electroless deposition technique. It was also found that the surface conductivity of the PA6 nanofibres was significantly improved by the Cu deposition. Thus, it is proposed that the combination of electrospinning and electroless deposition will provide a new approach to producing the functional nanofibres for various applications.[42, 43]

7.4.6 *Optoelectronic nanofilms*

Organic polymers possessing a conjugated p-electron system have fascinated scientists all over the world because of their electronic and physical properties. Recently, these nanostructured polymers have received great attention because of their potential application in electronic displays, electrode materials in batteries, semiconductors, sensors, capacitors, molecular electronic circuit elements and secondary batteries. In 2006, Kim *et al.*[44] reports on the self-assembly of conducting nanofilms of poly(3,4-ethylenedioxy-thiophene) prepared by using vapor-phase polymersation. The results show that the surfaces of nanofilms were extremely uniform and exhibited a compact and tight morphology. They were observed to exhibit good electro-conductive and optical properties that might be attributed to the growth of a thin film with nanoscale thicknesses and high-integrity surface structures. Thus, this nanofilm is capable of being applied to organic optoelectronic films for electrical devices such as organic displays and semiconductors.

In addition, by the spinning coating technology, Oh *et al.*[45] also fabricate nanofilms of poly(methyl silses-quioxane-co-thylenylsilsesquioxane) (PMSSQ–BTMSE) and poly-methylsilses-quioxane (PMSSQ) were prepared from the respective soluble prepolymer. Investigation on the optical, dielectric and thermal properties suggested that the PMSSQ–BTMSE and its nanoporous films on the nanometer scale are candidate materials for use as low and ultra-low dielectric interlayers in the fabrication of advanced microelectronic devices.

Most recently, B. S. Kim *et al.*[46] synthesised high charge density star polymers with oppositely charged arm structures: poly[2-(dimethylamino)ethyl methacrylate] (PDMAEMA) star and poly(acrylic acid) (PAA) star polymers with crosslinked cores. By exploiting the electrostatic interactions between the polyelectrolyte arms, they have integrated the PDMAEMA star and PAA star polymers within alternating multilayer thin films using layer-by-layer (LbL) assembly to generate all-star polyelectrolyte LbL films. The prepared star/star multilayer films illustrate nonuniform and nanoporous structures. Furthermore, results show that the assembled star/star multilayer films exhibit distinct morphological changes by undergoing extensive structural reorganisation upon post-treatment under different pH conditions that do not lead to any changes with their linear compositional counterparts. It is hypothesised that these differences are due to the star polyelectrolyte's compact structure and decreased extent of entanglement and interpenetration, which lead to a low degree of ionic crosslinking compared to their linear counterparts.

7.4.7 *Bacterial cellulose nanofibres*

Bacterial cellulose (BC), an indigenous dessert food known as a 'nata-de-coco', is produced by Acetobacter xylinum and consists of ribbon-shaped nanofibres structured in a weblike network. Because the nanofibres are made up of a bundle of cellulose microfibrils which are aggregates of semicrystalline extended cellulose chains, bacterial cellulose nanofibres have excellent physical properties. Their Young's modulus is 138 GPa, their tensile strength is estimated to be at least two GPa, and remarkably, their thermal-expansion coefficient in the axial direction is as small as 0.1×10^{-6} 1/K.[47]

In 2005, Yano *et al.* discovered that bacterial cellulose nanofibres had very promising characteristics as a reinforcing material for optically transparent plastics. Because of the size effect, the composite in a variety of resin was optically transparent even at high fibre content such as 70%. Moreover, the nanocomposites exhibited remarkable characteristics, such as low thermal expansion and high strength because of the BC nanofibre reinforcement, while maintaining the ease of bending of pure polymers.[48]

7.4.8 *Adaptive composites*

Materials for adaptive composites need not be restricted to solid constituents, but can incorporate fluid networks. For example, Sottos and co-workers [49] have developed a self-healing system based on a three-dimensional micro-vascular network embedded in the substrate. The network autonomously delivers a healing agent to repair cracks in a damaged polymeric surface sheet.

In 2008, Jeffrey *et al.*[50] reported a family of polymer nanocomposites, which mimic the architecture and display similar chemistry responsive mechanic adaptability of sea cucumbers. This kind of adaptive nanocomposite includes one kind of rigid cellulose nanofibres exhibit a reversible reduction by a factor of 40 of the tensile modulus, for example, from 800 to 20 megapascals (MPa), upon exposure to a chemical regulator that mediates nanofibre interactions. Using a host polymer with a thermal transition in the regime of interest, they demonstrated even larger modulus changes (4200 to 1.6 MPa) upon exposure to emulated physiological condition.

To design a 'mechanical-active' small-diameter artificial vascular graft, Takehisa Matsuda *et al.* had fabricated a tubular scaffold made of elastomeric poly(L-lactide-co-ecaprolactone) fabrics at different wall thicknesses using an electrospinning (ELSP) technique. In this kind of system, mechanical-responsive adaptive functionality was achieved, and the average outer diameter of the aorta, which periodically enlarged and shrank by responding to the applied pulsatile stress, appeared to increase with rotation rate.[51]

In addition, self cleaning fabrics containing TiO_2 nano-particles have been developed based on their photocatalytic ability of oxidising dirt and other contaminants. It is reported that the textile substrates like cotton, wool, polyamide and polyester were pre-treated with RF-plasma, MW-plasma or vacuum-UV irradiation in such a way that negatively charged TiO_2 chelating groups, such as carboxylic groups, are introduced by the pre-treatment methods. Such TiO_2-treated materials showed self cleaning activity under daylight environment. Furthermore, microbial protective clothing has been prepared by coating military garments with silver nanoparticles. These coatings in addition make the fabric self-wound healing, water- and dirt-repellent. Research is also going on to prepare the chemical and biological protective clothing with self-decontamination and drug-delivery characters. The decontaminants like nanometal oxides (MgO, Al_2O_3) along with activated carbon and antibiotics in the form of nano-coatings are being explored for this purpose.

Besides, after Li *et al.* first reports the observation of magetic hysteresis by producing the magnetic ceramic nanofibres of nickel ferrite with electrospinning method, Wang *et al.* reports the production of field-responsive superparamagnetic composite nanofibre via the electrospinning of polymer-stabilised magnetite nanoparticle suspensions in PEO and PVA solutions. It was reported that the nanofibre was superparamagnetic at room temperature, and responded to an externally-applied magnetic field by deflecting in the direction of increasing field gradient.[52]

7.4.9 *Others*

In addition to the adaptive and functional nanostructure mentioned above, there are also another group of special responsive and adaptive porphyrin nanostructures based on porphyrin, including nanofibre, nanotubes and nanospheres, by ionic self-assembly of two oppositely charged porphyrins in aqueous solution.[53] It was reported that these nano-structures are responsive and adaptive to environmental stimuli, such as chemicals and light, and it was observed that the nanospheres with a diameters of about 120 nm are stable structures between pH 2 and 12. In addition, the nanotubes are photocatalytic and capable of reducing metal ions that deposit selectively onto tube surfaces to form novel nanotube-metal composite systems. Moreover, the nanofibre bundles are highly fluorescent and emit intense red light when excited with blue light.

In addition, as the development of supramolcular chemistry, metal-responsive nanofibre which responses to the metal ions like Cu^{2+}, Ni^{2+}, Zn^{2+} or Co^{2+} and Cu^{2+} was achieved in a peptide assembly. It was reported that the metal responsiveness was acquired in the peptide assembly and nanofibre formation by attachment of his residues on the both termini of the β-sheet peptide.[54]

7.5 Applications of Nanofibres, Nanofilms of Adaptive Polymers

7.5.1 *Protective clothing applications*

Current protective clothing is mainly designed to contain charcoal absorbents and they show their limitations, particularly considering water permeability and extra weight-imposed to the article of clothing. Therefore, it is desirable to fabricate a lightweight and breathable fabric, which is insoluble in all solvents, permeable to both air and water vapor, highly reactive with nerve gases and other deadly chemical agents. Because of their great surface area to volume, fabrics composed of nanofibre are capable of neutralising chemical agents, but not influencing the air and water vapor permeability.

By layer-by-layer deposition during the electrospinning, nonwoven was fabricated to have high porosity but very small pore size. It is conceivable that this structure provides good resistance to the penetration of chemical harm agents in aerosol form.[55] Preliminary investigations have indicated that the electrospun nanofibres present both minimal impedance to moisture vapor diffusion and extremely efficiency in trapping aerosol particles compared with conventional textiles.[55,56] Thus, it shows strong promises as ideal protective clothing.

7.5.2. *Filtration applications*

Generally, in order to reach high filter efficiencies, pores in the filter material and the sizes of the channels should be adjusted to the fineness of the particles to be filtered. Furthermore, it is required to filter finer particles increasingly so that a transition from fibres with diameters in the micrometer ranges to fibres with diameters in the nanometer range.[57,58]

As one of the characteristic applications, nanofibrous nonwovens are used as coalescing and aerosol filters. Since tiny water droplets with dimensions in the nanometer range also result in the formation of ice crystals at high altitudes, coalescing filters are used to filter tiny water droplets from aviation fuel in aviation engineering.[59] It is also expected that nanofibre nonwovens can filter effectively the fine aerosol particles but its thicknesses do not cause an unacceptably high flow resistance [55,56,60] filters with thicknesses in the range of 10 mm and fibre coverages of about 1 $g.m^{-2}$ are suitable.

Actually, nanofibres are already used extensively in air filters. Unlike the nanofibre air filters, conventional air filters is mainly made from fibres with diameters in the micrometer range. They function by trapping particles floating in the air deep inside the filter and the whole filter material is involved in the retention of the particle. As a result, as the number of trapped particles increases, the pressure drop across the filter also increases and above a certain limit, the filter can no longer be used. The filter can be cleaned with a pressure blast, but a higher pressure drop remains with each cleaning process and eventually the filter has to be replaced by a new one. One strategy is to coat the crude filter material with a gossamer fleece composed of nanofibres. Thus, the particles can be captured on the surface, then the pressure drop remaining after each cleaning process can be reduced significantly and the filter lifetime can be prolonged to a factor of about ten. Another possible strategy for further improvement is by electrostatically charging the nanofibres.[55,57] During electrospinning, the nanofibres are often already charged, but this charge generally dissipates because of contact with an electrical ground. To obtain a permanent charge, corona polarisation or triboelectri polarisation was proposed. For example, polymers with different triboelectric properties were used to fabricate nanofibres having permanent charge.

In addition, it was reported that using chemically selective materials, such as the 'Smart Dust', could also increase the filter efficiency.[61] For the filter performance, the filtration efficiency is one of the most important concerns. Generally, it increases with decrease in fibre diameter, because tiny particles less than 0.5 mm can be easily trapped due to the very high surface area to volume ratio and resulting in high surface cohesion in the electrospun nanofibrous structured filters.

7.5.3 *Biomedical application*

7.5.3.1 *Delivery of drug and pharmaceutical composition*

It is always an important concern in medicine to deliver drugs or pharmaceuticals to patients in the most physiologically acceptable manner. Generally, the smaller the dimensions of the drug and the coating materials, the better the drug absorbed by human being. The principle for drug delivery with polymer nanofibre is that dissolution rate of a particular drug increases as surface area of both the drug and the corresponding carrier, if needed, increases. For example, Kenawy *et al.*[62] investigated delivery of tetracycline hydro-chloride based on the fibrous delivery matrices of poly(lactic acid), poly (ethylene-co-vinylacetate) and their blend. Bioabsorbable PLA nanofibre membranes were proposed to be used for loading an antibiotic drug Mefoxin. In this study, they demonstrated the preliminary efficiency of this nanofibre membrane compared with bulk film.

7.5.3.2 *Tissue engineering*

In the past decades, tissue engineering/biomaterials were developed greatly for the treatment of tissues or organs in malfunction in a human body, however, it is very difficult to design ideal scaffolds/synthetic matrices that can mimic the structure and biological functions of the natural extra cellurlar matrix (ECM). Recently, Laurencin CT *et al.* recognised that Human cells can attach and organise well around fibres with diameters smaller than those of the cells.[63] In this way, it was expected that scaffolds of nanofibre can provide an optimal template for cells to seed, migrate and grow. Thus, it is required to develop fibrous structures with fibre architectures which are beneficial for deposition and proliferation of cell for a successful regeneration of biological tissues and organs. The creation of reproducible and biocompatible three-dimensional scaffolds for cell in growth resulting in bio-matrix composites became of particular interest in tissue engineering for various tissue repair and replacement procedures. Recently, people have realised such scaffolds can be fabricated with synthetic biopolymers and/or biodegradable polymer nanofibres.[64,65] Furthermore, it is believed that converting biopolymers into nanofibres and networks which mimic native structures will ultimately enhance the utility of these materials as large diameter fibres do not mimic the morphological characteristics of the native fibrils.

7.5.3.3 *Wound dressing*

Nanofibres can also be used as wound dressing materials for the treatment of wounds or burns a human skin, as well as designed for haemostatic devices with some unique characteristics. For instance, applying an electric field, ultrafine fibres of biodegradable polymers can be directly sprayed/spun onto the injured skin as shown in Fig. 16. The resulting fibrous mat dressing can let wounds heal as it promotes the growth of normal skin and eliminates the formation of scar tissue.[55] For wound dressing, non-woven nanofibrous membrane mats are usually designed to have pore sizes ranging from 500 nm to 1 mm, small enough to protect the wound from bacterial penetration via aerosol particle capturing mechanisms and to have high surface area of 5-100 m^2/g, which is extremely efficient for fluid absorption and dermal delivery.

7.5.4 *Other functional application*

In addition to the application mentioned above, by electrospinning method, nanofibrous devices piezoelectric can also be produced from polymers having piezoelectric effect, such as polyvinylidene fluoride. Polymer nanofibres could also be used to develop functional sensors with the high surface area of nanofibres further improving the sensitivity. For instance, it was reported that poly(lactic acid co glycolic acid) (PLAGA) nanofibre films were employed as a new sensing interface for developing chemical and biochemical sensor applications.[65] Recently, highly sensitive optical sensors composed of fluorescent electrospun polymer nanofibre films were also reported to be produced.[66] Preliminary results indicate the sensitivities of nanofibre films to detect ferric and mercury ions and a nitro compound (2,4-dinitrotulene, DNT) are two to three orders of magnitude higher than that obtained from thin film sensors.

In addition, nanoscale tubes became important materials in many industry fields. It was proposed that electrospun ultrafine fibres can be used as templates to develop various nanotubes.[67,68] The tube material is usually coated on the nanofibre template, and the nanotube is formed once the template is removed through thermal degradation or solvent extraction. Thus, the template nanofibre must be stable during the coating as well as degradable or extractable without destructing the coating layer. For instance, by using PLA [poly(l-lactide)] nanofibres as the template, Bognitzki *et al.*[67] obtained polymer, composite of polymer and metal, and metal nanotubes respectively through chemical vapor deposition (CVD) coating and physical vapor deposition (PVD) coating and then thermal degradation. The wall thickness of the resultant tubes was in the range of 0.1-1 mm. Later, thinner nanotubes were also achieved with the similar procedure using both PA and PLA nanofibre as templates by Hou *et al.*[68]

References

1. Ondarcuhu, T. and Joachim, C., Drawing a single nanofibre over hundreds of microns. *Europhys Lett,*, 1998. 42(2): pp. 210-220.
2. Liu, M., Bian, F., and Sheng, F., FTIR study on molecular structure of poly(N-isopropylacrylamide) in mixed solvent of methanol and water. *European Polymer Journal*, 2005. 41(2): pp. 283-291.
3. Martin, C. R., Membrane-based synthesis of nanomaterials. *Chem Mater*, 1996. 8: pp. 1739-1746.
4. Ma, P. X. and Zhang, R., Synthetic nano-scale fibrous extracellular matrix. *J Biomed Mat Res.*, 1999. 46: pp. 60-72.
5. Liu, G. J., Ding, J. F., Qiao, L. J., Guo, A., Dymov, B. P., and Gleeson, J. T., Polystyrene-block-poly (2-cinnamoylethyl methacrylate) nanofibers-Preparation, characterization, and liquid crystalline properties. *Chem-A European J,*, 1999. 5: pp. 2740-2749.
6. Deitzel, J. M., Kleinmeyer, J. D., Hirvonen, J. K., and Tan, N. C. B., Controlled deposition of electrospun poly(ethylene oxide) fibers. *Polymer*, 2001. 42(19): pp. 8163-8170.
7. Reznik, S. N., Yarin, A. L., Theron, A., and Zussman, E., Transient and steady shapes of droplets attached to a surface in strong electric fields. *J. Fluid Mech.*, 2004. 516: pp. 349-377.
8. Cloupeau, M. and Prunetfoch, B., Electrostatic Spraying of Liquids in Cone-Jet Mode. *Journal of Electrostatics*, 1989. 22(2): pp. 135-159.
9. Geever, L. M., Cooney, C. C., Lyons, J. G., Kennedy, J. E., Nugent, M. J. D., Devery, S., and Higginbotham, C. L., Characterisation and controlled drug release from novel drug-loaded hydrogels. *European Journal of Pharmaceutics and Biopharmaceutics*, 2008. 69(3): pp. 1147-1159.

10. Dalton, P. D., Klinkhammer, K., Salber, J., Klee, D., and Moller, M., Direct in vitro electrospinning with polymer melts. *Biomacromolecules*, 2006. 7(3): pp. 686-690.
11. Martin, C. R., Membrane-based synthesis of nanomaterials. *Chemistry of Materials*, 1996. 8(8): pp. 1739-1746.
12. Lee, S. and Obendorf, S. K., Developing Protective Textile Materials as Barriers to Liquid Penetration Using Melt-Electrospinning. *J. Appl. Polym. Sci.*, 2006. 102: pp. 3430- 3437.
13. Doshi, J. and Reneker, D. H., Electrospinning Process and Applications of Electrospun Fibers. *Journal of Electrostatics*, 1995. 35(2-3): pp. 151-160.
14. Hoffman, A. S., Stayton, P. S., Bulmus, V., Chen, G. H., Chen, J. P., Cheung, C., Chilkoti, A., Ding, Z. L., Dong, L. C., Fong, R., Lackey, C. A., Long, C. J., Miura, M., Morris, J. E., Murthy, N., Nabeshima, Y., Park, T. G., Press, O. W., Shimoboji, T., Shoemaker, S., Yang, H. J., Monji, N., Nowinski, R. C., Cole, C. A., Priest, J. H., Harris, J. M., Nakamae, K., Nishino, T., and Miyata, T., Really smart bioconjugates of smart polymers and receptor proteins. *Journal of Biomedical Materials Research*, 2000. 52(4): pp. 577-586.
15. Zhuo, H. T., Hu, J. L., Chen, S. J. and Yeung, L. Y., Preparation of polyurethane nanofibres by electrospinning. *Journal of Applied Polymer Science*, 2008. 109(1): pp. 406-411.
16. Demir, M. M., Yilgor, I., Yilgor, E., and Erman, B., Electrospinning of polyurethane fibers. *Polymer*, 2002. 43(11): pp. 3303-3309.
17. Bergshoef, M. M. and Vancso, G. J., Transparent nanocomposites with ultrathin, electrospun nylon-4,6 fiber reinforcement. *Advanced Materials*, 1999. 11(16): pp. 1362-1365.
18. Reneker, D. H., Yarin, A. L., Fong, H., and Koombhongse, S., Bending instability of electrically charged liquid jets of polymer solutions in electrospinning. *Journal of Applid Physics*, 2000. 87: pp. 4531-4547.
19. Hohman, M. M., Shin, M., Rutledge, G., and Brenner, M. P., Electrospinning and electrically forced jets. I. Stability theory. *Physics of Fluids*, 2001. 13(8): pp. 2201-2220.
20. Theron, A., Zussman, E., and Yarin, A. L., Electrostatic field-assisted alignment of electrospun nanofibers. *Nanotechnology*, 2001. 12: pp. 384- 390.

21. Chen, Z., Mo, X., He, C., and Wang, H., Intermolecular interactions in electrospun collagen-chitosan complex nanofibers. *Carbohydrate Polymers*, 2008. 72(3): pp. 410-418.
22. Um, I. C., Fang, D. F., Hsiao, B. S., Okamoto, A., and Chu, B., Electro-spinning and electro-blowing of hyaluronic acid. *Biomacromolecules*, 2004. 5(4): pp. 1428-1436.
23. Whitesides, G. M. and Grzybowski, B., Self-assembly at all scales. *Science*, 2002. 295(5564): pp. 2418-2421.
24. Kato, E. and Murakami, T., Study of the volume phase transition of poly(N-isopropylacrylamide) gels in aqueous solutions of pyridine by depolarized Rayleigh and Raman scattering. *Polymer*, 2002. 43(21): pp. 5607-5613.
25. Li, D., Wang, Y., and Xia, Y., Electrospinning of polymeric and ceramic nanofibers as uniaxially aligned arrays. *Nano Lett.*, 2003. 3: pp. 1167- 1171.
26. Zussman, E., Yarin, A. L., and Weihs, D., A micro-aerodynamic decelerator based on permeable surfaces of nanofiber mats. *Experiments in Fluids*, 2002. 33: pp. 315-320.
27. Greiner, A. and Wendorff;, J. H., Electrospinning : A fascinating method for the preparation of ultrathin fibers. *Angewandte Chemie* 2007. 46(30): pp. 5670-5703
28. Kimizuka, N., self-assembly of supramolecular nanofibers. *Advanced Polymer Science*, 2008. 219: pp. 1-26.
29. Zhang, X.-Z., Jo Lewis, P., and Chu, C.-C., Fabrication and characterization of a smart drug delivery system: microsphere in hydrogel. *Biomaterials*, 2005. 26(16): pp. 3299-3309.
30. Krause, S., Dersch, R., Wendorff, J. H., and Finkelmann, H., Photocrosslinkable Liquid Crystal Main-Chain Polymers: Thin Films and Electrospinning. *Macromolue Rapid Communication,* 2007. 28: pp. 2062-2068.
31. Yang, Z., Huck, W. S., Clarke, S. M., A., T., and E.M., T., Shape memory nanoparticles from inherently non-spherical polymer colloids. *Nature*, 2005: pp. 486-490.
32. Vaia, R. and Baur, J., Adaptive composites. *Science*, 2008. 319(5862): pp. 420-421.
33. Shutava, T., Prouty, M., Kommireddy, D., and Lvov, Y., pH responsive decomposable layer-by-layer nanofilms and capsules on the basis of tannic acid. *Macromolecules*, 2005. 38(7): pp. 2850-2858.

34. Jin, S. P., Liu, M. Z., Chen, S. L., Bian, F. L., Chen, Y., Wang, B., Zhan, F. L., and Liu, S. X., Responsive properties and applications of the intelligent polymers and hydrogels. *Acta Physico-Chimica Sinica*, 2007. 23(3): pp. 438-446.
35. Zhao, Y., Yokoi, H., Tanaka, M., Kinoshita, T., and Tan, T. W., Self-assembled pH-responsive hydrogels composed of the RATEA16 peptide. *Biomacromolecules*, 2008. 9(6): pp. 1511-1518.
36. Kim, B. S., Gao, H. F., Argun, A. A., Matyjaszewski, K., and Hammond, P. T., All-Star Polymer Multilayers as pH-Responsive Nanofilms. *Macromolecules*, 2009. 42(1): pp. 368-375.
37. Meng Q. and J., H., A temperature-regulating fiber made of PEG-based smart copolymer. *Smart copolymerSolar Energy Materials & Solar Cells*, 2008. 92: pp. 1245-1252.
38. Quochi, F., Cordella, F., Orru, R., Communal, J. E., Verzeroli, P., Mura, A., Bongiovanni, G., Andreev, A., Sitter, H., and Sariciftci, N. S., Random laser action in self-organized para-sexiphenyl nanofibers grown by hot-wall epitaxy. *Applied physics letters*, 2004. 84(22): pp. 4454-4456.
39. Schiek, M., Lutzen, A., Koch, R., Al-Shamery, K., Balzer, F., Frese, R., and Rubahn, H. G., Nanofibers from functionalized para-phenylene molecules. *Applied physics letters*, 2005. 86(15): pp. 153107-153107-3.
40. Zhang, W., Huang, Z. H., Yan, E. Y., Wang, C., Xin, Y., Zhao, Q., and Tong, Y. B., Preparation of poly(phenylene vinylene) nano-fibers by electrospinning. *Materials Science and Engineering a-Structural Materials Properties Microstructure and Processing*, 2007. 443(1-2): pp. 292-295.
41. Wei, Q. F., Ye, H., Hou, D. Y., Wang, H. B., and Gao, W. D., Surface functionalization of polymer nanofibers by silver sputter coating. *Journal of Applied Polymer Science*, 2006. 99(5): pp. 2384-2388.
42. Tao, D., Wei, Q. F., Cai, Y. B., Xu, Q. X., and Sun, L. Y., Functionalization of polyamide 6 nanofibers by electroless deposition of copper. *Journal of Coatings Technology and Research*, 2008. 5(3): pp. 399-403.
43. Kato, H., Takemura, S., Ishii, A., Takarai, Y., Watanabe, Y., Sugiyama, T., Hiramatsu, T., Nanba, N., Nishikawa, O., and Taniguchi, M., Conducting polymer nanofilm growth on a nanoscale linked-crater pattern fabricated on an Al surface. *Journal of Vacuum Science & Technology A*, 2008. 26(4): pp. 824-831.

44. Kim, J. Y., Kwon, M. H., Min, Kwon, S., and Ihm, D. W., Self-Assembly and Crystalline Growth of Poly(3,4-ethylenedioxythiophene) Nanofilms. *Advanced Materials*, 2007. 19: pp. 3501-3506.
45. Oh, W., Hwang, Y., Park, Y. H., Ree, M., Chu, S. H., Char, K., Lee, J. K., and Kim, S. Y., Optical, dielectric and thermal properties of nanoscaled films of polyalkylsilsesquioxane composites with star-shaped poly(epsilon-caprolactone) and their derived nanoporous analogues. *Polymer*, 2003. 44(8): pp. 2519-2527.
46. B.Su., K., H.F., G., A.A., A., K., M., and T., H. P., All-Star Polymer Multilayers as pH-Responsive Nanofilms. *Macromolecules*, 2009. 42(1): pp. 368-375.
47. Ifuku, S., Nogi, M., Abe, K., Handa, K., Nakatsubo, F., and Yano, H., Surface modification of bacterial cellulose nanofibers for property enhancement of optically transparent composites: Dependence on acetyl-group DS. *Biomacromolecules*, 2007. 8(6): pp. 1973-1978.
48. Yano, H., Sugiyama, J., Nakagaito, A. N., Nogi, M., Matsuura, T., Hikita, M., and Handa, K., Optically transparent composites reinforced with networks of bacterial nanofibers. *Advanced Materials*, 2005. 17(2): pp. 153-155.
49. Toohey, K. S., Sottos, N. R., Lewis, J. A., Moore, J. S., and White, S. R., Self-healing materials with microvascular networks. *Nature Materials*, 2007. 6(8): pp. 581-585.
50. Capadona, J. R., Shanmuganathan, K., Tyler, D. J., Rowan, S. J., and Weder, C., Stimuli-responsive polymer nanocomposites inspired by the sea cucumber dermis. *Science*, 2008. 319(5868): pp. 1370-1374.
51. Inoguchi, H., Kwon, I. K., Inoue, E., Takamizawa, K., Maehara, Y., and Matsuda, T., Mechanical responses of a compliant electrospun poly(L-lactide-co-epsilon-caprolactone) small-diameter vascular graft. *Biomaterials*, 2006. 27(8): pp. 1470-1478.
52. Wang, A., Singh, H., Hatton, T. A., and Rutledge, G. C., Field-responsive superparamagnetic composite nanofibers by electrospinning. *Polymer*, 2004. 45(16): pp. 5505-5514.
53. Wang, Z. C., Medforth, C. J., and Shelnutt, J. A., Porphyrin nanotubes by ionic self-assembly. *Journal of the American Chemical Society*, 2004. 126(49): pp. 15954-15955.

54. Matsumura, S., Uemura, S., and Mihara, H., Metal-triggered nanofiber formation of His-containing beta-sheet peptide. *Supramolecular Chemistry*, 2006. 18(5): pp. 397-403.
55. Gibson, P., Schreuder-Gibson, H., and Rivin, D., Transport properties of porous membranes based on electrospun nanofibers. *Colloids and Surfaces a-Physicochemical and Engineering Aspects*, 2001. 187: pp. 469-481.
56. Schreuder-Gibson, H. L., Gibson, P., Senecal, K., Sennett, M., Walker, J., and Yeomans, W., Protective textile materials based on electrospun nanofibers. *Journal of Advanced Materials,*, 2002. 34(3): pp. 44-55.
57. Tsai, P. P., Schreuder-Gibson, H., and Gibson, P., Different electrostatic methods for making electret filters. *Journal of Electrostatics*, 2002. 54: pp. 333- 341.
58. Qin, X. H. and Wang, S. Y., Filtration properties of electrospinning nanofibers. *Journal of Applied Polymer Science*, 2006. 102: pp. 1285-1290.
59. Abdel-Ghani and M. S., D., G. A. , Simulation of non-woven fibre mats and the application to coalescers. *Chemical Engineering Science*, 1985(40): pp. 117-129.
60. Hajara, M. G., Metha, K., and Chase, G. G., Humidity, temperature, and polymer. nanofibers on drop coalescence in glass fiber media. *Seperation and Purification Technology* , 2003. 30: pp. 79- 88.
61. Schmedake, T. A., Cunin, F., Link, J. R., and Sailor, M. J., Standoff Detection of Chemicals Using Porous Silicon "Smart Dust" Particles. *Advanced Materials.*, 2002. 14: pp. 1270-1272.
62. Kenawy, R., G.L, B., K., M., J., L., D.G., S., Sanders E.H., and G.E., W., Release of tetracycline hydrochloride from electrospun poly(ethylene-co-vinylacetate), poly(lactic acid), and a blend. *J. Control Release*, 2002. 81(1-2): pp. 57-64.
63. Buchko, C. J., Chen, L. C., Shen, Y., and Martin, D. C., Processing and microstructural characterization of porous biocompatible protein polymer thin films. *Polymer*, 1999. 40(26): pp. 7397-7407.
64. Huang, L., McMillan, R. A., Apkarian, R. P., Pourdeyhimi, B., Conticello, V. P., and Chaikof, E. L., Generation of synthetic elastin-mimetic small diameter fibers and fiber networks. *Macromolecules*, 2000. 33(8): pp. 2989-2997.

65. Kwoun, S. J., Lec, R. M., Han, B., and Ko, F. K. *Polymer nanofiber thin films for biosensor applications.* in *In: Proceedings of the IEEE 27th Annual Northeast Bioengineering Conference*. 2001.
66. Lee, S. H., Ku, B. C., Wang, X., Samuelson, L. A., and Kumar, J., Design, synthesis and electrospinning of a novel fluorescent polymer for optical sensor applications. *Materials Research Society Symposium Proceedings*, 2002. 708: pp. 403-408.
67. Bognitzki, M., Hou, H. Q., Ishaque, M., Frese, T., Hellwig, M., Schwarte, C., Schaper, A., Wendorff, J. H., and Greiner, A., Polymer, metal, and hybrid nano- and mesotubes by coating degradable polymer template fibers (TUFT process). *Advanced Materials*, 2000. 12(9): pp. 637-640.
68. Hu, X., Zhou, J., Zhang, N., Tan, H., and Gao, C., Preparation and properties of an injectable scaffold of poly(lactic-co-glycolic acid) microparticles/chitosan hydrogel. *Journal of the Mechanical Behavior of Biomedical Materials*, 2008. 1(4): pp. 352-359.

Chapter 8

Cosmetics Applications of Adaptive and Functional Polymers

This chapter gives a brief description of the definition, development history and classification of cosmetics, and introduces the advantages and beautiful prospects of adaptive and functional polymers in the cosmetics field. We also further discuss the properties and cosmetic applications of five types of typical adaptive and functional polymers in this chapter, which are: hydrogels, cyclodextrins, polysaccharides, shape memory polymers and nanopolymer particles.

Keywords: cosmetic, adaptive, functional, polymer

8.1 Introduction

An old proverb says, ‘Beauty is in everyone’s heart’. From more than 4,000 years ago to modern society, people have never stopped their pursuit of beauty. Cosmetics are substances which can be used to beautify the human body. Their development is closely linked with the development of human science and technology.

Adaptive and functional polymers are widely studied and have become increasingly popular in cosmetic applications in recent years. Polymers can be designed to respond to external stimuli in a controlled and predictable manner by chemical synthesis. These adaptive polymers have complex and diverse structures, and when stimulated by outside environments, their molecular, aggregation and even appearance structures can be changed. Hence, they respond well to a small stimulus and show many special properties, such as photosensitivity, biocompatibility, thermal responsiveness and so on.

The application of adaptive and functional polymers in the cosmetics field can produce smart and multifunctional cosmetics. At the same time it also reflects developments in human knowledge and the future of cosmetics.

8.2 Definition, History and Classification of Cosmetics

8.2.1 *Definition of cosmetics*

The US Food and Drug Administration(FDA) which regulates cosmetics in the United States defines cosmetics as items that are intended to be applied to the human body for cleansing, beautifying, promoting attractiveness or altering the appearance without affecting the body's structure or functions.[1] This broad definition also includes any material intended for usage as a cosmetic.

Most cosmetics are mainly composed of two parts: substrate and supplements. Substrate is the ‘backbone’ of cosmetics. Ideal cosmetic substrates should have the following five basic conditions:[2] 1) ability to form a solid protective layer on the skin surface. This layer should both be able to block sand, dust and harmful substances (including bacteria) from the external body and provide nutrition to the skin; 2) breathability and inability to be rubbed off easily; 3) good compatibility with additives in cosmetics; 4) easy removal by water; 5) do not have side effects with skin and cannot penetrate the skin. Supplements consist of surfactants and/or other additives. Their proportion in cosmetics is lower than the substrate, but they also play essential role in cosmetic quality. Surfactants can modify the interface between various phases and make opposing phases coexist steadily. They can create energy barriers between the droplets of emulsion to prevent flocculation and coalescence. Additives play complex roles in cosmetics. Some of them can give cosmetics an aroma or good appearance, while others are used to help cosmetics in antisepsis or antioxidation. Furthermore, some additives can even endow cosmetic special functions, for example, chitosan can give cosmetics an antibacterial function.

8.2.2 *Development history of cosmetics*

8.2.2.1 *Original cosmetics*

The appearance of the first generation of cosmetics can be dated back to 4,000 years ago. These original cosmetics were naturally pure with no chemical treatments invovled. People just applied pure plants extracts, animal oils or fats (their main ingredients were various kinds of fatty acid glycerides) directly onto their skin. Two thousand years ago, the ancient Egyptians used flowers, animal oils, mineral oils and even cosmetic brushes in religious ceremonies, mummy preservation as well as personal care in the royal family. The ancient Romans also used natural cosmetics such as camphor, musk, sandalwood, lavender and clove oil on skin, hair, nails and lips. About 1,500 years ago, Naples was a famous commercial centre for cosmetic trading in the ancient Roman empire.[3] During ancient periods, cosmetics were also widely developed in China. A book named ‘Shen Nong's Herbal Classic’ which published in the Qin Dynasty, indicated that Angelica dahurica can be used to make a type of facial mask to nourish and moisturise the skin, and change the colour of the skin.[4]

8.2.2.2 *Emulsion cosmetics*

'Emulsion cosmetics' are based on an O/W emulsifying technique which appeared in the mid-18th century. Compared with their predecessors, emulsion cosmetics have two advantages. On the one hand, they allow the simultaneous use of lipophilic and hydrophilic ingredients in required amounts. On the other hand, they have desirable properties in many aspects, such as hand feeling, skin compatibility and ease of application. However, the emulsion system also has a major disadvantage which limits further applications. Their formation is accompanied by a large increase in interfacial energy and thus, they are always thermodynamically unstable. However, in the early 19th century, the introduction of dielectric surfactant and adoption of Hydrophile-Lipophile Balance (HLB) value evaluation system in cosmetics solved this key problem. From then on, cosmetic production gradually changed from a family type of shop into a professional industry.[5]

8.2.2.3 *Functional cosmetics*

Symbolising the third development stage of cosmetics is the appearance of functional cosmetics in the mid-20th century. Functional cosmetics employ advanced aliphatic alcohols as their substrates and contain many kinds of animal and plant extracts as additives (such as saponins, fruit acids, deep-sea fish protein and so on). These substrates can combine well with skin and form an isolation layer on the surface which can prevent skin from dehydration. Thus, they give comfort to our bodies. Functional cosmetics also solve skin problems such as whitening, removing acne, eliminating spots and reducing freckles, so they have become popular very quickly.[6]

8.2.2.4 *Bionic cosmetics*

Functional cosmetics have dominated cosmetics for nearly 40 years during the last century. However, because of the advancements of aliphatic alcohols with low molecular weight (only a few hundreds), they do not have enough intensity and can slide off skin very easily. The isolation layer also greatly weakens the absorption of functional additives on skin in functional cosmetics. Hence in the latter part of the century, scientists started to develop the latest generation of cosmetics, which is called 'bionic cosmetics'. Bionic cosmetics use synthetical or natural polymer materials as their substrate and contain some biological substances as their additives, such as ceramides, DNA, epidermal growth factors and so on'. Bionic cosmetics have many amazing effects, such as breast enhancement, body-building abilities, anti-aging and skin repairing. They represent the developing trend of cosmetics in the 21st century.[7] Some scientists even predict that in the future, this type of cosmetic may be used to rejuvenate the skin of the elderly.[8]

8.2.3 *Classification of cosmetics*

There are many classification methods in cosmetics and they can be divided into six types by their purpose:[9]

(1) Skin care products, which are used to clean and moisturise the skin and/or enhance the skin's metabolism, and so on.
(2) Hair care products, which include hair care, shampoo and shaving supplies. They provide hair with a natural, healthy, beautiful appearance as well as modify and fix hair.
(3) Oral care products, which are used to clean teeth and the oral cavity, prevent anti-caries and eliminate bad breath.
(4) Nail care products, with functions such as cleaning, painting and protecting nails.
(5) Beautifying cosmetics, which are used to modify features, promote the aroma and colour effect of other cosmetics, and thus increase the feeling of beauty for the whole body.
(6) Special purpose cosmetics, which include products for other areas of the body such as hair dyes, electric hair curlers, depilators, breast enhancers, spot removers, smell eliminators, sunscreens and so on.

The Japanese scholar Sadakazu divided cosmetics into eight categories by their functions and location for usage. They are: skin care, hair care, nail care, oral care, cleansing, foundation use, beauty and fragrance cosmetics.[10]

8.2.4 *Adaptive and functional polymers in cosmetics*

Adaptive and functional polymers are sensitive to external and small stimulations, such as light, temperature, pH value and so on. Their performance includes many special chemical and physical properties, such as photosensitivity, biocompatibility, thermal responsiveness, etc. In the author's opinion, they will become increasingly popular in cosmetic applications for three key reasons, as described in the following.

8.2.4.1 *Adaptive and functional polymers with unique advantages*

Nowadays, the cosmetics industry is facing a revolutionary shift. This shift is pushing the modern cosmetics industry towards the use of new ingredients.[11] The use of once favourably acceptable ingredients in cosmetics is rapidly becoming unpopular. These traditional ingredients appear to have many drawbacks, such as lack of stability, poor biocompatibility with the human body and so on. Furthermore, some of their solvents even create environmental issues. However, adaptive and functional polymers have many advantages in the making of cosmetics. Since they are macromolecular compounds, they show no skin infiltration and irritation, and thus they are very safe to use on our bodies. Due to their unique three-dimensional structures, they can form strong and stable repulsive barriers between droplets in emulsion to improve cosmetic stabilisation. Furthermore, functional ingredients can be embedded into adaptive polymers and then control-released in a suitable situation. In this way, cosmetics will work more effectivly and become more multifunctional.

8.2.4.2 *Increasingly maturing technology in polymer industry*

The second factor which causes an increase in the use of adaptive and functional polymers in the cosmetics field is the increasingly maturing technology in the polymer industry. Since the 1930s, many thousands of polymers have been prepared, but only less than a hundred of them reach the phase of large-scale industrial production in the following fifty years due to lack of technology. Hence, before the year 1990 there were only a few polymers that were used in the cosmetics industry. However, in recent years, great improvements in polymer technology, which include the production of biodegradable polymers and adaptive polymers at an industrial scale, demonstrate that the use of adaptive and functional polymers is feasible and will be increasingly popular in the future.[12]

8.2.4.3 *Consumer expectations*

The third reason comes from consumers. In modern society, there are more improvements in the quality of life and increasingly more consumers show great interest in multifunctional and smart cosmetics. There is an expectation that one cosmetic can perform various functions and perhaps cosmetics can even change their functions in different situations automatically.[13] Obviously, in this case, cosmetics producers will place more attention on new cosmetic developments to satisfy consumer expectations. Thus, adaptive and functional polymers will become attractive materials in making cosmetics due to their unique properties.

8.3 Cosmetic Applications of Adaptive and Functional Polymers

Adaptive polymers can respond to slight changes in the external environment in a dramatic way. Functional polymers can give products many useful abilities, such as having antibacteria, anticorrosive, emulsifying, and moisturising properties. Some of them can even give biological properties to cosmetics. These synthetic and natural polymers are potentially very useful for a variety of applications, including in the cosmetic sfield.

8.3.1 *Hydrogels*

Hydrogels are semi-solid materials that consist of a network of interacting polymers or colloidal particles dispersed in water.[14] Many chemicals can be used to prepare hydrogels, such as polyurethanes, polysaccharides, acrylamide and so on.[15] Hydrogels have excellent film forming, moisture adsorbing, desorbing and elastic properties. They can form stable suspensions or emulsions in aqueous solutions and some of them even have biological activities, such as antibacterial activity. Consequently, they are widely used in the cosmetics and biomedical fields.[16,17]

Load and release are very important properties of hydrogels in cosmetics applications. Poor stability and barrier properties of the skin limit the use of some bioactive materials in cosmetics. However, since hydrogels have three-dimensional matrices, bioactive materials can be loaded into them to improve stability and then released at a suitable site to take effect.[17]

Recently, stimuli-responsive hydrogels have also attracted much attention. This type of hydrogel can perform reversible and reproducible volume change that are even stimulated by small alterations of the external environment, such as temperature, pH or electricity.[18] Stimuli-responsive property can give hydrogels a novel controlled release property which can be used in cosmetics delivery systems. For example, research shows that pH-responsive alginate hydrogel beads can slowly release active ingredients due to a lowering pH, so that they have potential application as a delivery vehicle for drugs and cosmetics.[19] Among all stimuli-responsive hydrogels, temperature-sensitive hydrogels are the most commonly studied environmentally sensitive polymer systems,[20] and among all temperature responsive polymers, poly(*N*-isopropylacrylamide) (PNIPAAm) is probably the most extensively used one. PNIPAAm can form a thermally reversible hydrogel which shows a lower critical solution temperature (LCST) at about 33°C in water.[21] When the temperature falls below the LCST, the polymer chain of PNIPAAm is hydrated to an expanded form, which makes it soluble in water. When the temperature rises above the LCST, the PNIPAAm undergoes a reversible phase transition into an insoluble form and dehydrates into a compact form (see Fig. 8.1).

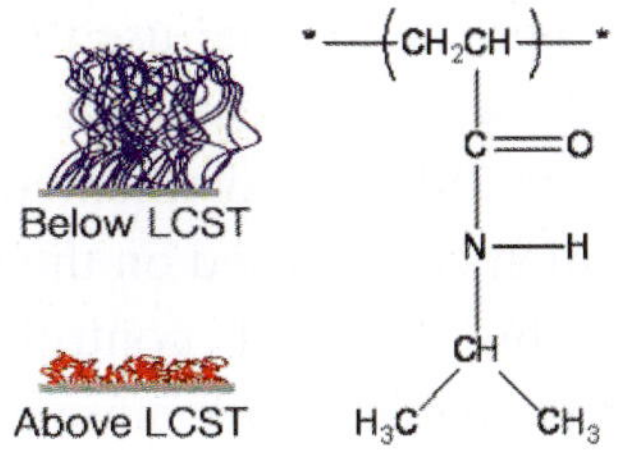

Fig. 8.1. Phase transition of PNIPAAm.[22]

Since the trigger temperature PNIPAAm hydrogel is very close to human body temperature, it can be used as a nutrient substance carrier in personal care products. First, the nutriments or active ingredients are encapsulated into the hydrogel and when the hydrogel is heated by the human body, nutrient substances will be control-released from the hydrogel [23] (see Fig. 8.2). To overcome the heavy syneresis and weak mechanical strength of pure PNIPAAm hydrogels, polyurethane (PU) can be used to obtain a PNIPAAm/PU copolymer hydrogel (see Fig. 8.3) and achieve a mild release of active ingredients.

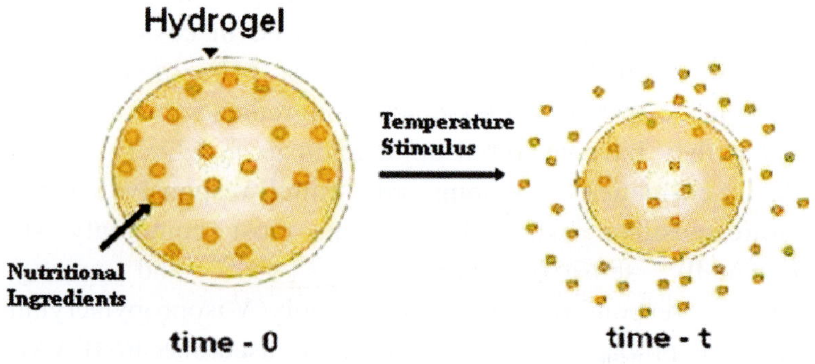

Fig. 8.2. Controlled release behaviour of reversible temperature sensitive hydrogels.[23]

According to Liu's studies,[24] PNIPAAm/PU copolymer hydrogels can be used to make facial masks and nonwoven fabrics can support the PNIPAAm/PU copolymer hydrogels as a facial mask substrate. Compared with traditional facial masks, the hydrogel facial mask has many advantages. It not only has a soft surface, good hand-feeling, and is aesthetically pleasing, but can also be used repeatedly. Furthermore, it can amazingly provide a semitransparent appearance and embed nutritional ingredients at 20°C, and then turn opaque, releasing nutritional ingredients which can be used on the human body (about 36-37°C) (see Fig. 8.4). This kind of controlled release can make nutritional ingredients in cosmetics work more efficiently.

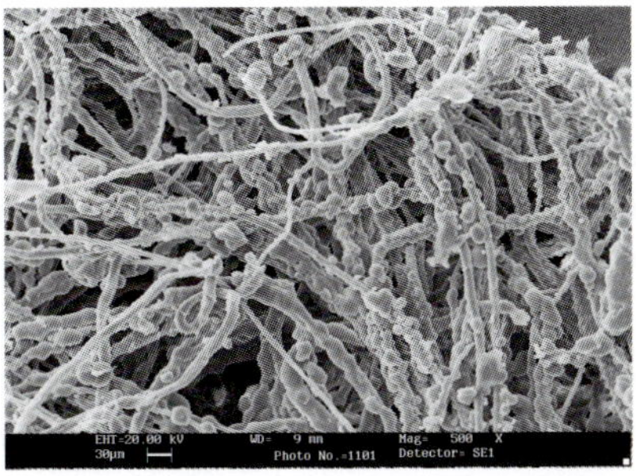

Fig. 8.3. Image of PNIPAAm/PU copolymer hydrogel.[24]

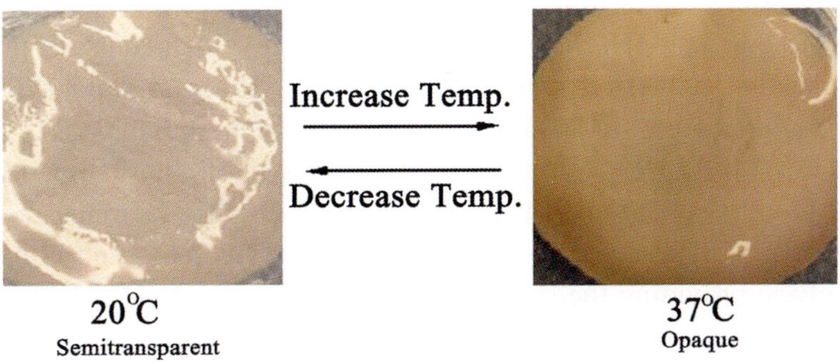

Fig. 8.4. PNIPAAm/PU hydrogel facial mask.

Controlled release systems have undergone revolutionary advancements in recent years. The introduction of stimuli-responsive hydrogels to controlled release systems significantly strengthens the release efficiency of active ingredients. The exploitation and use of these polymeric hydrogels in ingredient delivery systems will open up new areas of future cosmetics applications.

8.3.2 *Polysaccharides*

Polysaccharides are natural polymers. They are complex carbohydrates consisting of many (more than ten) smaller monosaccharide units linked by glycosidic bonds in branched or unbranched chains. They can display different behaviours once dissolved in water owing to their molecular structure. Since they exhibit good biocompatibility, biodegradability, nontoxicity and biological activities, they are widely used as thickening, emulsifying, stabilising, or flocculating agents in industrial fields.[25] The use of polysaccharides in cosmetics has been almost as long as the use of cosmetics themselves. It may be because they are a natural resource and can be easily extracted from animals or plants.[26] Some of the bioactivities of polysaccharides, such as antioxidation and antibacterial properties, make them potential lipid protectors in cosmetics formulations. Furthermore, they can also be helpful in the protection of human skin against ultraviolet (UV) induced damage.[27] Hyaluronic acid and chitosan are two of the most widely used polysaccharides in the cosmetics field.

Hyaluronic acid (HA) is common among many organisms and present in connective tissues of skin, cartilage, bone, synovial fluid (SF) and extracellular matrix (ECM). It is the most important functional natural polymer in living organisms.[28] It is composed of alternating (1-4)-β linked D-glucose and (1-3)-β linked N-acetyl-D-glucose amine residues (see Fig. 8.5). The repeating chains in HA are hydrated and coil upon themselves, providing the substance with elasticity and viscosity. At low concentrations and in solutions, HA adopts a conformation of an expanded coil. At higher concentrations, HA forms an entangled molecule that is able to capture a large amount of water.

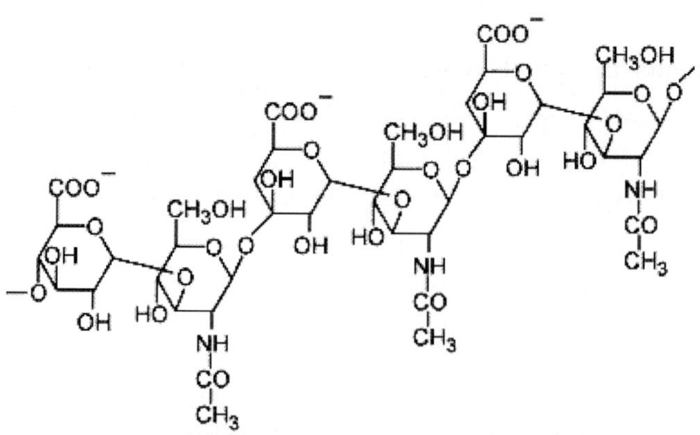

Fig. 8.5. Molecular structure of hyaluronic acid.

Since HA is nontoxic and has good biocompatibility and biodegradability, it is an ideal component in cosmetics. Apart from its moisturising function, much research have shown that HA acts as a free radical scavenger and can be used in facial cosmetics surgery as a 'rejuvenation' agent.[29] Rohrich reported that nonsurgical facial rejuvenation through soft-tissue augmentation with bioengineered HA has resulted in a major shift in the facial rejuvenation algorithm. HA fillers provide both patients and physicians with a biocompatible, easy-to-administer alternative that requires no pre-injection skin testing and produces reproducible, long-lasting results.[30]

Chitosan is deacetylated chitin. Apart from HA, chitin is the second most abundant natural polymer in the world. Chitosan are six-carbon sugar copolymers, which consist of β-D-glucosamine and N-acetyl-β-D-glucosamine generally in a 4:1 ratio. It has good water solubility, excellent biocompatibility and biodegradability, and furthermore, it has strong antimicrobial ability.

Chitosan has many useful properties. It can combine with surfactant to stabilise cosmetics emulsion, form a breathable film on the skin surface, and its amphoteric character makes this film both able to remove sebum and oils from skin and maintain moisture. Its complex ability also gives this film an ability to form complexation with heavy metals and remove them from skin effectively. Nowadays, chitosan is widely used in hair care for improving the suppleness of hair, reduction of static electricity or in skin care fields for maintaining moisture, treating acne and toning skin.[31]

One of the most important and widely used properties of chitosan is its antibacterial effect. The antibacterial function may be due to two mechanisms: one is that chitosan interacts with the membrane to alter cell permeability,[32] and even cause leakage of the intracellular components,[33,34] or involve binding with DNA to inhibit RNA synthesis.[35,36] Another mechanism is the absorption mode. Chitosan moleculars may create a thin membrane on the surface of the bacteria and cutoff the nutrient transportation port to cell 30, absorb the bacteria and prevent the cell movement, or interfere in the synthesis of the bacteria membrane. The bacteria would die due to the melting of the cell membrane or nutrient deficiency.[37]

Liu [38]reported that chitosan modified PNIPAAm/PU hydrogel based facial masks have strong antibacterial activity while pure PNIPAAm/PU hydrogels do not have such a property (see Fig. 8.6). Further studies showed that chitosan also does not have a negative impact on the controlled release behaviour of PNIPAAm/PU hydrogels. The release amount of nutritional ingredients in chitosan modified PNIPAAm/PU hydrogels is also similar with original pure PNIPAAm/PU hydrogels. Furthermore, the release amount of nutritional ingredients could be controlled by varying the temperature (see Fig. 8.7).

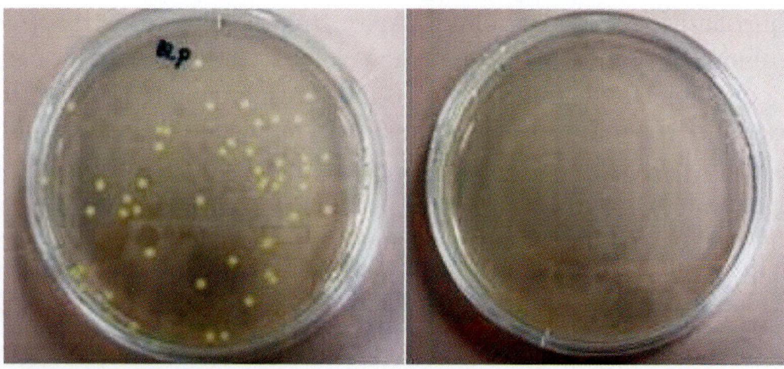

Fig. 8.6. PNIPAAm/PU hydrogel with chitosan modifications (left) and without modifications (right) [37]

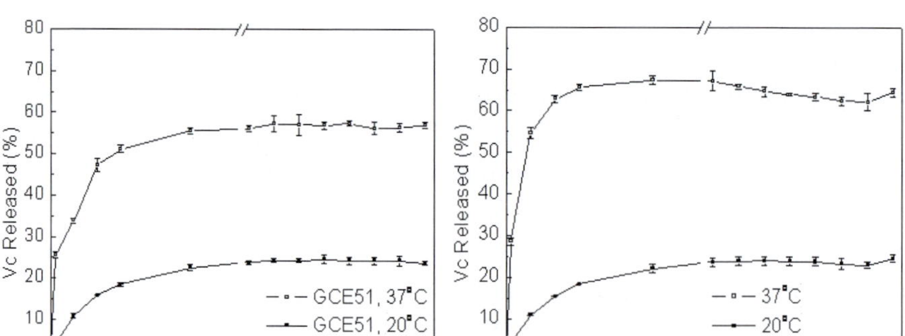

Fig. 8.7. Release amount of Vc modified by chitosan(right) or not modified by chitosan (left) [38]

Another very important property of chitosan is pH-sensitivity. It can form pH-sensitive complex with other polymers, and thus this complex polymer can swell at low pH and shrink at high pH to achieve controlled release. This property has many prospects in cosmetics applications for active ingredient delivery.[39]

Chemical modifications of chitosan to make chitosan derivatives are very popular in recent years. Carbohydrates can be grafted onto the chitosan backbone at the C-2 position to improve water solubility and stability. Alkylated chitosans show better surface activities than pure chitosan and the stability of the interfacial film offers great potential applications in the cosmetics field.[40] Chitosan also can graft with cyclodextrins (CDs) to form a CD-grafted chitosan complex. This complex shows long-lasting antibacterial activity as well as a wider antimicrobial range than pure chitosan because CDs can increase the dissolution of chitosan in water to enhance its antibacterial ability and slowly release chitosan to prolong its antibacterial duration. This CD-grafted chitosan complex can be used in cosmetics field to make effective antibacterial multifunctional cosmetics.[41]

Aside from HA and chitosan, there are also many other polysaccharides which are used in the cosmetics field. For example, Xanthan is an anionic polysaccharide made commercially by bacterial fermentation. It is a complex polysaccharide comprised of a primary chain of β-D-glucose and a branching trisaccharide side chain comprised of β-D-mannose. When used in cosmetics preparations, it not only endows cosmetics antioxidation and anti-inflammation effects, but can also perform better than traditional additives (such as Na-alginate and carboxymethyl cellulose(CMC)) in maintaining emulsion stabilisation.[42]

8.3.3 *Cyclodextrin*

CDs are cyclic oligosaccharides consisting of 6 to 8 α- (1,4) linked D(+)-glucopyranose units which are linked through R-1,4- glycosidic bonds (see Fig. 8.8). Due to the chair conformation of the glucopyranose units in CDs, its molecules display a truncated cone shape with a relatively hydrophobic central cavity surrounded by an external hydrophilic surface. The relatively hydrophobic interior cavity gives CDs the ability to host other hydrophobic molecules. In contrast, the hydrophilic exterior is sufficient enough to impart CD complexes with water solubility. Due to this arrangement, CDs can increase the solubilisation of poor water-soluble ingredients and improve the stabilisation of poorly stable compounds by entrapping them to form complexes.[43] The mechanisms of controlled degradation of such complexes are based on pH changes in the solution, leading to the cleavage of hydrogen or ionic bonds between the host and guest molecules. Thus, CD complexes can be used to achieve controlled release of biologically active compounds.[44]

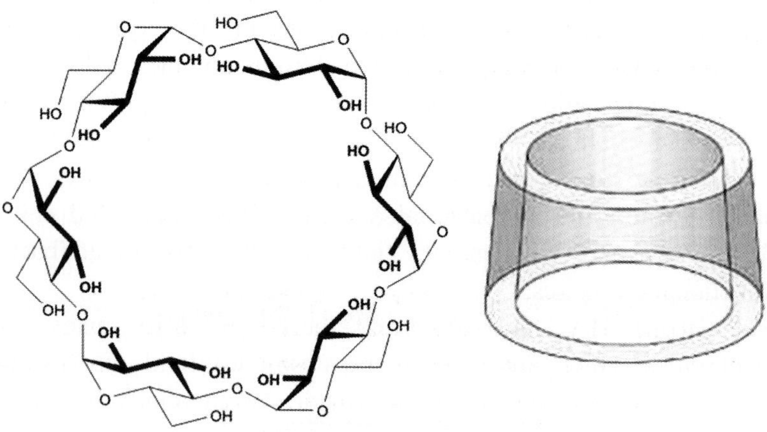

Fig. 8.8. Structure of Cyclodextrin

Nowadays, several CD-based formulations are commercialised by the pharmaceutical and cosmetics industries,[45] because CDs have many advantages in cosmetics application, such as: 1) improving the water solubility of lipophilic materials; 2) protecting guest molecules against decomposition, oxidation, hydrolysis or loss by evaporation; 3) controlling the release and skin absorption of active ingredients and 4) preventing interactions between various formulation ingredients.

So far, most of research in CDs and CD-based polymers focus on their embedding and releasing abilities. Samia Scott's research showed us that CD-polysaccharide associative gels can perform a good loading in benzophenone (widely used as sunscreen in the cosmetics field) and tamoxifen (an anticancer agent). The release rate was governed mainly by the dissociation of the complex molecules.[46] When CDs are combined with polymer based silicone, the submicron particles produced in that way allow the encapsulation of antifungal substances up to a 1:1 stoichiometry with respect to the CD. A simple oil-in-water emulsion cannot achieve such an encapsulation.[47] β -CDs can also be employed as a modified group to graft with PAAc and then synthesised with PAAm by using the sequential interpenetrating polymer network (IPN) method to form a thermoresponsive hydrogel (IPN(PAAc-g-β-CD/ PAAm) hydrogel). The experimental data showed that this IPN hydrogel can provide a positive delayed release of the active ingredient pattern and exhibit enhanced loading and enhanced controllable response rate in comparison with normal IPN(PAAc / PAAm) hydrogels (see Fig. 8.9).[48]

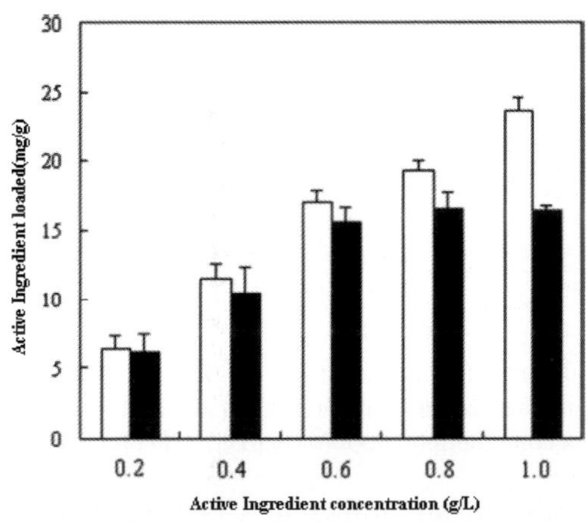

Fig. 8.9. Amounts of active ingredient loaded into the IPN hydrogels at 25°C: (□) PAAc-g-β-CD/PAAm and (■) PAAc/PAAm.[48]

Perfuming cosmetics products is an important part of the cosmetics industry. However, since fragrance materials are always poor in water solubility or insoluble in water and usually exist in a liquid state, the perfuming process may be difficult. Furthermore, the amount of fragrance materials in the product rapidly decreases during storage due to their volatility and poor stability. However, according to Ulya's study, by using CDs to form inclusion complexes of linalool and benzyl, this can significantly increase the water solubility. The interaction of the guest with CDs produces a higher-energy barrier to decrease the volatility of these two fragrance materials. Furthermore, the controlled release of linalool and benzyl can be achieved by preparing inclusion complexes of these fragrance materials with CDs, and the stability of these compounds can be increased by complexation (see Fig. 8.10).[49] Consequently, long-lasting fragrances are produced.

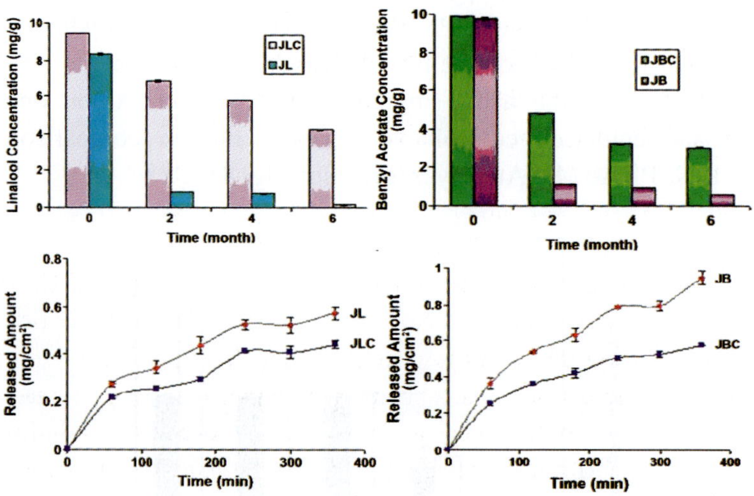

Fig. 8.10. Release and concentration conditions of benzyl acetate and linalool. JL (or JB) indicates formulations that contain uncomplexed linalool (or benzyl); JLC(or JBC) represent formulations containing linalool (or benzyl): CD (1:1) complex.[49]

In conclusion, the strong ability of CDs to complex fragrances can be combined with CD grafted thermo-responsive polymer gels to make smart controlled release perfuming cosmetics products. Since the LCST of this type of CD grafted thermo-responsive polymer gel is close to human body temperature, such a device is capable of releasing fragrances when heated by the human body.

8.3.4 *Shape memory polymers*

Shape memory polymers are polymeric smart materials which have the ability to return from a deformed state (temporary shape) to their original (permanent) shape induced by an external stimulus,[50] such as electric[51] or magnetic fields,[52] light [53] or a change in pH or temperature. The most widely used shape memory polymers today are polyurethanes (a type of temperature response polymer). Shape memory polyurethanes (SMPUs) are physically crosslinked thermoplastics. They have excellent performance and a lower cost. Nowadays, some shape memory polymers have already been used in textiles, and also have huge prospects in cosmetics applications.

There are several patents on SMPU applications in the cosmetics field for hair setting. When hair is treated by SMPU, it is able to be restored to the original parent geometry by application of temperature stimuli when the hair is deformed into a temporary configuration.[54] We performed an experiment according to a test method which is mentioned in US patents for further research on SMPU shape memory properties. Our research found that SMPU performance in hair settings is good and the shape recovery rate of hair is closely related to the content of SMPUs in hair setting solutions. The experiment consisted of two steps. In the first step, the hair was curled into a permanent shape in its original length and then a treatment composition, which contained shape memory polymers, was evenly sprayed onto the hair surface. Then, the SMPU solution was volatilised by heating it in the oven to shape the hair. In the second step, the hair was stretched with 37 g of weight for 30 min to achieve a deformed shape, and then heated at 75°C in an oven for recovery.

We repeated the experiment ten times for each sample of hair, and calculated the shape recovery rate each time (See Table 8.1 and Fig. 8.11).

Table 8.1. Shape recovery rate after deformation cycle.

	1st	2nd	3rd	4th	5th	6th	7th	8th	9th	10th
Sample 1 (0.03g/g Hair)	82.5	82.9	74.3	71.3	63.5	60.4	52.7	49.8	45.7	40.8
Sample 2 (0.045g/g Hair)	88.9	85.2	72.2	74.4	69.4	62.3	58.6	55.9	52.1	48.8
Sample 3 (0.06g/g Hair)	92.5	88.5	81.5	78.5	73.9	70.3	68.3	65.2	63.6	58.6
Sample 4 (No Chemical)	13.2	8.3	2.4	0.5	0	0	0	0	0	0

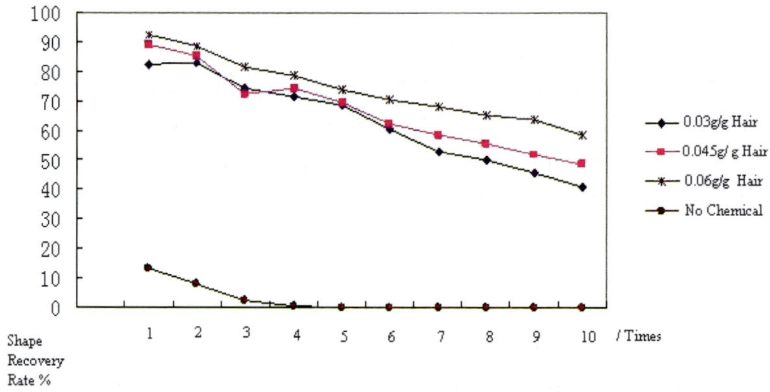

Fig. 8.11. Shape recovery rate after deformation cycle

From the experiment data, we found that the hair has a good shape recovery rate after treatment with SMPUs. More SMPUs added into the hair meant a better shape recovery rate.

8.3.5 *Nanopolymer particles*

Nanopolymer particles are polymers or copolymers having dispersed in nanoparticles. They are a subset of nanomaterials, and are defined as single particles with a diameter below 100 nm, although their agglomerates may be larger.[55] Nowadays, with in depth research in nanotechnology, people have began to recognise that nanopolymers have a larger surface area, stronger adsorption ability as well as surface reaction features in comparison with normal sized polymers. Some nanopolymers have great potential in the chemical industry, biomedicine and cosmetics applications, such as polymer/multiwall carbon nanotube composites, nanogels and dendrimers.

Recent studies reveal that some adaptive and functional polymer properties can be reinforced by using multiwall carbon nanotubes to make polymer/multiwall carbon nanotube composites. The SMP-MWNT composite is prepared by mixing SMP and multiwall carbon nanotubes (MWNT) in a solution together. Then, the mixed solution is stirred at high-speed in room temperature for 12 hours.[56]

Research data shows that by adding MWNT into the SMP, the recovery ratio increases from 83% (pure SMP) to 91% (SMP with 1.0 wt % MWNT). The recovery force also increases due to the interaction between the MWNTs and SMP, which helps in storing the internal elastic energy during loading and shape fixing.[57] Water vapor permeability (WVP) studies of various contents of SMP-MWNT composites also reveal that the WVP increases with an increase in the MWNT content. This could be attributed to the pathway effect of MWNTs by forming a passage where water molecules could diffuse rapidly (see Fig.8.12).[58] In comparison with pure SMPs, SMP-MWNT composites have improved thermal sensitivity, better shape recovery rates and mass transfer properties. They may find more applications than pure polyurethanes in the cosmetics field, such as hair setting.

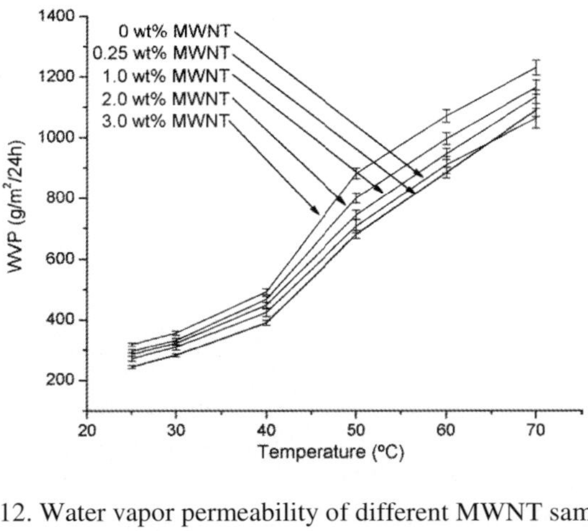

Fig. 8.12. Water vapor permeability of different MWNT samples [58]

Reinforcement of chitosan film with carbon nanotubes was also tested. The experiment showed that this composite exhibits a large increase of the tensile modulus with incorporation of only 0.8% of MWNT which may reinforce the film-forming ability of chitosan. Other advantages of chitosan in such materials are not only its biodegradability and antibacterial activity, but also the hydrophilicity introduced by adding polar groups that are able to form secondary interactions which may give chitosan more surface activities. These advantages are promising in the cosmetics and pharmaceutical fields.[59] Crosslinked polymeric nanogel particles have a porous structure dictated by the nature of the polymer, extent of crosslinking and size of the network domain. The characteristic porous structure and affinity of the polymer chains for particular moieties are key factors that make nanogel particles effective as carriers for specific attributes.[60]

Properties and cosmetics applications

Nanogel particles are currently finding increasing applications in personal care.[61] The properties of the core in the nanogel could be modulated by a change in hydrophobicity or hydrophilic units, and thus the capture of a guest molecule and its release could be controlled by external stimulus.[62] Recent research also shows that the advantages of nanogels include homogeneous distribution in the formulation, prolonged active ingredient activity, protection of ingredients against adverse conditions and the ability to release encapsulates at desired targets with desired results. Therefore, nanogels offer considerable advantages for cosmetics delivery.[60]

Chitosan-ovalbumin nanogels are green and steady nanogels which can be used in cosmetics applications. During the process of preparation, ovalbumin gelled forming nanospheres and chitosan chains are supposed to be partly trapped in the nanogel core, and while being heated, the rest of the chitosan chains should form the shell of the nanogels (see Fig. 8.13). This nanogel has excellent stability, and it does not change the size distribution after long-time storage nor dissociate in the pH range of 2-10.5. The chitosan shell structure also gives the nanogels the appealing biomedical properties of chitosan and pH-dependent properties. Another advantage of this system is that the production of the nanogels is convenient and green and no separation procedure is needed during the nanogel preparation. Such properties make chitosan-ovalbumin nanogels good candidates for cosmetics applications.[63]

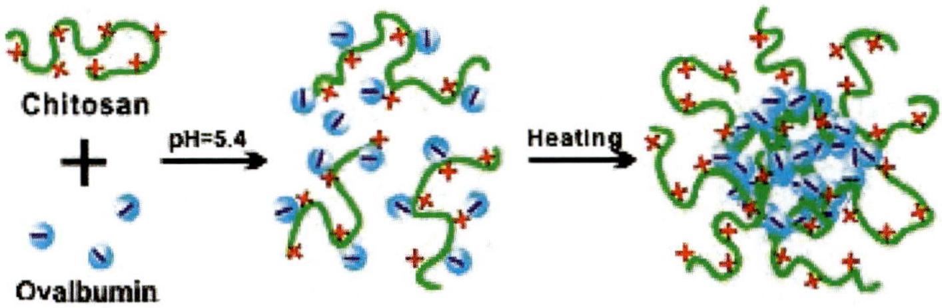

Fig.8.13. Chitosan-ovalbumin nanogels.[63]

Dendrimers are novel three-dimensional, hyperbranched, monodisperse nanometric macromolecules obtained by a reiterative sequence of reactions.[48] Dendrimers have a large number of 'tunable' surface groups and an interior that provides space as well as a microenvironment suitable for host-guest chemistry. Due to certain structural advantages, dendrimers can be achieved with a precise molecular weight that is necessary for reproducible pharmacokinetic data, high density of surface functionalities that may allow a higher payload, and a well-defined structure that provides a handle to tune toxicity and release properties. Poly(amidoamine) (PAMAM) and poly(propyleneimine) (PPI) are two commercially available dendrimers which have been widely studied for biological applications (see Fig. 8.14).[64]

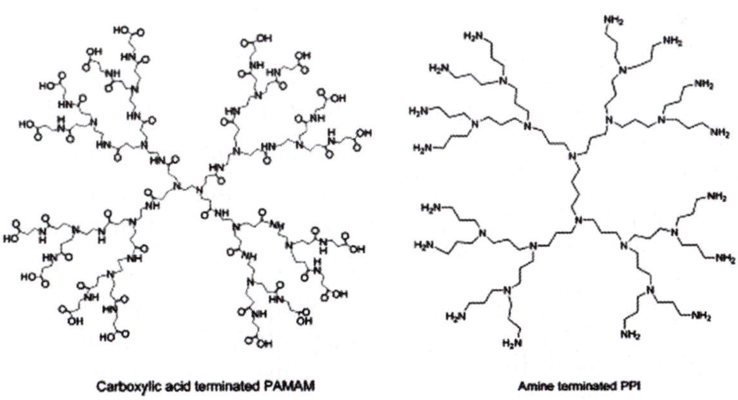

Fig. 8.14. Structures of PAMAM and PPI Dendrimers

Today, the delivery and release properties of dendrimers have gained substantial attention. Amphiphilic dendrimers can embed lipophilic or hydrophilic active ingredients in their interior space and release these ingredients in a suitable environment. Abhay's studies on PAMAM dendrimer revealed that dendrimers not only enhance the solubility of hydrophobic molecules, but also control the delivery of the bioactive compounds.[65] This target delivery property of dendrimers may have prospects in cosmetics applications to administer the delivery and release of active ingredients in cosmetics. Some amphiphilic dendrimers also possesses certain gram positive antibacterial activity and minimal eukaryotic cell toxicity. This indicates that these dendrimers will be a potential safety antibacterial agent in the future in many areas, such as medicine and cosmetics.[65] A recent study of dendrimer mediated solubilisation has been found to be superior to CD mediated solubilisation. PPI dendrimers are useful in solubility enhancement of not only acidic and basic, but also amphoteric ingredients. Their solubilisation ability is clearly regulated by pH and the chemical nature of dissolved ingredients. This property of dendrimers also has potential to be used in cosmetics preparations to improve stability and rheology.[65]

8.4 Glamourous Prospects in the Cosmetics Field

Adaptive polymer materials are auto-responsive materials. They can change their properties through internal or external feedback mechanisms. Their advanced design ideas are described as a major leap in the history of materials science. However, at present, only a minimal amount of adaptive and functional polymers are used in cosmetics to give cosmetics better performance in stability, safety and biocompatibility. Nevertheless, we still believe that in the future, as our research in adaptive and functional polymeric materials are increasingly expanded, they will have a wider use range in the field of cosmetics. Some possibilities of adaptive and functional cosmetics in the future can include the following.

8.4.1 *Multifunctional cosmetics*

Nowadays, if formulators can create products that excite consumers by performing more than one function, their products are more likely to succeed in the marketplace. Multifunctional cosmetics are a single cosmetics formulation which can provide more than one performance benefit. It can fulfill various requirements of customers and provide people with much convenience in just one use. Multifunctional cosmetics are already popular in our daily lives, such as two-in-one shampoos and multiuse skin care products. With the immense development in polymer science, more powerful multifunctional polymers will be able to endow more functions onto cosmetics in the future. The function of these polymers can range from acting as a preservative to acting as an emulsifier, and from acting as a moisturiser to acting as an emollient. Furthermore, some of them can even give biological properties to cosmetics such as having antioxidant and antibacterial properties. Some of the adaptive polymers can work as other functional ingredient carriers and release ingredients in a suitable environment to give more functions to a single cosmetic. It is undoubtedly that applications of adaptive and functional polymers to the cosmetics field will become more and more popular in the future.[14]

8.4.2 *Smart cosmetics*

8.4.2.1 *Adaptive hydrogel cosmetics*

Aside from temperature-sensitive hydrogels that we have mentioned above, some adaptive polymer gels have pH-sensitive properties. Shinichi reported that P(MAA-co-EGMA) nanoparticle gels show pH-responsive release behaviour when they are used to load functional materials for cosmetics applications.[62] Some polymeric hydrogels have photosensitive properties. When the light intensity is higher than a certain value, the polymer texture will change and then block the light from passing through.[67] Such unique polymers can be used to prepare 'smart' sunscreens in the future. Some adaptive polymer gels even have biochemical sensitivity. When they come into contact with the skin, they can release an appropriate amount of bioactive ingredient according to the biochemical environment of the skin surface.[68] This type of polymer will make cosmetics more durable and effective.

8.4.2.2 *Shape Memory Cosmetics*

If we combine shape memory polymers onto a mask, the mask may 'memorise' the pre-shaped figure and modify the skin to an ideal status. Heat-sensitive shape memory materials can also be used in hair settings to give human hair a 'memory' function. When the hair is in a disarray, we can heat the hair with a hair dryer to enable recovery to the original shape.

Scientists predict that future SMP cosmetics will have highly bionic properties and should be built at a cell level. Their response areas will also enlarge, and not only limited to one-to-one correspondence.[69] New developments in biology can also be combined shape memory technology, and in that way, skin care products can self-adjust according to the skin surface environment to 'memorise' and recover the original beauty of our body.

8.4.3 *Nanoparticle cosmetics*

The concepts of 'nanotechnology' were first given by physicist Richard Feynman at an American Physical Society meeting on December 29, 1959. Then, in the last decade of the 20th century, nanotechnology underwent a rapid development, and nanoparticles were applied in the industry, medical fields and other fields. Compared with normal sized polymers, nanopolymer particles have a larger surface area, stronger adsorption as well as a surface reaction feature. This allows them to protect associated ingredients much better than normal sized particles. Furthermore, nanopolymer particles can facilitate active ingredients to go across critical and specific biological barriers and hit specific targets.[55]

Polymeric nano-microspheres are typical nanopolymer particles which may have broad prospects in making sunscreen products. They can work as excellent TiO2 particle carriers. TiO2 is a widely used UV-shielding material, and its UV absorption capacity lies on its size. Hence, smaller in size means stronger UV absorption capacity. However, ultrafine TiO2 shows poor dispersion, thus weakening the absorption capacity of UV radiation. Nevertheless, using propylene cyanide copolymer nano-microspheres to load TiO_2 can prepare a stable and well dispersed composite particle.[69] Furthermore, PMMA-TiO_2 polymer composite microsphere not only has a good effect in preventing UV radiation, but also does not stop visible light through to the skin.[71]

8.5 Summary

The development history of cosmetics is synchronised with the history of scientific progress of humans. Today, increasingly more modern high-tech is applied in cosmetics research and development. The cosmetics industry has already stayed away from the common chemical industry areas and is an independent discipline which is based on pharmaceuticals, relying on refinements of the chemical industry and combining medicine, biological engineering, life sciences and micro-electronics technology. In other words, the cosmetics industry is gradually developing into a multidisciplinary area. At present, increasingly more adaptive and functional polymers have been applied in the cosmetics field to make smart and functional cosmetics. This indicates the development trend of future cosmetics. Valuable properties of adaptive and functional polymers, such as stimuli-response, shape memory, film-forming ability, dispersibility, thickening property, moisturizing ability and biocompatibility, make them essential elements in future cosmetics. Furthermore, with in-depth studies on polymer techniques, designing the polymer structure and functions at the molecular or even atomic level will become possible. It means that polymers will become more smart and multifunctional, thus they will be more efficient in cosmetics applications.

Polymer science has injected new vitality into the development of cosmetics and will create new research areas in the cosmetics field. The introduction of polymer science into cosmetics field will also provide new ideas and research methods for the development of cosmetics. We can boldly predict that in the near future, smart and new functional cosmetics will arrive. Multifunctional, nanometer and smart cosmetics will become the mainstream of cosmetics in the coming century.

References

1. Reed, S. I., Cosmetics and Your Health, in US Department of Health and Human Services. 2007.
2. Goddard, E. D. and Gruber, J. V., Principles of Polymer Science and Technology in Cosmetics and Personal Care. 1999, New York : Marcel Dekker.
3. Walter, P., Martinetto, P., Tsoucaris, G., Breniaux, R., Lefebvre, M.A., Richard, G., Talabot, J. and Dooryhee, E., Making make-up in ancient Egypt. *Nature*, 2001. 397: pp. 483-484.
4. Li, X. D., Application of Water-soluble Polymers in Cosmetic Formulation. *Flavour Fragrance Cosmetics*, 2004. 5: pp. 36-39.
5. Becher, P. and Emulsion, D., Theory and Practice. 2001, Oxford: Oxford University Press.
6. Zhang, H. M., Chang, H. H. and Wei, W. L., Medicated cosmetic--the development direction of 21st century cosmetic. *Detergent & Cosmetics*, 2003. 4: pp. 1-3.
7. Alberto, C., Yuan, D. and Shu, L., Emerging technologies and the future of cosmetic science. *Detergent & Cosmetics*, 2006. 3: pp. 3-5.
8. Vargha-Butler, E. I. and Kiss, E., Wettability of biodegradable surfaces. *Colloid Ploymer Science*, 2001. 279: pp. 1160-1168.
9. Zhang, J. M., Application of surfactants in cosmetics. *Detergent & Cosmetics*, 2008. 10: pp. 43-48.
10. Sadakazu, K. Z., Practical knowledge of cosmetics. 1981, Tokyo: Tokyo New Economics News.
11. Kuhn, T., Structure of Scientific Revolutions, 2nd ed. 1970, Chicago: Universuty Press.
12. Doi, Y., Microbial Polymers. 1990, New York: VCH Publishers.
13. Randy, S. and Perry, R., Multifunctional cosmetics 2003: New York : Marcel Dekker.
14. Rujirat, S., Manop, S., Jochen, W. and David J. M., Adsorption of protein-coated lipid droplets onto gellan gum hydrogel surfaces. *Food Research International*, 2008. 41: pp. 237-246.
15. Rene, G., Therapeutic and Cosmetic Indications of Lotrafilcon A Silicone Hydro- gel Extended Wear Lenses. *Ophthalmologica*, 2004. 21: pp. 33-38.
16. Subhas, C. K., Biraja, C. D., Rupesh D. and David L. K., Natural protective glue protein, sericin bioengineered by silkworms: Potential for biomedical and biotechnological applications. *Progress in Polymer Science*, 2008. 33: pp. 998-1012.

17. Roberta, C., Sonia, T., Rita, M. and Nevio, P., A novel dextran hydrogel linking trans-ferulic acid for the stabilization and transdermal delivery of vitamin E. *European Journal of Pharmaceutics and Biopharmaceutics*, 2009. 72: pp. 232-238.
18. Edelman, E. R. and Tanaka, T., Experimental Methods in Polymer Science: *Modern Methods in Polymer Research & Technology*. 2000, New York: Academic Press. 547-589.
19. Liu, H. X., Wang, C.Y., Gao, Q.X., Liu, X. X. and Tong, Z., Fabrication of novel core-shell hybrid alginate hydrogel beads. *International Journal of Pharmaceutics*, 2008. 351 pp. 104-112.
20. Bromberg, L. E. and Ron, E. S., Temperature-responsive gels and thermogelling polymer matrices for protein and peptide delivery. *Advanced Drug Delivery Reviews*, 1998. 31: pp. 197-221.
21. Heskins, M. and Guillet, J. E., Solution properties of poly (N-isopropylacrylamide). *Journal of Macromolar Science-Chemistry*, 1968. 8: pp. 1441-1455.
22. Michael, C. V. Progress in thermoresponsive polymer research. *Biomaterial of the Month-Archives*, 2008.
23. Joseph K. and Robert L. Controlled release of bioactive agents. *Trends in Biotechnology*, 1984. 2: pp. 47-51.
24. Hu, J. L., Liu, B.H. and Liu, W.G., Temperature/pH Dual Sensitive N-isopropylacrylamide/Polyurethane Copolymer Hydrogel-grafted Fabrics. *Textile Research Journal*, 2006. 76: pp. 853-861.
25. Lapasin, R. and Pricl, S., Rheology of Industrial Polysaccharides:Theory and Applications. Blackie Academic & Professional, Glasgow, 1995. 23: pp. 236-237.
26. Zhu, L., Yu, L. L. and Li, Z. J., Surfactants' applictions in cosmetics. *China Cosmetics Review*, 2007. 18: pp. 80-86.
27. Hagen, T. and Reinhard, H. H., The examination of polysaccharides as potential antioxidative compounds for topical administration using a lipid model system. *International Journal of Pharmaceutics*, 2005. 298: pp. 153-163.
28. Grigorij, K., Robert, S. and Peter, G., Hyaluronic acid: a natural biopolymer with a broad range of biomedical and industrial applications. *Biotechnology Letter*, 2007. 29: pp. 17-25.
29. Mary, P. and Lupo, M. D., Hyaluronic Acid Fillers in Facial Rejuvenation. *Seminars in Cutaneous Medicine and Surgery*, 2006: pp. 122-126.

30. Rohrich, R., Ghavami, A. and Crosby, M., The Role of Hyaluronic Acid Fillers (Restylane) in Facial Cosmetic Surgery: Review and Technical Considerations. *Plastic and Reconstructive Surgery* 2007. 120: pp. 41-54.
31. Marguerite, R., Chitin and chitosan: Properties and applications. *Progress of Polymer Science*, 2006. 31: pp. 603-622.
32. Helander, I. M., Nurmiaho-Lassila, E. L. and Ahvenainen, R., Chitosan disrupts the barrier properties of the outer membrane of Gramnegative bacteria. *International Journal of Food Microbiology*, 2001. 71: pp. 235-244.
33. Liu, H., Du, Y. and Sun, L., Chitosan kills bacteria through cell membrane damage. *International Journal of Food Microbiology*, 2004. 95: pp. 147-155.
34. Young, P. N., Lee, S. and Meyers, S. P., Antibacterial activity of chitosans and chitosan oligomers with different molecular weights. *International Journal of Food Microbiology*, 2002. 74: pp. 65-72.
35. Yang, J. M., Su, W. Y. and Leu, T. L., Evaluation of chitosan/PVA blended hydrogel membranes. *Journal of Membrane Science*, 2004. 236: pp. 39-51.
36. Martino, A., Sittinger, M. and Risbud, M. V., Chitosan: A versatile bio-polymer for orthopaedic tissue-engineering. *Biomaterials*, 2005. 26: pp. 5983-5990.
37. Yu, S. H., Mi, F. L., Tsai, C. H. and Peng, C. K., Miscibility, mechanical characteristic and platelet adhesion of 6-O-carboxymethylchitos- an/polyurethane semi-IPN membranes. *Journal of Membrane Science*, 2006. 276: pp. 68-80.
38. Liu, B. H., Hu, J.L. and Meng, Q.H., Nonwoven Supported Temperature-Sensitive Poly(N-isopropylacrylamide)/Polyurethane Copolymer Hydrogel With Antibacterial Activity. *Journal of Biomedical Materials Research Part B: Applied Biomaterials*, 2008. 89B: pp. 1-8.
39. Zang, Y., Guan Y. and Zhou, S., Single component chitosan hydrogel microcapsule from a layer-by-layer approach. *Bio-macromolecules*, 2005. 6: pp. 2365-2369.
40. Cristel, O. and Serge, L., Studies of the association of chitosan and alkylated chitosan with oppositely charged sodium dodecyl sulfate. *Colloids and Surfaces A: Physicochemical Eng.*, 2008. 330: pp. 201-206.

41. Khaled, E. T., Mohamed A. G., and Safaa E. R., Novel method for preparation of β-cyclodextrin/grafted chitosan and it's application. *Carbohydrate Polymers*, 2006. 63: pp. 385-392.
42. Amar, R. and Jayashree, M. N., Emulsification of Vegetable Oils using a Blend of Nonionic Surfactants for Cosmetic Applications. *Journal of Dispersion Science and Technology*, 2007. 28: pp. 1077-1080.
43. Loftsson, T., Jarho, P. and Jarvinen, T., Cyclodextrins in drug delivery Expert Opinion on Drug Delivery, 2005. 2: pp. 335-351.
44. Challa, R., Ahuja, A. and Khar, R.K., Cyclodextrins in drug delivery: an updated review. *Pharmceutical Science and Technology*, AAPS, 2005. 6: pp. 329-357.
45. Buschmann, H. J. and Schollmeyer, E., Applications of cyclodextrins in cosmetic products: a review. *Journal of Cosmetic Science*, 2002. 53: pp. 185-191.
46. Samia, D. M., Patrick, C. and Kawthar B. M., Cyclodextrin and Polysaccharide-Based Nanogels: Entrapment of Two Hydrophobic Molecules, Benzophenone and Tamoxifen. *Biomacromolecules*, 2009. 10: pp. 547-554.
47. Ahlem, N., Souhaira, H. and Hélène, P.L., Emulsions of β-cyclodextrins grafted to silicone for the transport of antifungal drugs. *Materials Science and Engineering*, 2008. 28: pp. 705-715.
48. Wang, Q. F., Li, S. M., Wang, Z.Y. and Liu, H. Z., Preparation and Characterization of a Positive Thermoresponsive Hydrogel for Drug Loading and Release. *Journal of Applied Polymer Science*, 2008. 111: pp. 1417-1425.
49. Ulya N., T., S. and Nilüfer, T., Use of Cyclodextrins as a Cosmetic Delivery System for Fragrance Materials: Linalool and Benzyl Acetate. *Pharmaceutical Science and Technology*, AAPS, 2007. 8: pp. E1-E9.
50. Lendlein, A. and Kelch, S., Shape-memory polymers. *Angewandte Chemie International Edition Eng.*, 2002. 41: pp. 2034-2057
51. Liu, Y., Review of electro-active shape-memory polymer composite. *Composites Science and Technology*, 2009. pp. 69: pp. 2064-2068.
52. Mohr, R., Initiation of shape-memory effect by inductive heating of magnetic nanoparticles in thermoplastic polymers. *Proceedings of the National Academy of Sciences*, 2006. 103: pp. 3540-3545.
53. Lendlein, A. and Kelch, S., Light-induced shape-memory polymers. *Nature*, 2005. 434: pp. 879-882.

54. Lendlein, A. and Kelch, S., Method for producting shape memory effects in hair in connection with hydrophobic active ingredients, in US, 2008/0311050A1[P]. 2008: USA.
55. Gerhard, J. N., Jurgen, L. and Christele, R., Grey Goo on the Skin? Nanotechnology, Cosmetic and Sunscreen Safety. *Critical Reviews in Toxicology*, 2007. 37: pp. 251-277.
56. Subrata, M. and Hu, J. L., Shape Memory Studies of Functionalized MWNT-reinforced Polyurethane Copolymers. *Iranian Polymer Journal*, 2006. 15: pp. 135-142.
57. Meng, Q. H., Hu, J. L. and Zhu, Y., Shape-Memory Polyurethane/ Multiwalled Carbon Nanotube Fibers. *Journal of Applied Polymer Science*, 2007. 106: pp. 837-848.
58. Meng, Q. H., Hu, J. L. and Subrata, M., Thermal sensitive shape recovery and mass transfer properties of polyurethane/modified MWNT composite membranes synthesized via in situ solution prepolymerization. *Journal of Membrane Science*, 2008. 319: pp. 102-110.
59. Tang, C. Y., Xiang, L. X. and Su J. X., Largely improved tensile properties of chitosan film via unique synergistic reinforcing effect of carbon nanotube and clay. *Journal of Physical Chemistry B*, 2008. 112: pp. 3876-3881.
60. Somasundaran, P., Somil, C., Mehta, L. R. and Soma. C., Nanotechnology and Related Safety Issues for Delivery of Active Ingredients in Cosmetics. *Materials Research Society Bulletin*, 2007. 32: pp. 779-786.
61. Li, Q. L. and Latreche, D., Polymeric Particles and Fragrance Delivery Systems, in EP, 1443058[P]. 2004: Europe.
62. Shinichi, Y., Makoto, S., Tatsuya, E. and Yotaro, M., Preparation and Characterization of a pH-Responsive Nanogel Based on a Photo-Cross-Linked Micelle Formed From Block Copolymers with Controlled Structure. *Langmuir*, 2009. 25: pp. 5258-5265.
63. Yu, S. Y., Hu, J. H., Pan, X. Y., Yao, P. and Jiang, M., Stable and pH-Sensitive Nanogels Prepared by Self-Assembly of Chitosan and Ovalbumin. *Langmuir*, 2006. 22: pp. 2754-2759.
64. Umesh, G., Hrushikesh, B. A. and Narendra, K. J., Polypropylene Imine. Dendrimer Mediated Solubility Enhancement: Effect of pH and Functional Groups of Hydrophobes. *Journal of Pharmaceutial Science*, 2007. 10: pp. 358-367.

65. Abhay, A., Abhay, S. C., Prakash, V. D. and Narendra K. J., Poly(amidoamine) (PAMAM) Dendritic Nanostructures for Controlled Sitespecific: delivery of acidic anti-inflammatory active ingredient. *Pharmaceutical Science and Technology. AAPS*, 2005. 6: pp. E536-E542.
66. Song, Q. H. and Song, Z. B., Development of Surfactants and the Future of Cosmetics. *Fine Chemicals*, 2002. 19: pp. 11-13.
67. Hu, Z. B., CO2 laser-controlled transmission of visible light in N-isopropylacrylamide gel. *Polymer Gels Networks*, 1995. 3: pp. 267-279.
68. Robert, P., Temperature-sensitive aqueous microgels. *Advance in Colloid and Interface Science*, 2000. 85: pp. 1 -33.
69. Zhan, J. J., Chen, Z. H. and Pan, Y., Progress in Shape Memory Polyurethane. *China Polymer Bulletin*, 2006. 8: pp. 21-28.
70. Hua, D. B., The Nanotechnology and polymer nanomaterials composites. *Modern Plastics Processing and Applications*, 1999. 11: pp. 52-56.
71. Yasugi, K., Nagasaki, Y. and Kato, M., News from the Controlled Release Society. *Journal of Control Release*, 1999. 12: pp. 89-94.

Chapter 9

Medical Applications of Adaptive Polymers

This chapter summarises different medical applications and the unique medical performance of adaptive polymer materials that use several different types of adaptive polymers. The content emphasises the progress and frontiers of adaptive polymeric products for medical applications, including design requirements and specific examples of tissue engineering, diagnosis, medical treatments and medical devices. These adaptive polymers include polymer gels, and shape memory, active, molecular imprinted and modified polymers, as well as other polymers with adaptive properties.

Keywords: adaptive polymer; tissue engineering; diagnosis; medical treatment; medical device

9.1 Introduction

Polymers are applied in a large number of medical applications,[1] such as tissue engineering, drug delivery or implantation of medical devices.[2]

Adaptive polymers are distinctly different from other polymers, and their physical and chemical properties are sensitive to changes in the environment.[3] Adaptive polymers not only ensure the basic functions of parts, such as structure, shape and optical, etc., but are also smart and can be adapted to their environment to attain optimisation of their behaviour. In other words, depending on changes in external conditions (such as pressure, temperature, adsorbed little molecules, electric field, optical wavelength, magnetic field, pH value and so on),[4] adaptive polymeric materials are able to change their properties, such as conductive, mechanical, surface, composition, structure, or their functions.[5]

Adaptive polymer materials cover a broader range of materials. Aside from the possession of a smart structure, each has functionality that is attributed to functional polymer materials. Mostly, adaptive polymeric materials are embedded in systems whose intrinsic properties are able to favorably change and meet their performance needs. A universal adaptive polymer does not exist and some routes coexist to obtain extrinsic or intrinsic adaptive polymers. In recent years, adaptive polymers which are used in medical fields are growing at a steady rate. These polymers are generally inexpensive and versatile, and their qualities are used for many bulk applications.

9.2 Aims and Requirements of Medical Adaptive Polymers

Adaptive polymers that are used in the medical field need to meet certain performance requirements of medical material, or modified to meet medical standards.[6] These performance requirements include bio-compatibility, functionality, nontoxicity, mechanical properties and a number of other properties. These requirements are discussed in more detail in the following sections.

9.2.1 *Biocompatibility*

Biocompatibility refers to the ability of an adaptive polymer material to perform with appropriate host response in specific applications.[7] Biocompatibility has been considered one of the most important items for estimating a material for application in human organisms. It can be evaluated by '*in vitro*' and '*in vivo*' modes. Some effective methods have been found, such as cellular reaction, cell adhesion and the healing process.[8] These methods are generally evaluated through cell systems, experimental, histological and host responses that involve carcinogenic, immunogenic, and thrombogenic responses. The complexity of host responses is a result of a series of steps that include many interdependent mechanisms of material tissue interactions. These interactions determine the final properties of adaptive polymers in a biological environment. In the application of implanted devices, the primary goal is minimising and adjusting interactions between material and tissue. In the application of biodegradable materials, the situation is the opposite, due to the degradation by-products which are able to strongly interact with living systems.

9.2.2 *Functionality*

Multicomponent adaptive polymers, showing more complex structures than homopolymers, naturally have more properties for medical applications, caused by special prepared methods or following modification strategies. Thus, adaptive polymer materials offer many properties that can be chosen according to functional necessities. Adaptive polymer products with properties that enhance their functionality plays a crucial role in quality, and have unlimited potential for innovative[9] new products. Nowadays, the modification of adaptive polymers has emerged as an effective means of controlling the adaptive polymers in medical applications. In fact, the ability of the functional groups which can be separated to certain boundaries can be utilised for the preparation of homogeneous polymeric materials with adaptive surfaces.[6] Then, the adaptive surface characteristics of the polymer materials will response to environment variation. It is particularly important where the location of functionalities is on the polymer backbone, which can migrate to the surface of the polymer.[10]

9.2.3 *Nontoxicity*

In all of the medical applications for adaptive polymers, it is acceptable to have direct contact between the material and tissues. Therefore, considering the final medical application of these adaptive materials, adequate safety tests for specific adaptive polymer matrix are necessary.[11] Since there is a necessary requirement for nontoxicty, toxicology for medical materials has evolved into a sophisticated science. Unless the material is specifically engineered for special requirements, for example, a 'smart bomb' drug that can target cancer cells and destroy them,[12] all medical materials should be safe. Many low molecular weight polymers exhibit some level of physiologic activity and cell toxicity. It is reasonable to emphasise that the material should not release anything from its mass unless specifically designed to do so. Therefore, when a new medical material is under development, toxicity is also examined along with the methods to evaluate the safety of the design standards.

9.2.4 *Mechanical properties*

Hardness, strength, toughness, plasticity, brittleness, elasticity, ductility and malleability are mechanical properties that are used as measurements of how adaptive polymers behave under a load. These characteristics are described in terms of the types of force or stress that the adaptive polymer must withstand and how these are resisted. In fact, medical materials have mechanical requirements which originate from the physical and chemical properties of the adaptive polymer. The requirements include physical properties, mechanical performance and mechanical durability. Mechanical performance is the most important.[12] Following that, mechanical durability needs to be taken into consideration. Finally, the physical properties of medical materials will also address other aspects of performance. To achieve these requirements, these elements must be considered, including design principles from physics, chemistry, mechanical engineering, chemical engineering and materials science.

9.2.5 *Other properties*

In addition to the above requirements which must be satisfied, adaptive polymers also need to meet other practical requirements in medical applications. A number of essential requirements, including physical characteristics, such as cost, geometry, clarity, size, colour of adaptive polymeric materials, and bioresponses, such as degradation rate and ease of processing, determine which material is required for special applications.

9.3 Adaptive Polymeric Materials for Medical Applications

Adaptive polymers have many significant applications in the medical field as delivery systems of therapeutic agents,[13] tissue scaffolds,[14] artificial organs,[15] molecular switches,[16] bioseparation devices,[17] sensors or actuator systems.[18] In this chapter, we will mainly focus on introducing recent research on adaptive polymers that are used in tissue engineering, diagnosis, medical treatments and medical devices.

9.3.1 *Adaptive polymeric materials for tissue engineering*

In recent years, there have been an abundance of research carried out in various aspects of tissue engineering with different adaptive polymers.[9,19] In this study, electroactive polymers (EAPs) that are used in artificial muscles[20,21] and biodegradable polymer scaffolds[14] have priority in research.

9.3.1.1 *Electroactive polymer for artificial muscles*

A natural muscle is a contractile organ. It consists of fibres which 'actuate' force and motion in response to nerve stimulation. Animals use muscles to change the chemical energy of adenosine triphosphate (ATP) into mechanical energy. Artificial muscles are synthetic materials, but they behave just like biological muscles. Artificial muscles are able to match or exceed the remarkable behaviour of natural muscles, which are needed for artificial limbs and damaged organs, insect-like air vehicles or humanoid robots. Artificial muscles should be similar in resilience and in their ability to produce large actuation strains. They are made up of special materials which dramatically swell and shrink under chemical or electrical stimuli.

There are several different types of polymer artificial muscle. EAPs, which are human made actuators that most closely emulate muscles, emerged in the last fifteen years, and exhibit large strains in response to electrical stimulation (Fig. 9.1). Due to this response, EAPs are named 'artificial muscles'.[21] Artificial muscles prepared with EAPs are a rapidly emerging actuation technology with significant differences from traditional electromagnetic actuators and other emerging new technologies, such as shape memory alloys and piezoelectric technology. A comparison of electroactive polymer artificial muscles and electromagnetic actuator muscles reveals that actuator muscles have a significant advantage, which is energy density. That is to say, more energy is created per unit mass from the actuator itself.

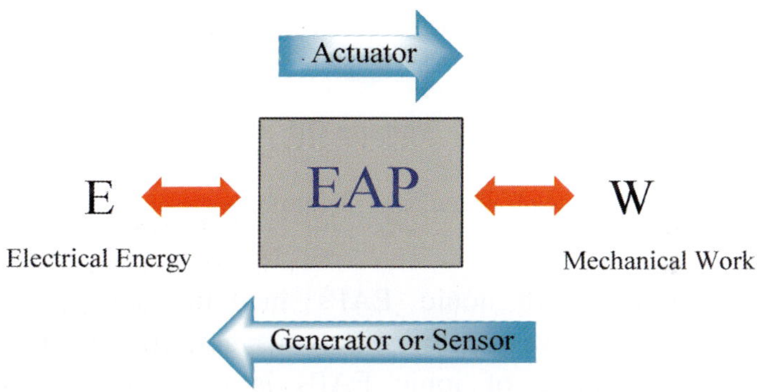

Fig. 9.1 EAP converts electrical energy to mechanical work and vice versa.

There are two kinds of EAPs: electronic and ionic. Each type has its own advantages and disadvantages.[22] Electronic EAPs rely on the motion of electrons as opposed to ionic electroactive polymers which rely on the motion of ions.

Common types of electronic EAPs include ferroelectric polymers, dielectric EAPs or electrostatically strited polymers, electrostrictive graft elastomers and liquid crystal elastomers. An example of an electric electroactive polymer is the dielectric elastomer, which is a film in which thin carbon-based electrodes sandwich a soft plastic, such as silicone or acrylic (Fig. 9.2).[23] Electricity arouses the electrodes to come together and squeeze the plastic, which expands to three times in comparison to a normal area in about half a second. Just like human muscles, actuators made of dielectric elastomers exert up to 30 times as much force. Electronic EAPs exhibit rapid response (in milliseconds) and high mechanical energy density, and then, it can be operated for a long time in room conditions. Recent developments allow for low voltages (~20 MV/meter) in ferroelectric EAPs. Moreover, electronic EAP produces mostly monopolar actuation due to associated electrostriction effects that are independent of the voltage polarity.

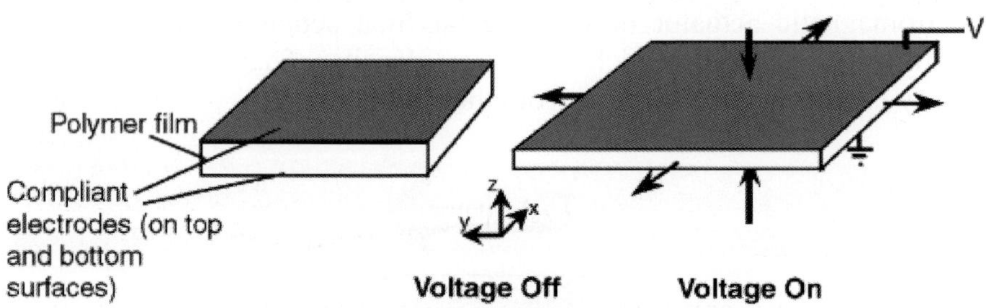

Fig. 9.2 Principle operation of dielectric elastomer actuation.[23]

Common types of ionic EAPs include ionic polymer-metal composites, conductive polymers, polymer gels and electrorheological fluids.[22] An example of ionic EAPs is the ionic polymer metal composite (IPMC) (Fig. 9.3). It consists of two metal-foil electrodes that sandwich wet, teflon-like plastic soaked with lithium ions.[24] Twelve volts cause the lithium ions to be positively charged and migrate toward the negatively charged foil layer, finally bending the ionic polymer-metal composites due to the bulking up of that side of the actuator. Advantages of ionic EAPs include natural bi-directional actuation that depends on the voltage polarity, low voltage, and some have the unique capability of bi-stability. However, they also have disadvantages, such as low electromechanical coupling efficiency, and need to use electrolytes and encapsulation.

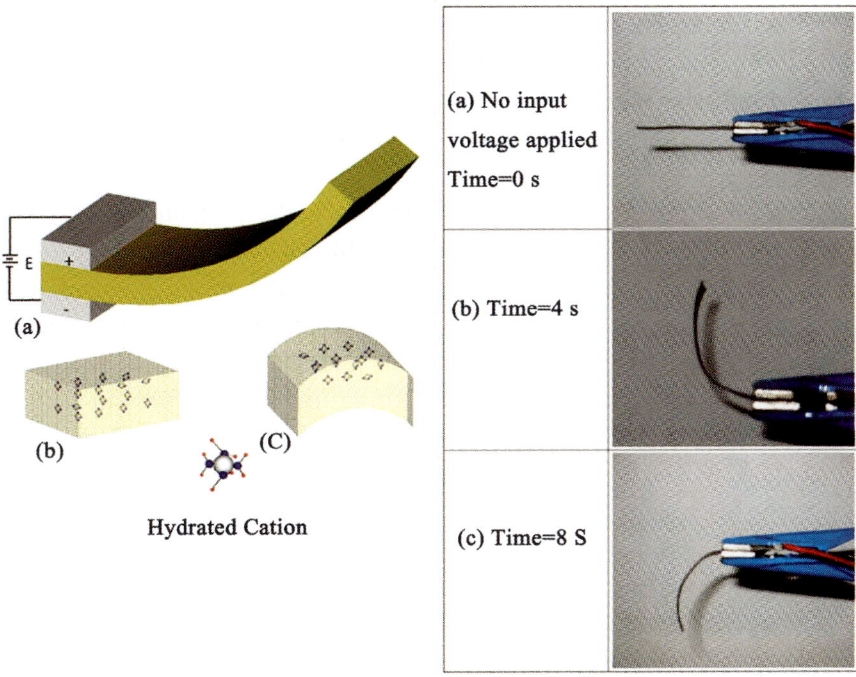

Fig. 9.3 The principle operations of IPMC: left side (a) bending of IPMC with applied voltage, (b) IPMC section with no applied voltage, and (c) IPMC section with applied voltage. Right side: typical IPMC placed under an applied AC voltage step function input of 3.0 V: (a) no voltage applied, (b) positive polarity, and (c) negative polarity.[24]

Novel uses for artificial muscles include: (a) smart implants with tiny perforations that contain a pharmaceutical, plugged by artificial muscles, and (b) an implant with tiny sensors which sense blood concentrations of certain chemicals.

9.3.1.2 *Biodegradable scaffold for tissue engineering*

With the increasing need for donor organs and tissues to treat various diseases, engineering tissues may be a promising method in fulfilling this demand. Scaffold imitates the structure and function of extracellular matrix (ECM) and therefore, plays a key role in replacing ECM. It is the core of tissue engineering (TE) to look for seed cells with strong regeneration capacity and biological materials adapted to cell survival.[25] Using bioscaffolds, healthy tissues can be grown to replace diseased or damaged cells or tissues. The growth of several types of tissues have been attempted, and these include smooth muscles, bones, heart valves, and cartilage.[26]

These bioscaffolds act as a temporary structure on which the cells can grow. Over time, bioscaffolds degrade as the cells form their own tissue matrix. The bioscaffolds can be used to grow tissues *in vitro*, outside the body, or *in vivo*, inside the body. For *in vitro* applications, healthy cells are seeded from the patient to the scaffold, and the tissue is grown in a simulated physiological environment. During this process, growth factors may be incorporated into the scaffold to enhance cell growth and viability. The healthy tissue that was grown can then be implanted into the patient to replace the diseased region. For *in vivo* applications, the damaged cells are first removed, and then the scaffold is directly implanted into the patient. The scaffolds may also be implanted to fill voids that are present in the tissue. The surrounding healthy cells will grow on the scaffold, and the scaffold is replaced by the ECM as the material degrades.[26]

Several requirements of the scaffolds are necessary for the growth of tissues. As the cells grow and form their own structure, the scaffold should degrade so that it does not impede tissue growth. Thus, the degradation rate should be similar to the growth rate of the cells. Furthermore, as the material degrades, it cannot release any toxins nor have any significant negative effects on cell growth or the human body. The morphology of the bioscaffold, which includes properties such as relative density, pore density, average pore size and porosity, is also important to tissue growth. The scaffold should have a large surface-area-to-volume ratio and be highly porous in such a way that nutrients and waste products can be circulated to and from the cells. Finally, depending on the application and the types of tissues being grown, the bioscaffolds may be under different stress and loading conditions, and constitutive models have been developed to predict the scaffold behaviour under loads. A failure of the material may cause improper tissue growth or tearing of the scaffold; thus, the bioscaffold should have sufficient mechanical strength to withstand the stresses.

Therefore, selection and construction of scaffold materials are key factors for tissue engineering. Many scaffolds are synthetic polymers, including poly(lactide) acid, poly(glycolide) acid, polyanhydride, polycaprolactone, polycarbonate, polyorthoester and poly(DL-lactide-co-glycolide) acid (PLGA). Simple polymers can provide architectural support for neo-tissue development. Unfortunately, these simple polymers are not able to adequately mimic the complex interactions between progenitor cells and adult stem, and also cannot participate in the formation of functional tissue. Future advances in TE and regenerative medicine will depend on the development of 'smart' biomaterials.[19] It is possible that smart polymers offer promise for revolutionary improvements in TE scaffolds, and a major goal is to impart smart biomaterials with specific properties beyond the physical properties of polymers. Research has found that proteins can be conjugated either randomly or in a site-specific manner, and introduce a reactive amino acid at a particular position.[27] It has been shown that the induction of a change in the conformational state of the smart polymer can control the activity of a protein, if a conjugation site is introduced near its ligand-binding domain.[28] This may allow delivery of cells to a desired location, and recovery of specific cells.

Hydrogels are an attractive scaffold material because their structure is similar to the ECM of many tissues.[29] Smart hydrogels are able to form scaffolds for two reasons. First, their interior environment is aqueous. Moreover, they can release cells at an appropriate place with a suitable stimulus.[9]

Recently, biomedical engineering advances have led to the development of the concept of tissue or organ printing.[30] Organ printing can be defined as layer-by-layer additive robotic biofabrication of three-dimensional functional living macrotissues and organ constructs that use tissue spheroids as building blocks.[31] Layered deposition of cells and cell aggregates use various rapid prototyping (RP) techniques, which confer reproducible control over cell placement, and slow cell distribution within the scaffold, and then yields a defined scaffold structure (Fig. 9.4).[32] Also, this technique would enable fabrication of vascular beds and allow larger tissue constructs to survive after implantation. Moreover, printed tissues can be used as a tool to research cell–cell and cell–matrix interactions in pharmacology by mimicking the microenvironment of the cells.

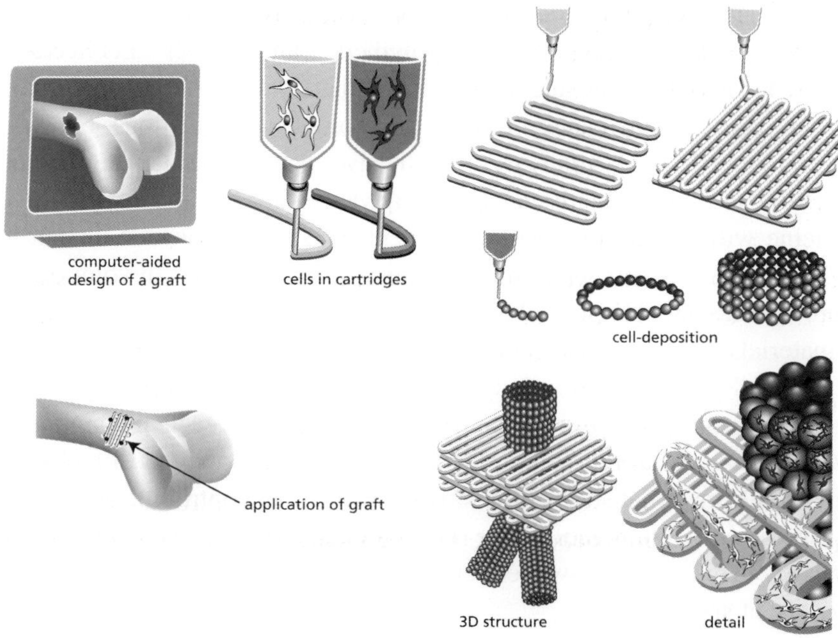

Fig. 9.4 Concept of organ printing.[32]

9.3.2 *Adaptive polymeric materials for diagnosis and medical treatments*

The study and application of adaptive polymers that are used in medical treatment and diagnosis mainly concentrate on drug delivery,[33] smart polymer nanoparticles for diagnostics,[34] adaptive polymeric materials in biosensors[35] and adaptive polymeric materials for eye related materials.[36]

9.3.2.1 *Adaptive polymers in drug delivery*

The use of adaptive polymers also include polymeric materials as drug delivery devices.[5] A number of adaptive polymers are potentially useful for this purpose, including hydrogels,[13] amphiphilic polymers,[37] molecular imprinting polymers[38] and other adaptive polymers.[9]

Hydrogels have been used as carriers for delivery of drugs, peptides or proteins in pharmaceutical applications. They have been used to control release systems or as carriers in swellable and swelling-controlled release devices.[39] The swelling and release behaviour of these gels may be dependent on drug concentration, temperature, pH, and ionic strength (Fig. 9.5). An approach of using hydrogels that are sensitive to glucose is the immobilisation of glucose oxidase on a pH-sensitive hydrogel.[39] Glucose oxidase acts as a glucose sensor as it produces gluconic acid by an enzymatic reaction with glucose. The gluconic acid that is produced lowers the pH of the medium, resulting in a release of insulin because of significant changes in swelling. An alternate route through phenyborate–poly(vinyl alcohol) polymers was discussed by Hisamitsu *et al.*[40] Kumashiro *et al.* proposed a delivery mechanism based on temperature range and enzyme activity.[41] The synthesised tem-perature-responsive hydrogels only allow enzyme-triggered polymer degradation below a higher critical solution temperature and above a lower critical solution temperature. They anticipated that this technique can control the release of drug molecules, which depend on both enzyme selectivity and changes in body temperature.

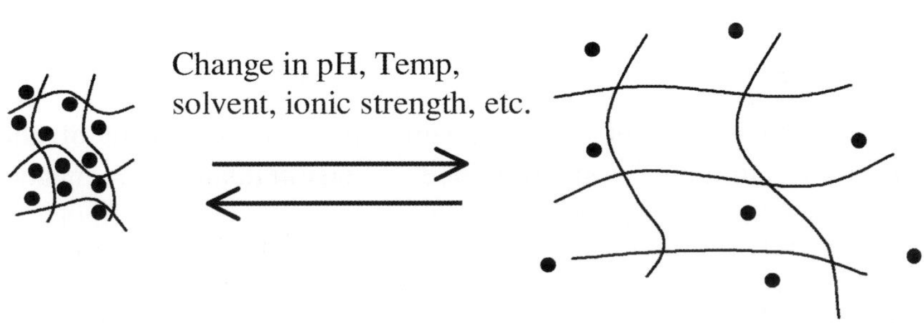

Fig. 9.5 Intelligent, stimuli-responsive hydrogels. Modulated release of drug (circles).[39]

Amphiphilic copolymers are well developed as precursors for the preparation of micellar drug carriers. Poly(*N*-isopropylacrylamide) (PNIPAAm) is one of the most extensively studied thermo-sensitive polymers that exhibits a lower critical solution temperature (LCST) at around 33°C in an aqueous solution. Over the past decade, considerable efforts have been devoted to design and prepare PNIPAAm-based thermo-sensitive polymeric micelles as delivery vehicles for controlled drug release.[42] The pH- and thermo-sensitive 'schizophrenic' micellization behaviour of a polypeptide hybrid double hydrophilic diblock copolymer (DHBC) based on PNIPAAm-*b*-poly(L-glutamic acid) (PNIPAAm-*b*-PLGA) was thoroughly investigated.[43] pH sensitive poly-(NIPAAm/MAA) nanoparticles were embedded along with glucose oxidase and catalase in an ethylcellulosebased membrane. The rate of insulin release was modulated by glucose concentration and changed with volume changes in the embedded nanoparticles. When polymer dimension was reduced to a nanometre, the response lag time was short and led to a faster response. Unfortunately, the permeability cycles (to test reusability) were not reproducible because the design did not 'allow sufficient time for removing the diffusants from the membrane'.[44] PNIPAAm-*b*-PLGA was able to self-assemble into micelles with PNIPAAm-cores at alkaline pH and elevated temperatures, and micelles with PLGA-cores at acidic pH and room temperature (Fig. 9.6). 'Schizophrenic' micellization of PNIPAAm-*b*-PLGA led to the facile location of peptide sequence either within micellar cores or coronas. Moreover, studies revealed that the formation of α-helix secondary structures at low pH considerably affects the pH-induced micellisation kinetics of PNIPAAm-*b*-PLGA. The incorporation of polypeptide sequences into DHBCs could endow them with structural versatility, tunable spatial arrangement of chain segments within self-assembled nanostructures, enhanced biocompatibility, and broader applications in the biomedical field.

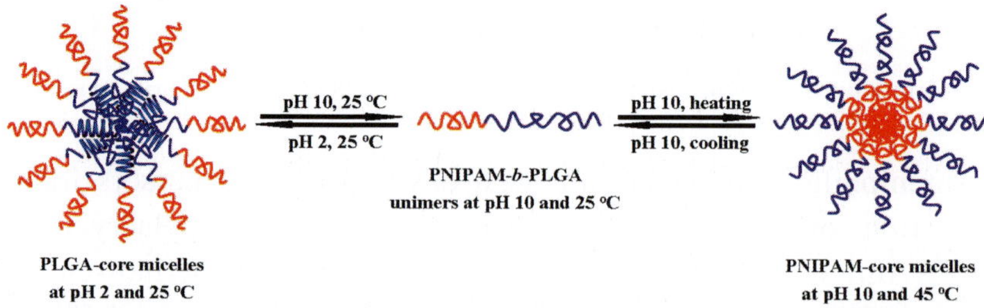

Fig. 9.6 Schematic illustration of thermo- and pH-responsive micellisation of PNIPAM65-*b*-PLGA110 associated with coil-to-helix transitions.[43]

Molecular imprinting is a rapidly evolving technique for preparing synthetic receptors.[45] The objective of producing drug delivery systems that are targeted to specific sites and tissues has also necessitated the creation of systems with highly specific recognition capabilities. The technique of molecular imprinting allows for the preparation of synthetic polymers with specific binding sites for a target molecule.[38] Molecular imprinting polymers (MIPs) can be obtained if the target is present and acts as a molecular template during the polymersation. The functional groups become fixed in defined positions in the polymer network following polymersation with a high degree of crosslinking. Following that, the functional groups (template) are removed by solvent extraction or chemical cleavage, and leave cavities which are complementary to the template in terms of shape, size and arrangement of the functional groups. These highly specific receptor sites can recognise the target molecule with high specificity.

MIPs have been largely used to improve the separation processes of target moleculars and show great potential as biosensors and enzyme mimics. If MIPs are used in drug delivery, it would be able to create fascinating functions, such as adhesion, targeting and responsive delivery (Fig. 9.7). This is possible by creating specificity to compounds on cell or tissue surfaces. For example, surface imprinting can be used for cell separation. From another point of view, MIPs also can be used to extract excess or toxic compounds from the blood or body fluids.[38]

The ability of MIPs to provide various functions entails several steps as described in the following. The molecularly imprinted polymer reaches the desired site and captures the undesirable compound, and then the linkages holding the polymer at the specific site degrade over time until the undesirable compound is totally degraded and the polymer moves with the fluid. Many researchers admit that MIPs can be used as antibody-mimics. In the pharmaceutical field, antibodies have been used as drug targeting devices that are coupled to drug carriers, such as liposomes or polymeric micelles, because they can selectively recognise (pathogenic) cells or tissues. Therefore, it is possible to synthesise drug targeting devices by creating imprints of targets on cells or tissue surfaces.[45]

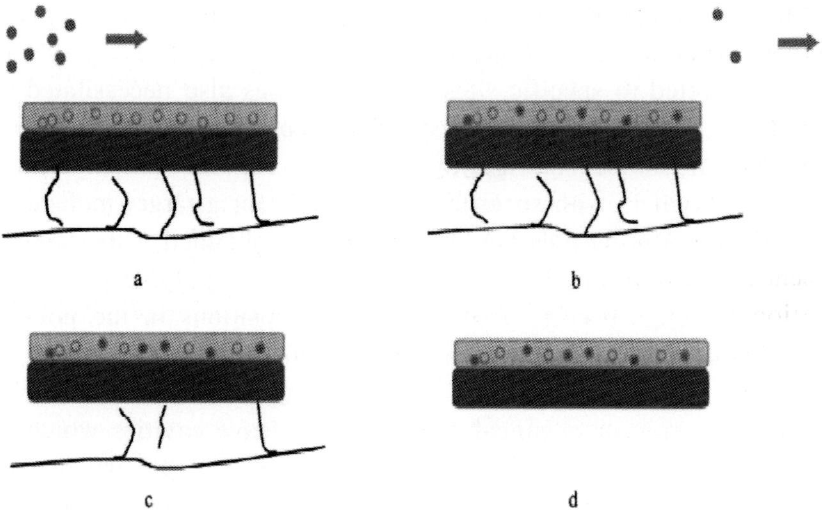

Fig. 9.7 Example of target action in biological systems with MIPs. Using the molecular imprinting technique, undesirable compounds can be removed from blood and other body fluids. Different constituents in the system can provide functions, such as adhesion and degradation. (a) The polymer reaches the desired site and adheres to the surface through the adhesive layer. (b) The molecularly imprinted side captures the undesirable compound. (c) The linkages holding the polymer at the specific site degrade over time. (d) The polymer moves with the fluid because the linkages that are holding it on the surface are totally degraded.[38]

9.3.2.2 *Smart polymer nanoparticles for diagnostics*

The unique properties of nanoscale particles have been exploited to develop new biotechnology applications by coupling the particles to biomolecules.[46] Nanoengineered particles have been used in various experimental and clinical conditions because they can reach specific molecular targets on diseased cells. Their medical applications involve a large deal of research which revolve around malignant diseases, and diagnostic and therapeutic applications.[47] Polymers have often been used as building blocks for nanoengineered particles.[34] Studies have shown that polymeric micelles that are composed of amphiphilic block copolymers represent a promising role as diagnostic agents.[48] Development in the fields of nanotechnology and microfluidics has created new opportunities for smart polymers.[34] Smart polymers have been conjugated to biomolecules, such as streptavidin and DNA, to form unique hybrid molecules.[49] These have been used to develop unique diagnostic assays, because they can reversibly form controlled nanoparticles.[50] Hoffman *et al.* tried to utilise smart polymer nanoparticles to enhance the capturing or detecting of rare proteins and DNA from non-invasive samples to increase the accuracy of diagnostic tests.

They recently developed two nano and microscale technologies for diagnostic applications.[34] The first is a reversible particle system that uses stimuli-responsive polymer-protein conjugates. They found that conjugates of streptavidin and temperature-responsive PNIPAAm rapidly form stable and uniformly sized mesoscale particles above the LCST of the polymer. The size of these particles is dependent on concentration, molecular weight of the polymer used and formulation parameters, such as the heating rate.[51]

The second is a stimuli-responsive bioanalytical bead system.[34] Latex beads were dualconjugated with PNIPAAm and an affinity ligand to confer temperature-responsiveness onto the beads. Above the LCST of the PNIPAAm, the bead surface becomes hydrophobic and the modified beads aggregate and adhere to the walls of microfluidic channels. They have been used to develop a reversible microfluidic affinity chro-matography matrix for the upstream processing of complex fluids and immunoassays. The use of this system in an immunoassay format has been demonstrated by using a model digoxin antibody-antigen system.[52] The digoxin capture system is immobilised in a locally heated region of a microfluidic channel, and a competition immunoassay is performed with digoxin and a fluorescent analogue (Fig. 9.8). The small molecules are found to bind to the immobilised antibody, and by measuring the amount of the unbound fluorescent analogue, the concentration of digoxin is determined. Both technologies can be used in a wide variety of formats, including microfluidic-based micro-total analytical system devices and simple, rapid field tests.

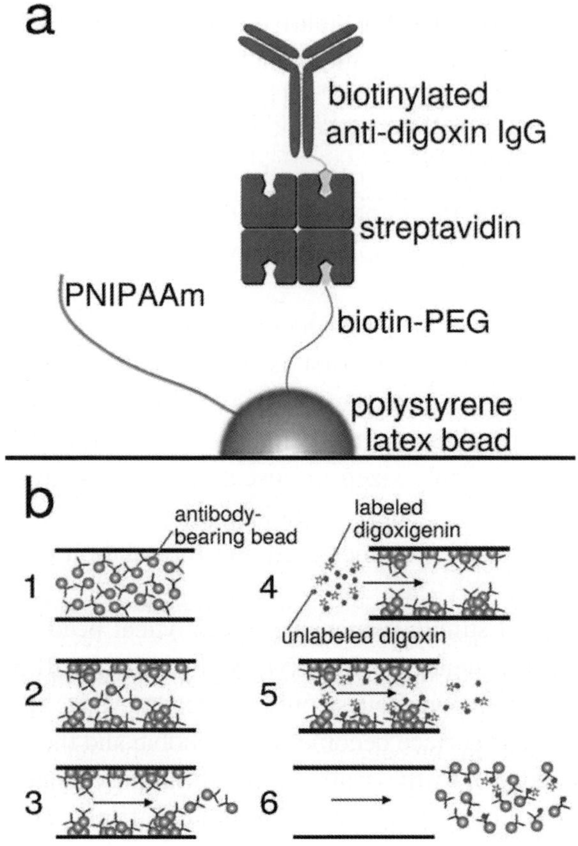

Fig. 9.8 Smart bead immunoassay system. Panel a: The assembled smart bead construct. A 100 nm diameter latex nanobead bead is surface-conjugated with biotin-PEG and PNIPAAm. Streptavidin is bound to the exposed biotin, providing a binding site for the biotinylated anti-digoxin IgG. Panel b: A schematic illustration of the experimental protocol. Suspended smart beads are loaded into the PET microfluidic channel (1). The temperature in the channel is then increased from room temperature to 37°C, resulting in aggregation and adhesion of the beads to the channel wall (2). Flow is initiated (the presence and direction of flow is indicated by an arrow in this diagram), washing unabsorbed beads out of the channel (3). A mixture of fluorescently labeled digoxigenin (at a fixed concentration) and digoxin (at varying concentrations) is flowed into the channel (4). Components of this mixture that fail to bind the immobilised antibodies are washed through (5). Since digoxin and labeled digoxigenin compete for antibody binding, a higher concentration of digoxin means that a greater amount of labeled digoxigenin will flow through at this step. Finally, the temperature in the channel is reduced, and the aggregation/absorption process reversed as antigen-bound beads leave the channel with the flow stream (6).[52]

9.3.2.3 *Adaptive polymeric materials in biosensors*

Nowadays, biomedical diagnostic research promotes the progress of evaluative techniques. Most of the improvements are due to biosensor technology, which has played an important role in improving materials that are used to sense, respond to signals, and carry signals. For biosensor applications, the sensing element is typically natural bioreceptors, such as enzymes, nucleic acids/DNA, antibody/antigen, and cellular structures/cells, due to their evolved high affinity and specificity. Biomimetic sensing elements can be advantageous over their biological counterparts, because they can be designed to mimic biological recognition pathways, and at the same time, they also exhibit other abiotic properties, such as greater stability in harsh environments.The development of long-term implantable biosensors has been seriously impeded by the deleterious effects of biofouling and lack of poor in vivo biocompatibility. Hence, most clinical applications have been restricted to academic studies rather than routine clinical monitoring. The selection of materials and fabrication techniques ultimately affect ample biosensor function and the performance of a biosensor. Consequently, ways to solve biocompatibility and biofouling problems will inevitably focus upon development in biosensor design with the technology of new materials. Success in this area will enable the widespread use of biosensors in long-term clinical monitoring and *in vivo* diagnosis.The use of intelligent polymer systems in sensors provides the possibilities of combining sensing, transduction of signals, and response in the same independent device as well as controlling biological events based on the signal by delivery of drugs or other means. Hydrogels and conducting EAPs as adaptive polymers are currently receiving widespread attention in biosensor development.[35]

Holtz *et al.* developed a hydrogel-based photonic crystal which acts as a glucose sensor for *diabetes mellitus*.[53] This material is crystalline polymer spheres, and polymerise within a hydrogel which swells and shrinks reversibly in certain analytes. Alexeev *et al.* also developed a photonic crystal glucose-sensing material, which consists of a crystalline colloidal array polymerised within a polymer network of a poly-acrylamide-poly(ethylene glycol) hydrogel with pendent phenylboronic acid groups.[54] The swelling event of polymer can increase the separation between the immobilised nanospheres, and shift the Bragg peak of diffracted light to longer wavelengths, and finally, induce a red-shift in the optical properties of the polymer. This hydrogel can be applied in contact lenses or ocular inserts to indirectly detect blood glucose levels via tear fluid. In this modified system, boronic acid derivatives are polymerised within a polyacrylamide-PEG. Shrinkage of hydrogels will cause a blue-shift because of crosslinks between glucose and the derivatives. The patient themselves can determine their blood glucose levels via a colour chart. Yang *et al.* demonstrated the use of small-molecule hydrogels for enzyme sensing.[55] In their work, β-lactam-containing conjugates with β-lactamase cleave the scissile β-lactam amide bonds, and release a potent hydrogelator. Hydrogel formation is exploited in the biological sensing of β-lactamases, bacterial enzymes that cause antibiotic inactivation in resistant bacterial strains. This approach provides a low cost and easy way that could be used to screen for inhibitors in this class of enzymes.

Conductive EAPs can be synthesised under mild conditions, and it is easy to incorporate a range of biological moieties (antibodies, enzymes and even whole living cells) into the polymer structure.[56] The unique electronic properties allow direct and indirect communication with the biochemistries and produce a range of analytical signals. Hence, the electrically conducting polymers with numerous features can enable rapid electron transfer in the fabrication of biosensors and excellent materials for immobilisation of biomolecules.[57] Nowadays, biosensors are applied in medical diagnosis, such as lactate, glucose, ethanol, fructose, cholesterol, urea, etc. Shen *et al.* described a conducting polyaniline-polyisoprene which was effectively used to immobilise a glucose oxidase forming glucose biosensor.[58] The biosensor has remarkable long-term stability and high permselectivity, which can determine H_2O_2 in electroactive interferent ascorbic acid. Gambhir *et al.*[59] recently co-immobilised urease and glutamate dehydrogenase on electrochemically prepared polypyrrole/polyvinyl sulphonate for the preparation of a urea biosensor. Kumar *et al.*[60] utilised doped dodecylbenzene sulphonate on polypyrrole films for immobilisation of cholesterol oxidase. These films were used to estimate cholesterol via ferricyanide mediation.

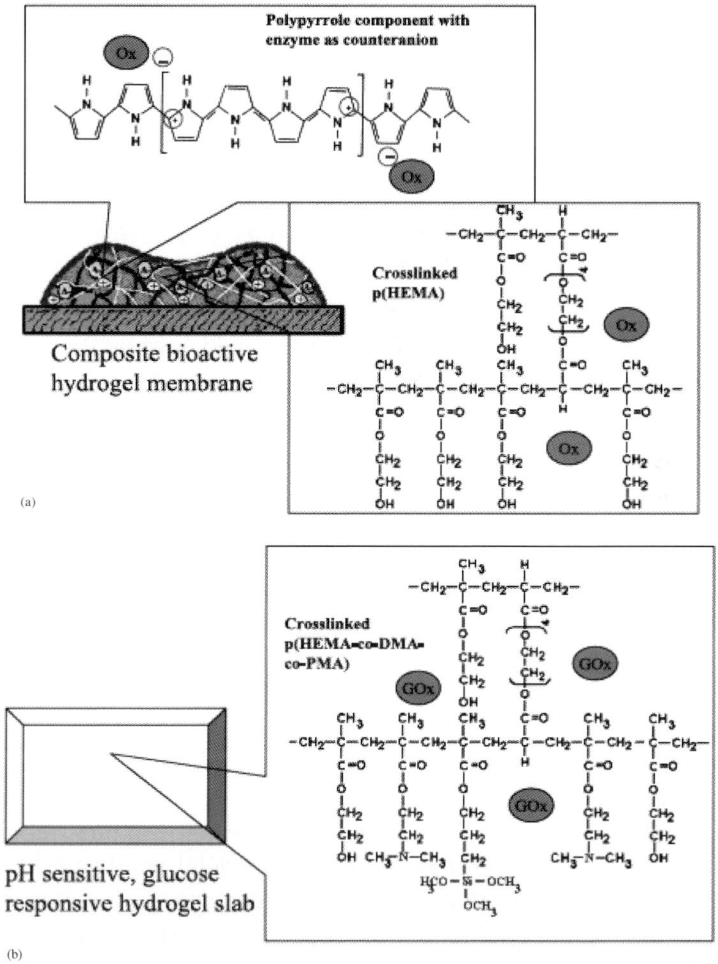

Fig. 9.9 (a) Schematic representation of the three components of a novel hydrogel composite matrix*/the cross-linked hydrophilic hydrogel, the positively charged electroactive PPy component and the negatively charged (net) oxidase enzyme. (b) Partial structure of the hydrogel network used for the fabrication of engineered devices for the controlled release of insulin.

Brahim *et al.* studied electroactive hydrogel composites that were integrated with hydrogels and conducting electroactive polymers. Enzymes were entrapped within biosensor matrices and could be control released through chemical stimulations[35] (Fig. 9.9). Enhanced polymer biosensing capabilities have been demonstrated in the fabrication of galactose amperometric, cholesterol and glucose biosensors. All of the biosensors show good performance, such as displaying extended linear response ranges (10^{-5}-10^{-2} M), rapid response times (<60 s), storage stabilities of up to one year, and excellent screening of physiological interferents, such as uric acid, ascorbic acid and acetaminophen. The polymeric biosensors will response to pH changes (neutral to acidic), when the crosslinked hydrogel components of these composite membranes are prepared with the amine containing dimethylaminoethyl methacrylate monomer. The ability to entrap glucose oxidase and form gluconic acid within these materials render them glucose-responsive. In this 'bio-smart' co-loaded insulin with glucose oxidase, there is a twofold increase in the release rate after the devices were immersed in glucose solutions. This demonstrates the potential of such systems to function as chemically-synthesised artificial pancreas. The polypyrrole (PPy) component of the composite film can enhance the permselective properties of the crosslinked p(HEMA) hydrogel.

9.3.2.4 *Adaptive polymeric materials for eye related materials*

Insertion of a punctum plug is an effective way to treat dry eyes.[61] Conventional plugs produce inherent problems which include irritation of the cornea and conjunctiva, epithelial laceration, granulation formation and frequent extrusion. Surgical punctal occlusion by cautery or laser also has several problems, such as recanalisation, deformation of the punctum and symblepharon formation.[62] The thermosensitive hydrophobic acrylic material of the SmartPlug™ can eliminate the possibility of ocular irritation from conventional plugs and surgical punctal occlusion. SmartPlug™ insertion is shown in Fig. 9.10.[62] This plug has two 'smart' properties. The first one is its transition from rigid solid to soft gel at around 30°C. The second 'smart' property is inducing a solid to gel transition when expanding the diameter from 0.4 mm up to 1 mm during the temperature changes. This expansion in diameter is accompanied with a length decrease in the SmartPlug™. Research suggest that SmartPlug™ is a safe and effective treatment for the dry eye syndrome.[62]

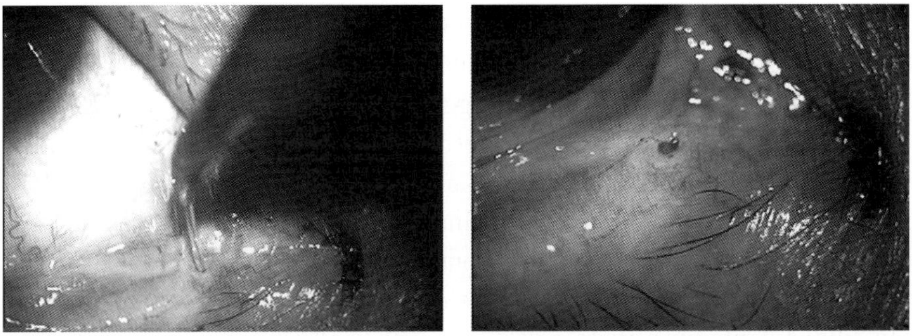

Fig 9.10 (Left panel) Smart Plug™ is inserted with centrally grooved forceps (Medennium Model 502), two-thirds of the Smart Plug™ is inserted into the punctum and one-third is left protruding from the punctum. (Right panel) After insertion, there is no protruding cap left externally in any of the plug insertions.[62]

Researchers are developing lenses that can alter focal length in response to changes in the environment.[63] These 'smart' contact lenses can adapt to their environment without human intervention. Now, scientists are starting to develop ‘adaptive lenses’ made from liquids or soft polymer materials that can change their focal length and response to external stimuli. Electrically tunable-focus liquid crystal lenses have promising applications in eyeglasses, mobile phone cameras, auto beam steering, and other machine visions.[64]

The rate of success in curing eye ailments with ophthalmic drugs mostly depends on achieving sufficient drug concentration on the cornea for an adequate period of time. However more than 90% of the typical delivery of drugs by eye drops is very inefficient, and worse, there is the inconvenience of frequent applications and, in some instances, this leads to serious side effects.[65] A number of researchers have proposed using smart hydrogel contact lenses for ophthalmic drug delivery and hope that they can increase the residence time of the drug in the eye and minimise side effects. It has been acknowledged that ophthalmic drugs have a much longer residence time in the post-lens tear film and a soft contact lens-based ophthalmic drug delivery system also can improve the bioavailability of ophthalmic drugs. The ‘smart’ hydrogels which react to disease-specific environmental triggers and/or chemical signals to effect drug release, are emerging as an ideal polymeric drug carrier with a loading capacity which is able to penetrate into and/or reside at the desired site and then release the active substances in a controlled manner.[66] Nowadays, some polymeric hydrogels have been investigated for soft contact lens-based ophthalmic drug delivery systems: (1) hydrogels for traditional contact lenses to absorb and release drugs[67]; (2) hydrogel use for piggyback contact lenses which integrate with drug plate or solution;[68] (3) modified hydrogels to fix drugs onto the surface of contact lenses;[69] (4) hydrogels for complex drugs that are dispersed in the contact lenses;[70] (5) ion ligand-containing hydrogels;[71] and (6) use of molecularly imprinted hydrogels for contact lens.[72] Moreover, prolonged drug release time can also be achieved when using ophthalmic inserts; solid devices which are placed in the eye, but the inserts must then be removed after usage. These problems can be resolved provided that drug delivery systems are based on biodegradable polymeric materials. Nanosystems with surface-segregated chitosan or poly-ethyleneglycol have been found to be relatively stable and also efficient at overcoming mucosal barriers (Fig. 9.11).[66,73] Drugs combined with biodegradable polymers will be released from the material into the eye in a predesigned manner. It is believed that the release of drugs will be constant over a long period, or triggered by the environment or a chemical signal. When they are no longer needed, the drug delivering polymer can be broken down naturally by the body.[66]

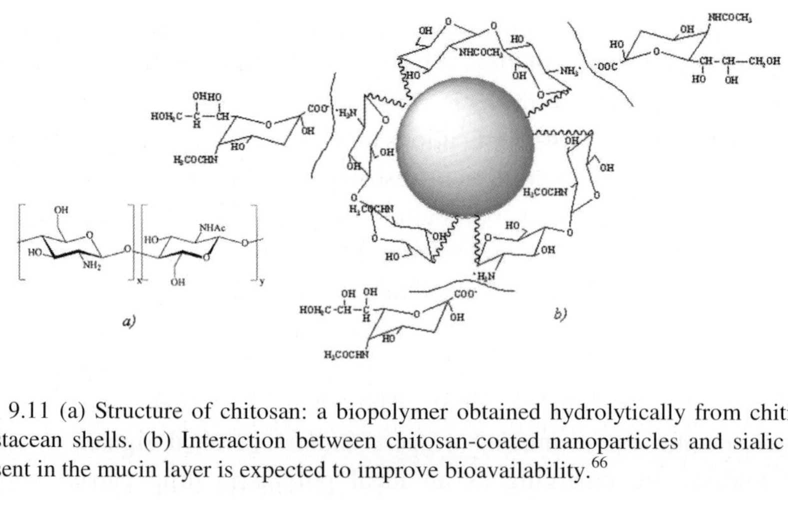

Fig. 9.11 (a) Structure of chitosan: a biopolymer obtained hydrolytically from chitin of crustacean shells. (b) Interaction between chitosan-coated nanoparticles and sialic acid present in the mucin layer is expected to improve bioavailability.[66]

9.3.3 *Adaptive polymeric materials for medical devices*

With the development of medical technologies and materials science, many adaptive polymeric materials are applied in medical devices.[74] Shape memory polymers are a class of adaptive polymers which have dual-shape capabilities under different conditions.[75] When they are exposed to an appropriate stimulus, they can change their shape in a predefined way from shape A to shape B. Shape memory research work started with the thermally induced dual-shape effect. It is important that shape memory polymers possess the ability to memorise a permanent shape, and the permanent shape is substantially different from their initial temporary shape. Large bulky devices which are prepared by shape memory polymers could potentially be introduced into the body in a compressed temporary shape by means of minimally invasive surgery. Then they will expand on demand to a permanent shape to fit as required when stimulated by the body's condition.[76] As a result, a complex mechanical transmutation could be performed automatically instead of manually by the surgeon. Shape memory polymers will initiate the transition from a temporary to permanent shape when there is an external stimulus, such as a temperature increase to the switching transition temperature. Moreover, aside from the dual-shape capability of shape memory polymers, they can be multifunctional, such as being biofunctional or biodegradable. Shape memory polymers constitute a group of high performance smart materials that have recently gained widespread attention. Their potential role in clinical applications has only become recognised in the last 5-6 years.[15] Presently, some SMP applications are already used in the medical field, while others are under development.

An attractive application of shape memory polymers is their utilisation in active medical devices.[15] Hayashi *et al.* investigated SMPs for several medical applications.[77] As a result of these investigations, SMP materials will soon be used to manufacture catheters which will remain stiff externally for accurate manipulation by the physician, but then will become softer and more comfortable inside the human body. Soft catheters will cause fewer arterial wall injuries than rigid catheters, and will be easier to manipulate and conduct in tortuous vessels. In addition, polyurethanes are known to be non-thrombogenic, so these catheters will not activate the coagulation cascade.

Due to their low thrombogenicity, polyurethanes are widely used in the construction of hematology related products and devices.[15] An example is a laser-activated device for the mechanical removal of blood clots (Fig. 9.12).[78] The device can be inserted into the blood vessel by minimally invasive surgery, and then the shape memory material coils into its permanent shape upon laser activation. Finally, the mechanical removal of the thrombus (blood clot) is enabled. Another example is shape memory material utilisation on obesity, which is one of the major health problems in developed countries. In most cases, overeating is the key problem, and it can be circumvented by curbing appetite. Shape memory polymers (SMPs) as biodegradable intragastric implants are put into the patient's stomach. After stimulation by gastric acid, the polymers inflate and provide the patient with a satiety feeling after a small amount of food has been eaten.[79]

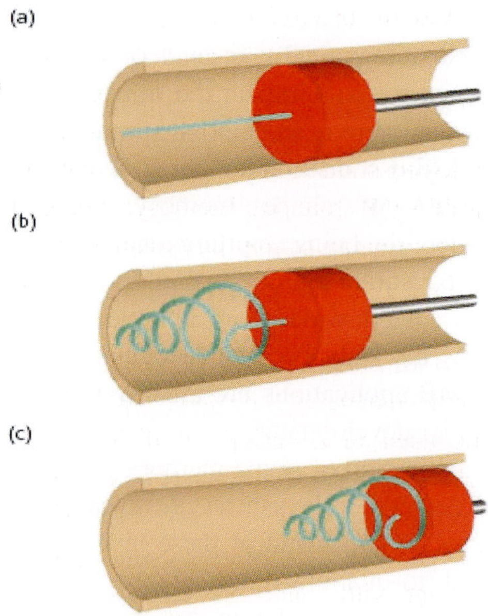

Fig. 9.12 Depiction of removal of a clot in a blood vessel using the laser-activated shape memory polymer microactuator coupled to an optical fibre. (a) In its temporary straight rod form, the microactuator is delivered through a catheter distal to the blood clot. (b) The microactuator is then transformed into its permanent corkscrew form by laser heating. (c) The deployed microactuator is retracted to capture the thrombus. (Reprinted with permission from 40. 2005, The Optical Society of America).[78]

The most recent developments in polyurethane-based SMP foams have widened their potential medical applications.[15] The SMP foams are outstanding, owing to their high shape recovery ratio in compression. For instance, they can be used for micro foldable vehicles, shape determination and microtags. Shape memory foams can be fitted for a hearing aid properly, because it is often proposed that they work as a measuring device to survey the shape of the human ear canal.[80] Polyurethane foam with a glass transition temperature (Tg) switching transition is commercially available. It has good recoverability after 83% compression. In addition, intracranial aneurysm can be a serious condition, because it can not be detected until the aneurysm ruptures, causing hemorrhaging within the subarachnoid space surrounding the brain. The typical treatment for large aneurysms is using platinum coils. However, the aneurysm eventually re-opens as a result of the bio-inertness of platinum in about 15% of the cases treated by platinum coils. Calomeri as a shape memory polymer was investigated as a candidate for aneurysm coils. The experiment demonstrated that typical hemodynamic forces do not hinder the shape recovery process when SMP coils are deployed inside a simulated aneurysm model.[81]

The usefulness of SMPs has also been investigated in wound closure.[76] A design for smart surgical suture has been examined. This suture is applied loosely in its temporary elongated shape which was obtained by elongating the fibre with controlled stress. When the temperature is raised above Tg, the suture shrinks and tightens the knot. Then, these sutures are made into degradable materials with good hydrolytic degradation. SMP materials could be used in a variety of different medical devices and diagnostic products as deployable elements of implants from vascular grafts to components of cardiac pacemakers and artificial hearts. Recently, Wache *et al.* conducted a feasibility study and preliminary developed a polymer vascular stent with shape memory as a drug delivery system.[82] Samples from thermoplastic polyurethane-based SMPs were manufactured by injection molding. The use of the shape memory polymer stent as a drug delivery system leads to significant reduction of restenosis and thrombosis. The manufacturing of SMP stents by injection molding, extrusion or dip-coating technologies makes economical sense for production purposes. Compared to the production of conventional metal stents, the production costs are reduced by more than 50%.[15]

In addition to shape memory polymers, other materials in the medical device field have a wide range of applications because of their behaviour and properties. For example, electro active polymers as a class of advanced materials show many new revolutionary properties. The EAP technology can be applied in medical devices in wide range, such as control of drug release, anchoring of devices and movement and properties of medical devices.[83]

9.4 Summary

The past decades have seen considerable efforts in the use of adaptive polymer materials for medical applications. An overview of these medical applications has been elaborately presented here. Presently, applications that use adaptive polymeric materials are already evident in the medical field, while others are still under development. Most of these applications are still being tested in laboratories worldwide with some at the infancy stage. Adaptive polymeric materials will become an important research area in the 21st century. Significant advancements are necessary before clinical usage or commercialisation can be realised. Listed below are some of the important points that should be taken into account for future research, development and applications on adaptive polymeric materials.

The high performance of adaptive polymers can determine the final medical application of polymer materials. From this review, it is evident that biocompatibility, functionality, nontoxicity, mechanical properties and other properties are very important characteristic factors of adaptive polymeric materials for different purposes. Therefore, it is important to search for new methods to prepare materials and explore the relationship between the structures and properties. A key point in the preparation of materials is to control the structural and compositional evolution for obtaining superior properties.

More attention should be given to modified technologies and response theories so that better biological compatibility and high performance of adaptive polymeric materials can be realised. For effective applications of adaptive polymers, functionalising the polymer surface or developing novel functional adaptive polymers are generally required.

Additional efforts should be made on assistive technologies, such as medical designs, information collection and clinic applications so as to obtain good behaviours and clear results, as well as improve medical applications of the adaptive polymeric materials.

Finally, investigation on ways to integrate the high performances of adaptive polymer materials or adaptive polymeric material systems requires properties of medical materials which are needed to create new medical applications for both medical devices and clinical diagnosis. Effective applications also depend on the level of understanding on the essential properties of previously used adaptive polymeric materials. The majority of the current investigations have made use of the advantages of thermal, electro or optic adaptive polymer materials. Other response features of these adaptive materials in terms of mechanical, physical, chemical and biological properties need further exploration.

In order to advance the medical applications of adaptive polymer materials from a perspective to a commercialised stage, much research which involve surgeons, physicists, chemists, biologists and material scientists are required. It is believed that more funding and efforts from research organisations, government and the industry in this field will shorten the distance between theory and practical utilisation stages. The authors hope that this review will act as a catalyst, stimulating interest and further research in such areas.

References

1. Hutmachera, D. W., Polymers for Medical Applications. *Encyclopedia of Materials: Science and Technology*, 2001: pp. 7664-7673.
2. Jagur-Grodzinski, J., Biomedical application of functional polymers. *Reactive and Functional Polymers*, 1999. 39(2): pp. 99-138.
3. Kalaitzidou, K. and Crosby, A. J., Adaptive polymer particles. *Applied Physics Letters*, 2008. 93(4): p. 041910.

4. Aguilar, M. R., Elvira, C., Gallardo, A., Vázquez, B., and Román, J. S., *Smart Polymers and Their Applications as Biomaterials.* 2007, Topics in Tissue Engineering , 2007. 3: pp.1-27.
5. Galaev, I. Y. and Mattiasson, B., 'Smart'polymers and what they could do in biotechnology and medicine. *Trends in biotechnology*, 1999. 17(8): pp. 335-340.
6. Luzinov, I., Minko, S., and Tsukruk, V. V., Adaptive and responsive surfaces through controlled reorganization of interfacial polymer layers. *Progress in Polymer Science*, 2004. 29(7): pp. 635-698.
7. Campoccia, D., Doherty, P., Radice, M., Brun, P., Abatangelo, G., and Williams, D. F., Semi-synthetic resorbable materials from hyaluronan esterification. *Biomaterials*, 1998. 19(23): pp. 2101-2127.
8. Mrue, F., Netto, J. C., Ceneviva, R., Lachat, J. J., Thomazini, J. A., and Tambelini, H., Evaluation of the biocompatibility of a new biomembrane. *Materials Research*, 2004. 7: pp. 277-283.
9. Roy, I. and Gupta, M. N., Smart polymeric materials: emerging biochemical applications. *Chemistry & Biology*, 2003. 10(12): pp. 1161-1171.
10. O'Rourke-Muisener, P. A. V., Koberstein, J. T., and Kumar, S., Optimal chain architectures for the molecular design of functional polymer surfaces. *Macromolecules*, 2003. 36(3): pp. 771-781.
11. Severian, D., *Polymeric biomaterials.* 1994: Dekker New York etc.
12. Ratner, B. D., Hoffman, A. S., Schoen, F. J., and Lemons, J. E., *Biomaterials science: an introduction to materials in medicine.* 1996: Elsevier Academic Press, ISBN 0-12-582463-7, 2004.
13. Kikuchi, A. and Okano, T., Pulsatile drug release control using hydrogels. *Advanced drug delivery reviews*, 2002. 54(1): pp. 53-77.
14. Holmes, N. S., A review of particle formation events and growth in the atmosphere in the various environments and discussion of mechanistic implications. *Atmospheric Environment*, 2007. 41(10): pp. 2183-2201.
15. Sokolowski, W., Metcalfe, A., Hayashi, S., Yahia, L., and Raymond, J., Medical applications of shape memory polymers. *Biomed Mater*, 2007. 2. (1):S23-27.
16. Shimoboji, T., Larenas, E., Fowler, T., Hoffman, A. S., and Stayton, P. S., Temperature-Induced Switching of Enzyme Activity with Smart Polymer? Enzyme Conjugates. *Bioconjugate Chem*, 2003. 14(3): pp. 517-525.

17. Roy, I. and Gupta, M. N., κ-Carrageenan as a new smart macroaffinity ligand for the purification of pullulanase. *Journal of Chromatography A*, 2003. 998(1-2): pp. 103-108.
18. Boczkowska, A. and Leonowicz, M., Intelligent materials for intelligent textiles. *FIBRES & TEXTILES in Eastern Europe*, 2006. 14(5): pp. 13-17.
19. Furth, M. E., Atala, A., and Van Dyke, M. E., Smart biomaterials design for tissue engineering and regenerative medicine. *Biomaterials*, 2007. 28(34): pp. 5068-5073.
20. Mirfakhrai, T., Madden, J. D. W., and Baughman, R. H., Polymer artificial muscles. *Materials Today*, 2007. 10(4): pp. 30-38.
21. Bar-Cohen, Y., Bionic humans using EAP as artifical muscles-reality and challenges. *ASAIO Journal*, 2004. 50(2): pp. 158-158.
22. Bar-Cohen, Y., Biomimetics using electroactive polymers(EAP) as artificial muscles-: A review. *Journal of advanced materials*, 2006. 38(4): pp. 3-9.
23. Pelrine, R., Kornbluh, R., and Kofod, G., High-strain actuator materials based on dielectric elastomers. *Advanced Materials*, 2000. 12(16): pp. 1223-1225.
24. Paquette, J. W., Kim, K. J., and Kim, D., Low temperature characteristics of ionic polymer–metal composite actuators. *Sensors & Actuators: A. Physical*, 2005. 118(1): pp. 135-143.
25. Tang, Z. P., Chen, X. Y., and Tang, R. H., Current application of scaffold materials for nerve tissue engineering. *Journal of Clinical Rehabilitative Tissue Engineering Research* 2008. 12(1): pp. 189-192.
26. Ye, W. J., Xin J. H., Li, P., Lee, K. D., and Kwong T., Durable antibacterial finish on cotton fabric by using chitosan-based polymeric core-shell particles. *Journal of Applied Polymer Science*, 2006. 102(2): pp. 1787-1793.
27. Hoffman, A. S., Bioconjugates of intelligent polymers and recognition proteins for use in diagnostics and affinity separations. *Clinical Chemistry*, 2000. 46(9): pp. 1478-1486.
28. Stayton, P. S., Shimoboji, T., Long, C., Chilkoti, A., Ghen, G., Harris, J. M., and Hoffman, A. S., Control of protein–ligand recognition using a stimuli-responsive polymer. Nature, 1995. 378: pp. 472-474.
29. Drury, J. L. and Mooney, D. J., Hydrogels for tissue engineering: scaffold design variables and applications. *Biomaterials*, 2003. 24(24): pp. 4337-4351.

30. Mironov, V., Boland, T., Trusk, T., Forgacs, G., and Markwald, R. R., Organ printing: computer-aided jet-based 3D tissue engineering. *Trends in biotechnology*, 2003. 21(4): pp. 157-161.
31. Mironov, V., Visconti, R. P., Kasyanov, V., Forgacs, G., Drake, C. J., and Markwald, R. R., Organ printing: Tissue spheroids as building blocks. *Biomaterials*, 2009. 30(12): pp. 2164-2174.
32. Fedorovich, N. E., Alblas, J., de Wijn, J. R., Hennink, W. E., Verbout, A. J., and Dhert, W. J. A., Hydrogels as extracellular matrices for skeletal tissue engineering: state-of-the-art and novel application in organ printing. *Tissue Engineering*, 2007. 13(8): pp. 1905-1925.
33. Qiu, Y. and Park, K., Environment-sensitive hydrogels for drug delivery. *Advanced drug delivery reviews*, 2001. 53(3): pp. 321-339.
34. Kulkarni, S., Malmstadt, N., Hoffman, A. S., and Stayton, P. S., Micro and Nanoscale Smart Polymer Technologies in Biomedicine. *BioMEMS and biomedical nanotechnology, biomedical & biological nanotechnology*, Springer, 2007. III, pp. 289-304.
35. Brahim, S., Narinesingh, D., and Guiseppi-Elie, A., Bio-smart hydrogels: co-joined molecular recognition and signal transduction in biosensor fabrication and drug delivery. *Biosensors and Bioelectronics*, 2002. 17(11-12): pp. 973-981.
36. Eljarrat-Binstock, E., Orucov, F., Aldouby, Y., Frucht-Pery, J., and Domb, A. J., Charged nanoparticles delivery to the eye using hydrogel iontophoresis. *Journal of Controlled Release*, 2008. 126(2): pp. 156-161.
37. Qu, T., Wang, A., Yuan, J., Shi, J., and Gao, Q., Preparation and characterization of thermo-responsive amphiphilic triblock copolymer and its self-assembled micelle for controlled drug release. *Colloids and Surfaces B: Biointerfaces*, 2009. 72(1): pp. 94-100.
38. Bures, P., Huang, Y., Oral, E., and Peppas, N. A., Surface modifications and molecular imprinting of polymers in medical and pharmaceutical applications. *Journal of Controlled Release*, 2001. 72(1-3): pp. 25-33.
39. Byrne, M. E., Park, K., and Peppas, N. A., Molecular imprinting within hydrogels. *Advanced Drug Delivery Reviews*, 2002. 54 (1): pp.149-161.
40. Hisamitsu, I., Kataoka, K., Okano, T., and Sakurai, Y., Glucose-responsive gel from phenylborate polymer and poly (vinyl alcohol): prompt response at physiological pH through the interaction of

borate with amino group in the gel. *Pharmaceutical research*, 1997. 14(3): pp. 289-293.

41. Kumashiro, Y., Ooya, T., and Yui, N., Dextran hydrogels containing poly (N-isopropylacrylamide) as grafts and crosslinkers exhibiting enzymatic regulation in a specific temperature range. *Macromolecular Rapid Communications*, 2004. 25(8): pp.867-872.
42. Wei, H., Cheng, S. X., Zhang, X. Z., and Zhuo, R. X., Thermo-sensitive polymeric micelles based on poly(N-isopropylacrylamide) as drug carriers. *Progress in Polymer Science*, 2009. In press.
43. Rao, J. Y., Luo, Z. F., Ge, Z. S., Liu, H., and Liu S. Y. "Schizophrenic" micellization associated withcoil-to-helix transitions based on polypeptide hybrid double hydrophilic rod-coil diblock copolymer. *Biomacromolecules,* 2007. 8: pp.3871-3878.
44. Zhang, K. and Wu, X. Y., Modulated insulin permeation across a glucose-sensitive polymeric composite membrane. *Journal of Controlled Release*, 2002. 80(1-3): pp. 169-178.
45. van Nostrum, C. F., Molecular imprinting: A new tool for drug innovation. *Drug Discovery Today: Technologies*, 2005. 2(1): pp. 119-124.
46. Kumar, A. P., Depan, D., Singh Tomer, N., and Singh, R. P., Nanoscale particles for polymer degradation and stabilization—Trends and future perspectives. *Progress in Polymer Science*, 2009. 34(6): pp. 479-515.
47. Kairemo, K., Erba, P., Bergst, K., and Pauwels, E. K. J., Nanoparticles in Cancer. *Current Radiopharmaceuticals*, 2008. 1: pp. 30-36.
48. Trubetskoy, V. S., Polymeric micelles as carriers of diagnostic agents. *Advanced drug delivery reviews*, 1999. 37(1-3): pp. 81-88.
49. Shimoboji, T., Ding, Z., Stayton, P. S., and Hoffman, A. S., Mechanistic Investigation of Smart Polymer? Protein Conjugates. *Bioconjugate Chem*, 2001. 12(2): pp. 314-319.
50. Hoffman, A. S., Stayton, P. S., Bulmus, V., Chen, G., Chen, J., Cheung, C., Chilkoti, A., Ding, Z., Dong, L., Fong, R., Lackey, C. A., Long, C. J., Miura, M., Morris, J. E., Murthy, N., and et al., Really smart bioconjugates of smart polymers and receptor proteins. *Journal of Biomedical Materials Research*, 2000. 52(4): pp. 577-586.
51. Kulkarni, S., Schilli, C., Muller, A. H. E., Hoffman, A. S., and Stayton, P. S., Reversible Meso-Scale Smart Polymer? Protein

Particles of Controlled Sizes. *Bioconjugate Chem*, 2004. 15(4): pp. 747-753.

52. Malmstadt, N., Hoffman, A. S., and Stayton, P. S., "Smart" mobile affinity matrix for microfluidic immunoassays. *Lab on a Chip*, 2004. 4(4): pp. 412-415.
53. Holtz, J. H. and Asher, S. A., Polymerized colloidal crystal hydrogel films as intelligent chemical sensing materials. *Nature*, 1997. 389: pp. 829-832.
54. Alexeev, V. L., Das, S., Finegold, D. N., and Asher, S. A., Photonic crystal glucose-sensing material for noninvasive monitoring of glucose in tear fluid. *Clinical Chemistry*, 2004. 50(12): pp. 2353-2360.
55. Yang, Z. M., Gu, H. W., Fu, D. G., Gao, P., Lam, K. J. K., and Xu, B., Enzymatic formation of supramolecular hydrogels, *Advanced Materials*, 2006. 16(16): pp. 1440-1444.
56. Wallace, G. G., Smyth, M., and Zhao, H., Conducting electroactive polymer-based biosensors. *Trends in Analytical Chemistry*, 1999. 18(4): pp. 245-251.
57. Gerard, M., Chaubey, A., and Malhotra, B. D., Application of conducting polymers to biosensors. *Biosensors and Bioelectronics*, 2002. 17(5): pp. 345-359.
58. Xue, H., Shen, Z., and Li, Y., Polyaniline–polyisoprene composite film based glucose biosensor with high permselectivity. *Synthetic Metals*, 2001. 124(2-3): pp. 345-349.
59. Gambhir, A., Gerard, M., Jain, S. K., and Malhotra, B. D., Characterization of DNA immobilized on electrochemically prepared conducting polypyrrole-polyvinyl sulfonate films. *Applied Biochemistry and Biotechnology*, 2001. 96(1): pp. 313-319.
60. Kumar, A., Chaubey, A., Grover, S. K., and Malhotra, B. D., Immobilization of cholesterol oxidase and potassium ferricyanide on dodecylbenzene sulfonate ion-doped polypyrrole film. *Journal of Applied Polymer Science*, 2001. 82(14): pp. 3486-3491.
61. Tai, M. C., Cosar, C. B., Cohen, E. J., Rapuano, C. J., and Laibson, P. R., The Clinical Efficacy of Silicone Punctal Plug Therapy. *Cornea*, 2002. 21(2): pp. 135-139.
62. Kojima, T., Dogru, M., Ishida, R., Goto, E., Matsumoto, Y., and Tsubota, K., Clinical Evaluation of the Smart Plug? in the Treatment of Dry Eyes. *American journal of ophthalmology*, 2006. 141(2): pp. 386-388.

63. Ren, H. and Wu, S. T., Tunable electronic lens using a gradient polymer network liquid crystal. *Applied Physics Letters*, 2003. 82: pp. 22-24.
64. Ren, H. and Wu, S. T., Adaptive liquid crystal lens with large focal length tunability. *Optics Express*, 2006. 14(23): pp. 11292-11298.
65. Xinming, L., Yingde, C., Lloyd, A. W., Mikhalovsky, S. V., Sandeman, S. R., Howel, C. A., and Liewen, L., Polymeric hydrogels for novel contact lens-based ophthalmic drug delivery systems: A review. *Contact Lens and Anterior Eye*, 2008. 31(2): pp. 57-64.
66. Barbu, E., Verestiuc, L., Nevell, T.G.., and Tsibouklis, J., Polymeric materials for ophthalmic drug delivery:trends and perspectives. *Journal of* Materials *Chemistry*, 2006.16: pp.3439-3443.
67. Peterson, R. C., Wolffsohn, J. S., Nick, J., Winterton, L., and Lally, J., Clinical performance of daily disposable soft contact lenses using sustained release technology. *Contact Lens and Anterior Eye*, 2006. 29(3): pp. 127-134.
68. Le Bourlais, C., Acar, L., Zia, H., Sado, P. A., Needham, T., and Leverge, R., Ophthalmic drug delivery systems-recent advances. *Progress in retinal and eye research*, 1998. 17(1): pp. 33-58.
69. Danion, A., Brochu, H., Martin, Y., and Vermette, P., Fabrication and characterization of contact lenses bearing surface-immobilized layers of intact liposomes. *Journal of Biomedical Materials Research Part A*, 2007. 82A(1): pp. 41-51.
70. Gulsen, D., Li, C. C., and Chauhan, A., Dispersion of DMPC liposomes in contact lenses for ophthalmic drug delivery. *Current Eye Research*, 2005. 30(12): pp. 1071-1080.
71. Sato, T., Uchida, R., Tanigawa, H., Uno, K., and Murakami, A., Application of polymer gels containing side-chain phosphate groups to drug-delivery contact lenses. *Journal of Applied Polymer Science*, 2005. 98(2): pp. 731-735.
72. Alvarez-Lorenzo, C., Hiratani, H., and Concheiro, A., Contact Lenses for Drug Delivery: Achieving Sustained Release with Novel Systems. *American Journal of Drug Delivery*, 2006. 4(3): pp. 131-151.
73. José Alonso, M., Nanomedicines for overcoming biological barriers. *Biomedicine & Pharmacotherapy*, 2004. 58(3): pp. 168-172.
74. El Feninat, F., Laroche, G., Fiset, M., and Mantovani, D., Shape memory materials for biomedical applications. *Advanced Engineering Materials*, 2002. 4(3): pp. 91-104.

75. Behl, M. and Lendlein, A., Shape-memory polymers. *Materials Today-Kidlington*, 2007. 10(4): pp. 20-29.
76. Lendlein, A. and Kelch, S., Shape-memory polymers. *Angewandte Chemie International Edition*, 2002. 41(12): pp. 2034-2057.
77. Hayashi, S., Properties and applications of polyurethane series shape memory polymer Int. *Prog. Urethanes*, 1993. 6: pp. 90–115.
78. Small Iv, W., Wilson, T., Benett, W., Loge, J., and Maitland, D., Laser-activated shape memory polymer intravascular thrombectomy device. *Optics Express*, 2005. 13(20): pp. 8204-8213.
79. Marco, D., *Biodegradable self-inflating intragastric implants for curbing appetite*. 2006, WO Patent 2006092789A2.
80. Huang, W. M., Lee, C. W., and Teo, H. P., Thermomechanical behavior of a polyurethane shape memory polymer foam. *Journal of Intelligent Material Systems and Structures*, 2006. 17(8): pp. 753-760.
81. Hampikian, J. M., Heaton, B. C., Tong, F. C., Zhang, Z., and Wong, C. P., Mechanical and radiographic properties of a shape memory polymer composite for intracranial aneurysm coils. *Materials science & engineering C. Biomimetic and supramolecular systems*, 2006. 26(8): pp. 1373-1379.
82. Wache, H. M., Tartakowska, D. J., Hentrich, A., and Wagner, M. H., Development of a polymer stent with shape memory effect as a drug delivery system. *Journal of Materials Science: Materials in Medicine*, 2003. 14(2): pp. 109-112.
83. Snyder, A. J., Tews, A. M., Frecker, M. I., Zhang, Q., and Runt, J. P., Biomedical Applications of Electroactive Polymers 17 Jun 2003 6 pages. *Biomimetics*, Spring 2009. ADA429656.

Chapter 10

Special Adaptive and Functional Polymers and Their Applications

10.1 Structure, Synthesis, Properties and Applications of Dendrimer Macromolecules

10.1.1 *Introduction*

Dendritic macromolecules (dendrimers and hyperbranched macromolecules) have become popular due to their unique structures. They distinguish themselves from the conventional random coil of linear polymers by a high degree of branching and high density of reactive terminal functional groups. The first dendrimer was presented by Buhleier *et al.*[1] in 1978 with an initial name called 'cascade' molecules. After seven years, several important papers about dendritic macromolecules were published.[2,3] Since then, the interest in dendrimer research has grown rapidly.

Dendrimers are well-defined hyperbranched macromolecules that emanate from a single core and ramify outward with each subsequent branching unit which has many internal voids and channels. A typical dendrimer is composed of three components as shown in Fig. 10.1: (1) a multifunctional central core (C), (2) layers of repeating branched units (generations) (B), and (3) a high density of surface groups also known as terminal groups (S). The macromolecule size can be precisely controlled because their synthesis process is a repetitive sequence of steps.

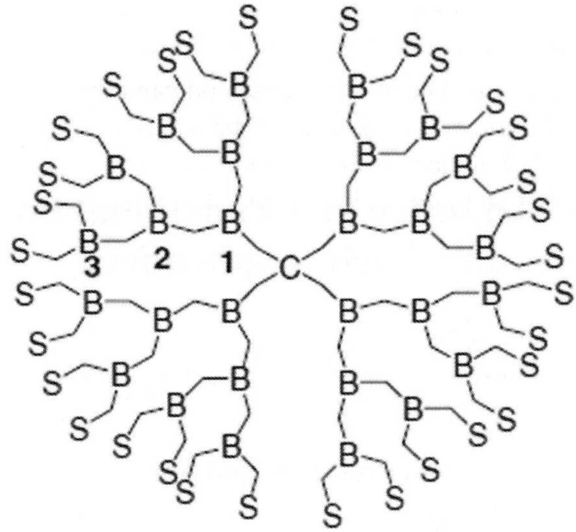

Fig. 10.1 Typical structure of a 3 generation dendrimer.[4]

10.1.2 *Synthesis of dendrimer macromolecules*

The structure of dendrimer molecules imposes strict requirements on the synthesis process. Chemicals must be reacted under optimal conditions and the reactive product has to be carefully purified, otherwise dendrimers with perfect branches and layers cannot be obtained.

Dendrimers can be synthesised by fully controlled approaches, such as the following growth methods: divergent, convergent, double-exponential, mixed-growth and other accelerated techniques. The selection of a synthesis route is determined by the kind of monomer used and molecular structures desired.

10.1.2.1 *Divergent synthesis*

In the divergent methods, a dendrimer grows outwards from the multifunctional core molecule layer by layer as shown in Fig. 10.2. The first generation is generated by the core molecule which is reacting with monomers that contain one reactive group and two dormant groups. Then, the periphery of the molecules is activated for reactions with more monomers. To obtain perfect dendrimers, the reaction process has to be fully controlled. This process is repeated until the required generations of dendrimer is obtained. The divergent method can create dendrimers with large sizes and generations. Recently, scientists have also synthesised heterogeneously functionalised dendrimers with several types of functionalities bound to the surface by the divergent method.

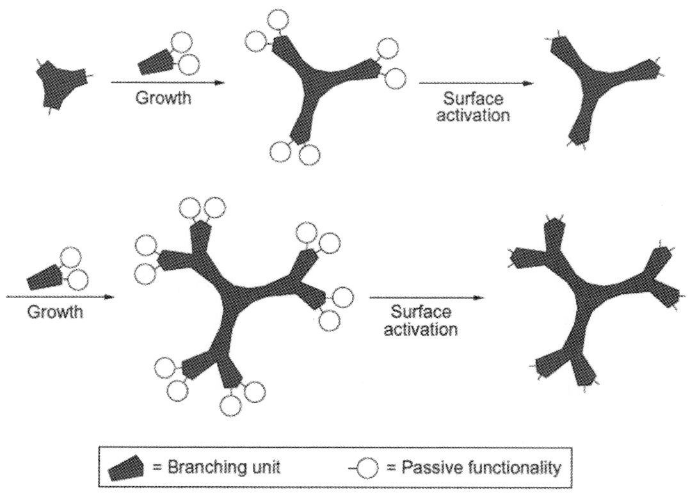

Fig. 10.2. Schematic representation of a dendrimer synthesis by the divergent approach.[5,6]

De *et al.*[7,8] reported the divergent synthesis process of a poly(propylene imine) dendrimer. This dendrimer is based on a skeleton of poly alkylamines with each nitrogen atom as a branching point. Repeated double alkylations of the amines with acrylonitrile by the Michael addition occur, which result in a branched alkyl chain structure. A new set of primary amines are produced by subsequent reduction. The amines can then be alkylated to provide further branching.

Although the divergent approach has been applied to synthesise various dendrimer molecules with great success, the divergent synthesis has problems with increases in dendrimer generation. This is because, firstly, the number of reaction points increase exponentially throughout the synthesis of the dendrimer. The rapid increase in the number of terminal groups together with the rapid increase of dendrimer molecular weight (MW) causes slow reaction kinetics. This makes it difficult to obtain dendrimers with high generation. In addition, side and incomplete reactions of terminal groups lead to numerous structure defects in high generation dendrimers, secondly, the purification of the desired product from reactants or 'deletion products' is hard due to the similarity in mass, size and properties of desired products and by products.

10.1.2.2 *Convergent synthesis*

The convergent synthesis method for dendrimers was first described in 1990 to prepare poly-benzylether that contained dendrimers.[9-11] The schema for the convergent growth route is presented in Fig. 10.3. The convergent method is different from the divergent synthesis method, as it proceeds inwards from the surface by gradually coupling monomers to form dendritic wedges (dendrons). When the dendrons are large enough, several surface units join together and end up at the core to give a complete dendrimer. In comparison with the divergent method, the convergent method is more powerful in producing regular and precise dendrimer macromolecules. In addition, the differences between any byproducts and the desired larger sized product are quite substantial, so that it is easier to purify the final product.

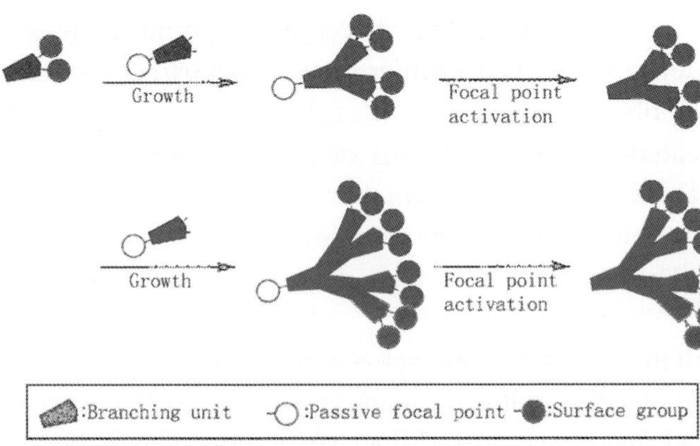

Fig. 10.3 Dendrimer synthesis by the convergent approach.[5]

The convergent method is highly flexible for synthesising asymmetric dendrimers or dendrimers with mixed structural elements while not completely coupling equal dendrons. The terminal groups, interior blocks and the focal point groups can be changed.[12] Different segments may be coupled together to create dendrimers with a heterogeneous morphology.[13,14] The problem of the convergent method is that the reactive focal group may be buried at the focal point of the dendrons, and as a result, the steric crowding decreases the reactivity of the focal group.

10.1.2.3 *Other approaches with improved efficiency*

Some other methods for improving the efficiency of activation or protection of monomers, condensation reactions, and purification by chromatographic separations, are also used, such as double-stage convergent growth, hypercore or branched monomer, double-exponential dendrimer growth and orthogonal coupling approaches.

In the double-stage convergent growth approach, the focal points of dendrons are attached in a divergent manner to the periphery of a dendron or dendrimer which can be prepared by convergent or divergent growth.[15] Both core dendritic molecules and dendrons are of lower generations. In this way, for example, the fourth generation dendritic aliphatic polyester starting with 2,2-bis(hydroxymethyl)propionic acid (1) can be synthesised with N,N'-dicyclohexylcarbodiimide (DCC) as a coupling agent in six steps only, involving two purifications by column chromatographic separations. If this dendritic macromolecule is prepared by the conventional divergent approach, a number of purification steps corresponding to the generation is needed.

Highly functionalised 'branched monomers' which are pre-branched analogs of the cores and monomers can be used in the convergent method to accelerate synthesis of dendrimer, which is called a hypercore (large dendritic core molecules) or branched monomer approach synthesis method.[15,16] The pre-branched oligomers then react together to give dendrimers in fewer steps and higher yields.

In the double exponential synthesis method, the number of repeat units on every dendrimer molecule increases following a double exponential function in terms of generation. The schematic representation of the double exponential growth of a dendrimer is presented in Fig. 10.4. This synthesis process employs a single trifunctional monomer of the type $Ap(Bp)_2$, which has orthogonally protected functional groups. The first step of the repetitive synthesis is selective removal of the protecting groups on Ap in one portion, and selective removal of the protecting groups on Bp in the second portion. The second step of the repetitive synthesis is the coupling of the two monoprotected intermediates in a proper stoichiometric ratio. The new monodendron maintains a single Ap group at the focal point, but contains X^2 Bp peripheral groups where x is the number of peripheral groups in the preceding generation. The dendrimers of higher generations can be obtained by repeated selective deprotection and coupling processes.

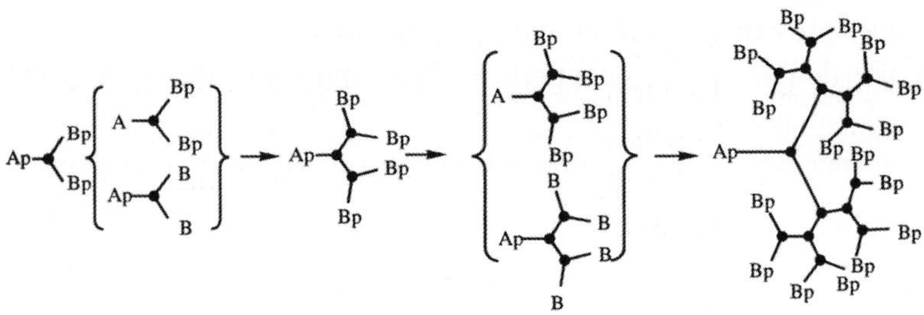

Fig. 10.4 The double exponential growth of a dendrimer.[17,18]

Two monomer approaches do not need laborious protection and activation processes. These approaches require the use of two AB_2 monomers, for example, AB_2 and CD_2. One of the two different functional groups of AB_2 react selectively with one of the two different functional groups of CD_2, such as A reacts with D and C reacts with B respectively. An example is a poly(ether uretane) dendrimer synthesised from 3,5-diisocyanatobenzyl chloride and 3.5-dihydroxybenzyl alcohol that can react pairwise in a one-pot sequential addition procedure.[19]

A less time consuming synthesis approach is based on an orthogonal coupling strategy. This orthogonal coupling reaction also eliminates protection and activation steps. Supermolecular self-assembling is another way to construct dendrimer macromolecules. Dendrons may contain hydrogen bonding or metal complex bonding which can create well-defined complexes with dendrimeric structures by self-assembling. Zimmerman's group built dendrimers by a self-assembly of complementary blocks.[20] Freshet dendrons that contained bis-isophtalic acid are applied. The dendrons in chloroform spontaneously forms hexameric aggregates through carboxylic acid-carboxylic acid hydrogen bonding. The stability of the hexamer is affected by many factors, such as solvent concentration, polarity and temperature. In a diluted solution, polar solvents, such as tetrahydrofuran, and dimethyl sulfoxid, and high temperature, the aggregate tends to break up into monomers.

10.1.3 *Properties and applications*

Dendirmers have many distinctive properties that are different from other macromolecules. Dendrimers have extremely low molecule weight dispersion. The dimension and M_W of dendrimers can be exactly controlled during the synthesis process. Dendrimer solubility is controllable by terminal groups. Hydrophilic groups terminated dendrimers can be solved in polar solvents, and hydrophobic groups terminated dendrimers can be solved in nonpolar solvents. In a dendrimer solution, dendrimer molecules form tightly packed balls, which result in low viscosity of the solution. In addition, with increasing dendrimer generations, the viscosity of the dendrimer solution reaches a peak value and then decreases. This property is completely different with that of linear macromolecule solutions, whose viscosity increase with increasing molecular weight. Due to the special properties of dendrimers, they have been widely used.

10.1.3.1 *Medical applications*

One of the pioneering applications of dendrimer macromolecules is biomedical application, which includes drug and gene delivery, magnetic resonance imaging (MRI) for diagnostics, antiviral and antibacterial materials and as drugs.

The special structures of dendrimers, which include high density functional groups on the surface and voids created by branches, make them good candidates for drug delivery. Drugs may be complexed in the interior voids created by branches, or covalently coupled on the dendrimer surface.[6,21-25]

Three driving forces are used for the encapsulation of guest molecules to the interior of dendrimers, which include physical encapsulation, multiple noncovalent chemical interaction and hydrophobic interaction.

10.1.3.2. *Nanoscale catalyst*

The high surface and high solubility of dendrimers with nanoscopic dimensions contribute to their good catalyst effect which are well documented in numerous papers.[26-29] Homogenous catalysts have good accessibility of active sites, but are often difficult to separate. Heterogeneous catalysts are easy to separate, but the kinetics of the reaction is constrained by low mass transport speed. Dendrimers as catalysts have both the advantages of homogenous and heterogeneous catalysts. The size and solubility of dendrimers or metallodendrimers can be easily changed and the catalytic sites at the core or periphery can be located. The soluble dendrimers may be removed from the reaction by ultrafiltration or dialysis.[30,31]

10.1.3.3 *Dendritic sensors*

Dendritic sensors can be used to detect inorganic or organic species. A poly(propylene amine) dendrimer periphery with 32 dansyl units is determined by Balzani *et al.*[32] If a co-ion is incorporated into the dendrimer, the fluorescence of all the dansyl units is quenched. Thus even at a low concentration, the co-ions can be detected because the dendrimer increases the sensitivity of the sensor. Valerio *et al.*[33] attached ferrocene groups to the periphery of dendrimers with different generations to sense small inorganic anions. The dendrimer sensors can sense and detect inorganic anions because anions change the redox potential of the ferrocene unit. Dendrimer sensors can also sense organic molecules. James *et al.*[34] used a dendrimer with eight boronic acids and eight anthracene groups to detect the binding of saccharide. The binding effect can be monitored by changing the fluorescence of the anthracene units.

10.1.3.4. *Energy 'funnels'*

Plants have photosynthesis with carotenoids that absorb and transfer solar energy to chlorophyll molecules. Chemists have investigated dendrimers that act with similar functions, which are called energy 'funnels'. A conjugated dendrimer with perylene chromophores as the core is fabricated.[35] The dendrimer branches have the function of absorbing and transferring the energy of light to the perylene core. The energy harvesting ability can be increased by increasing dendrimer generations. Poly(aryl ether) dendritic branches are also reported to be used to absorb infrared radiation and transfer the energy to the core to induce the photoisomerisation of a central azobenzene unit. The dendritic branches in dendrimers generate high steric bulk. They are proposed to be used to isolate chemical species from their environment.[36] In addition, dendrimer receptors have been developed, which are called dendrophanes. These dendrimers have specific recognition of guest molecules.[37]

10.1.4 ***Applications in textiles***

10.1.4.1 *Dye extraction*

Poly(propylene imine) dendrimer can be modified with polar end groups of fatty acid amides to produce an effective extractant of anionic dye molecules from water to solvent, such as toluene as shown in Fig. 10.5. The interior of the poly(propylene imine) dendrimer are tertiary amine groups which have high interaction with acid dyes at a sufficiently low pH. At a sufficiently low pH, the tertiary amine groups give an acid–base interaction with acid dyes. The dendrimer releases the absorbed dyes at a high pH value. The modified dendrimer can be used for the selective purification of anionic compounds.

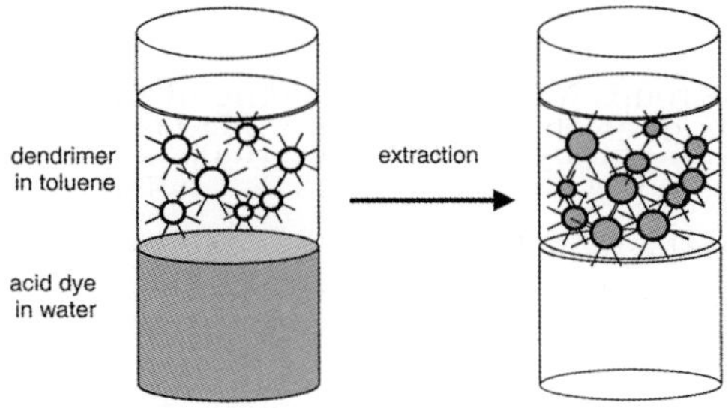

Fig. 10.5 Schematic extraction of an acid dye from water by a modified poly(propylene imine) dendrimer in toluene.[38,39]

The same mechanism can be used to extract polar ionic dye methyl orange from water into CO_2 using fluorinated dendrimers. The modified dendrimer is obtained by functionalising a four-generation hydrophilic dendrimer poly(propylene imine) dendrimers (DAB)-dendr-(NH2)32[40] with a 'CO_2-philic' shell, which is derived from heptamer acid fluoride of hexfluoropropylene oxide. The macromolecules are proposed to be used for remediating contaminated water, extracting pharmaceutical products, transporting reagents for reaction and targeted drug delivery.

10.1.4.2 *Dyeing of PP fibres*

The same method is used for the dyeing of hydrophobic fibres; poly(propylene) (PP) fibres. PP fibres are difficult to be dyed because they are hydrophobic. Alkylated dendrimer, polypropylene imine dendrimer and PP chips are blended before spinning.[41,42] A small amount, such as 4% of dendrimer, has limited influence on the processing and properties of PP fibres while it is efficient in improving the dyeing effects of PP fibres. In an appropriate dye solution, the dye is extracted to the fibre incorporated with alkylated dendrimer.[38]

Another strategy to use dendrimers for dyeing is by precipitating a polar shell from a dye solution for materials which are not compatible with polar dye molecules. A part of the dye can be caught by dendrimers.

10.1.4.3 *Dyeing of cotton*

Reactive dyes of cotton rely on a high pH, which is commonly over 10.5, and large amounts of electrolytes (NaCl or Na_2SO_4) to obtain satisfactory dyeing results. Large quantities of dyes are not fixed onto the cotton. Dendrimers with primary amino terminal groups can be reactive with dyes under neutral/acidic pH conditions to improve the dyeing properties. A dendrimer deriving from Am_{16}decanamide$_8$ (Astramol ™, DSM New Business Development) was used on cotton fabrics.[43] The dyeing was conducted by a competitive method. The dendrimer treated samples and untreated samples were dyed competitively in the same bath. It was found that dendrimer-treated cotton fabrics enhance colour strength with reactive dyes, even when dyeing has been conducted without both electrolyte and alkali. The dendrimer pre-treatment can save electrolytes and alkalis.

10.1.4.4 *Multi-functional finishing of textiles*

Dendrimers with different functional terminal groups can be employed to finish cotton or wool textiles to obtain multifunction properties.[44] The inorganic functional chemicals can react with dendirmer terminal groups through the sol-gel method. After the finishing process, a special inorganic/organic hybrid network comprising of nano dendirmers and functional inorganic chemicals are formed on the textile substrate. This special structure can bring textiles with functions of anti-UV irradiation, antibacteria, etc. The dendrimers with -COOH, -NH, etc. terminal groups may bring hydrophilic, antibacterial and antistatic properties to finished textiles simultaneously. The dendrimers with alkyl terminal groups provide textiles that are waterproof, have good hand feeling and soil repellency. The amphiphilic dendrimers and dendimers with siloxane groups bring good hand feeling to the textiles. Furthermore, because of the dendrimer nanostructure, functions of nanomaterial finishing are imposed onto textiles, which are time endurance and high washability while minimally influencing textile physical properties. In addition, the internal voids of dendrimer molecules and channels in the nanosphere offer a way to embed drugs or perfumes for controllable release in finished textiles.

10.1.5 *Summary*

Although dendrimers have been introduced for several decades, reliable and practical methods for dendrimer synthesis have been developed, and a great amount of applications for dendrimers have been proposed, especially in medicine, so far, their applications in the textile field are just in the initial stages. The high density terminal groups on dendrimers can be modified to prepare finishing agents with special functions. With the development of modern textiles with multifunction properties, we believe that dendrimers could play an important role in fulfilling the possibility of real multifunctional textiles.

10.2 Structure, Synthesis, Properties and Applications of Hyperbranched Macromolecules

10.2.1 *Introduction*

Although various methods have been used to reduce the tedious isolation and purification procedures during dendrimer or dendron synthesis procedures, the laborious and time consuming stepwise synthetic schemes limit the broad applications of dendrimers. Hyperbranched macromolecules can be synthesised in a single step by self-polymersation of AB_x multi-unctional monomers. They are easily synthesised on a large scale and sometimes can be regarded as alternatives of dendrimers.

Figure 10.6 shows the schematic representation of a hyperbranched macromolecule formed by an AB_2 monomer reaction in an uncontrolled growth manner. As shown in Fig. 10. 6, linear and dendritic subunits can be found in hyperbranched structures. Three different subunits are typically found in hyperbranched structures. Reaction of both B functionalities results leads to the dendritic unit. Reaction of only one B functionality leads to a linear unit; and if neither of the B functionalities reacts, this forms a terminal unit. Therefore, due to the incomplete reactions, a hyperbranched macromolecule can be considered as an intermediate between a perfect dendrimer and a linear macromolecule. Even though the overall composition of hyperbranched macromolecules and dendrimers is similar, the degree of branching is quite different.[17,45-47] Due to the relative ease of synthesis of hyperbranched macromolecules in comparison with dendrimers, hyperbranched macromolecules have greater application priority.

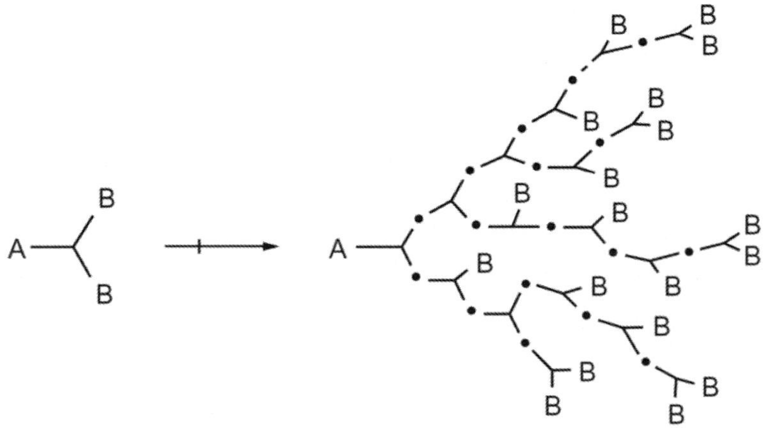

Fig. 10.6 The schematic representation of a hyperbranched macromolecule formed by AB_2 monomer reaction in an uncontrolled growth manner.[47]

10.2.2 *Synthesis of hyperbranched macromolecules*

The synthesis methods of hyperbranched macromolecules fall into seven main strategies: (1) step-growth polycondensation of monomer AB_x, (2) self-condensing vinyl polymerisation of AB_x monomers, (3) self-condensing ring-opening polymerisation of latent AB_x monomers, (4) proton-transfer polymersation of AB_x monomer, (5) polymersation of two types of monomers, (6) 'A_2 + B_3' methodology and (7) couple-monomer methodology.

10.2.2.1 *Step-growth polycondensation of monomer AB_x*

Many types of hyperbranched macromolecules are synthesised by the step-growth polycondensation of AB_x type monomers since it was first reported in 1990.[48,49] For the step-growth polycondensation of monomer AB_x of hyperbranched macromolecules, the prerequisites are: $x \geq 2$ and the functionality A reacts solely with the functionality B of another molecule. AB_x monomers must be synthesised prior to polymersation. This step-growth polycondensation obeys the normal step-growth polymersation characteristics. Some problems of step-growth polycondensation of monomer AB_x may occur during the synthesis process. First, gelation is liable to happen, which causes difficulty of purifying desired products. Second, side reactions such as crosslinking may occur during the synthesis process. Now, a broad range of hyperbranched polymers have been prepared by the step-growth polycondensation route, such as hyperbranched polyphenylenes, polyethers, polyesters, polyamides, polycarbonates and poly(ether ketone)s. Through polyadditions of the AB_x monomers that contain double or triple bonds, some hyperbranched macromolecules are also prepared, such as hyperbranched polyurethanes,[50-52] polycarbosilanes,[53,54] polyamides [55] and poly(acetophenone)s.[56,57] Polyphenylene is the first prepared hyperbranched macromolecule through this method.[58,59]

10.2.2.2 *Self-condensing vinyl polymersation*

The polymersation of AB* of functional vinyl monomers is significantly different from the step-growth polymersation of AB_2 monomers. The condensation reaction of AB_2-monomers directly generates hyperbranched macromolecules because the reactivity of the end groups is the same. For self-condensing vinyl polymersation, monomers need to have one vinyl group and one initiating moiety in the AB^* radical, cation or even a carbanion as the activated species. First, the B groups of the AB monomers are activated to produce the initiating B^* sites. B^* then initiates the propagation of the vinyl group A, which can form a dimer with a vinyl group, a growth site and an initiating site. The dimers function as an AB_2 monomer, and undergo further polymersation to generate hyperbranched macromolecules. In the process, controlled polymersation is preferred[60,61] in order to avoid gelation and crosslinking which are induced by dimerisation and chain transfer reactions.

The first hyperbranched macromolecule synthesised by this strategy is reported by Frechet *et al.*[62] The monomer is a 3-(1-chloroethyl)-ethenylbenzene monomer with both the polymerisable vinyl group of styrene and the latent initiating moiety of 1-hloroethylbenzene under cationic conditions. Figure 10.7 shows the preparation of the hyperbranched polystyrene by cationic polymersation in the presence of $SnCl_4$ and tetrabutylammonium bromide. The synthesis is conducted in dry dichloromethane at temperatures between -15 and -20°C. The approximately equimolar $SnCl_4$ to the AB* monomer results in the formation of hyperbranched macromolecules.

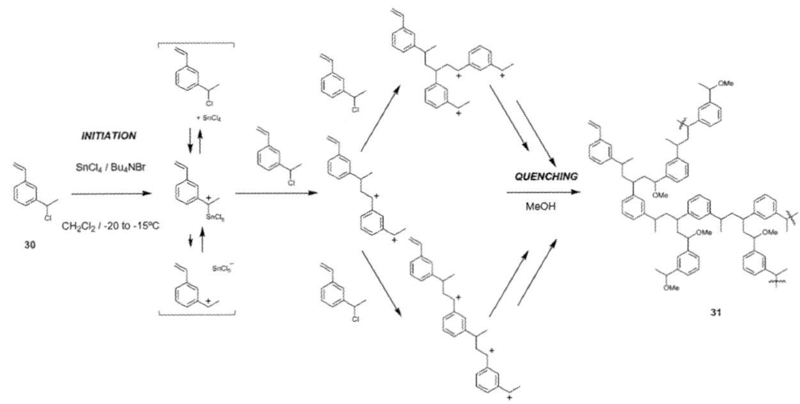

Fig. 10.7 Preparation of a hyperbranched polystyrene by self-condensing vinyl polymersation.[62]

In a similar fashion, through atom transfer radical polymersation (ATRP), hyperbranched poly(acrylatc)s and poly(methacrylate)s have been synthesised from monomers that feature methacrylate or acrylate functionalities in the presence of Cu (I) and a ligand.[63-65] The bulk polymersation of 2-(2-Bromopropionyloxy) ethyl acrylate is conducted in the presence of copper (I) bromide, copper (II) bromide and 4,4'-di-tert-butyl-2,2'-bipyridine at 100°C. For the synthesis of hyperbranched macromolecules, ATRP has the disadvantage of coupling or crosslinking when the macromolecule concentration is high. For this reason, chain transfer agents are used to inhibit the crosslinking reactions for the synthesis of hyperbranched methacrylate macromolecules.[66,67]

10.2.2.3 *Ring-opening multibranching polymerisation of latent AB_x monomers*

A third method for preparing hyperbranched macromolecules is the ring-opening multibranching polymersation method. In conception, when comparing ring-opening polymerisation with conventional step-growth polymerisation, it has the advantage where no low M_W compound needs to be removed. It benefits the formation of high M_W macromolecules.[49,68,69] The number of propagating chain ends increase with the polymerisation process because the reaction produces branch points. The reaction is promoted by proper initiators into the latent molecules. Hyperbranched macromolecules from cyclic carbamates, epoxide, oxctancs and caprolacroncs are reported.

Hyperbranched polyamines consisting of primary, secondary and tertiary amino moieties are synthesised by palladium-catalysed ring opening polymerisation.[68,69] An AB* monomer, 5,5-dimethyl-6-ethenylperhydro-1,3-oxazin-2-one, with bis-(dibenzylideneacetone)-palladium (Pd(dba)2) are used as the catalyst and bis (diphenylphosphino)ethane (dppe) as the ligand. Primary or secondary amines are employed as the initiating agents. The hyperbranched polyamine macromolecule growth by ring-opening multi-branching polymerisation is presented in Fig. 10.8. The primary amine initiating species attack the electrophilic site of the π-allyl palladium complex to produce a diamine that features primary and secondary amines, and then, both of these residues attack the π-allyl palladium complex to free two types of triamine, which are branched units and linear components.

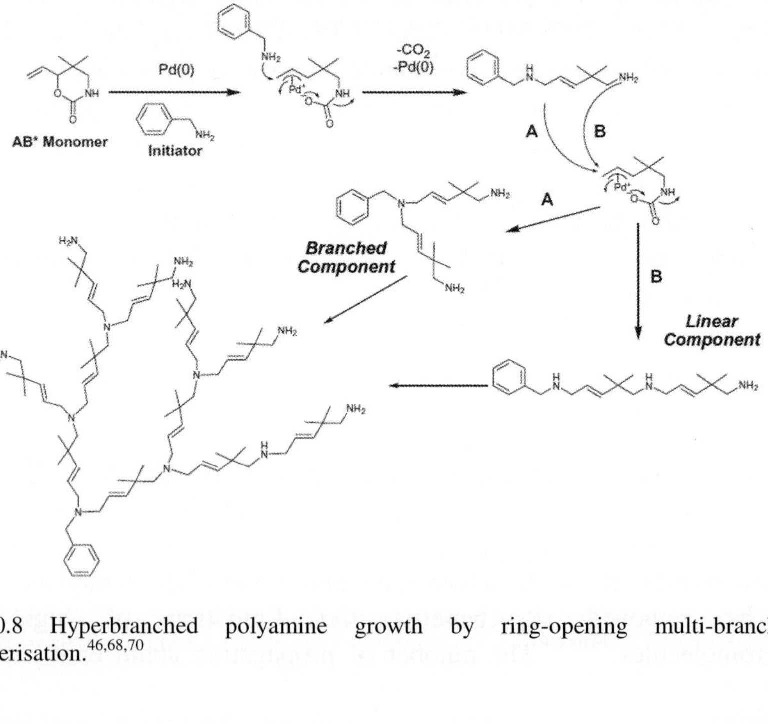

Fig. 10.8 Hyperbranched polyamine growth by ring-opening multi-branching polymerisation.[46,68,70]

10.2.2.4 *Other strategies for the synthesis of hyperbranched macromolecules*

10.2.2.4.1 The A_2 +B_3 route to hyperbranched macromolecules

Hyperbranched polyamides,[71] polycarbonates[72] and polyureas[72] are synthesised by the A_2 +B_3 route. Jikei *et al.*[71] reported the growth of the hyperbranched polyamide macromolecule of diamine (A_2) with trimesic acid (B_3). The A_2 reacts with the B_3 by polycondensation in the presence of condensation agents; triphenyl phosphite and pyridine.

10.2.2.4.2 Graft 'onto' and graft 'from' approaches

Comb or star shaped macromolecules can be produced by these two methods. Comb-burst poly(ethylenimine)-poly 2-ethyl-2-oxazoline) copolymers, and poly(ethylenimine) homopolymers are synthesised based on the polyoxazoline methodology. As shown in Fig. 10.9 by using nitroxide-mediated 'living' free radical polymerisation and ATRP, graft and dendrigraft macromolecules that use monomer styrene, methyl methacrylate and n-butyl methacrylate are prepared.[73]

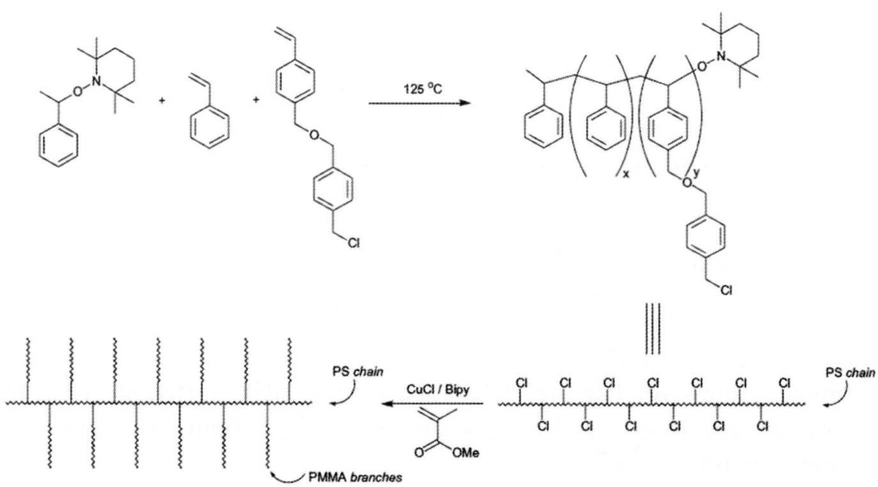

Fig. 10.9 Synthesis of a graft polymer by a graft 'from' polymerisation method.[73]

10.2.3 *Properties and applications of hyperbranched macromolecules*

Hyperbranched macromolecules have numerous potential applications due to their specific structures, such as no chain entanglements, branched shapes, and large number of functional groups within the molecules. The solubility, compatibility, reactivity, adhesion, electrochemical and luminescent properties of hyperbranched macromolecules are tailorable by modifying their function groups.[46,74] Figure 10.10 shows a wide variety of applications for hyperbranched macromolecules.

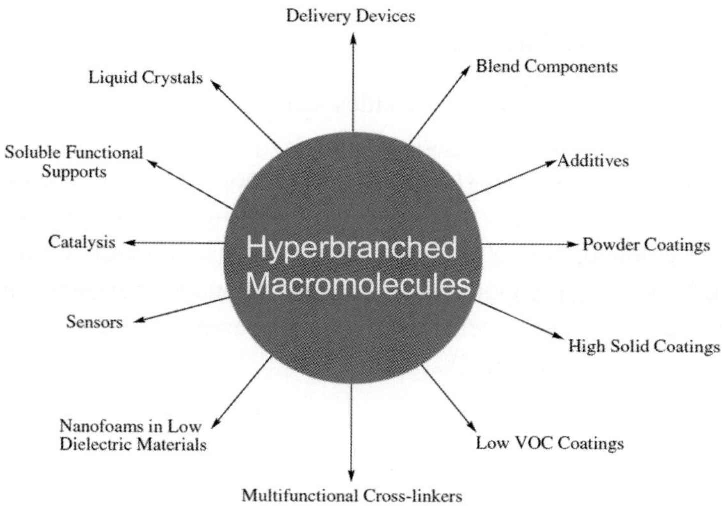

Fig. 10.10 Applications of hyperbranched macromolecules.[47]

10.2.3.1 *Coating*

Hyperbranched macromolecules have high solubility, low viscosity and are soluble in a wide range of solvents. Depending on the reactive terminal groups on hyperbranched macromolecules, different types of resins are developed, such as powder, high solid, flame retardant and barrier coatings for flexible packaging. For UV-curing, coating hyperbranched macromolecules based resins is usually finished with methacrylate or acrylate. The hyperbranched macromolecule based UV-curing resins have low viscosities and high curing rates more so than other kinds of macromolecules. For low temperature powder coatings, the resins with hyperbranched macromolecules have suitable melt viscosities.

10.2.3.2 *Medicine*

Hyperbranched macromolecules also have many applications in the medicine field. The most important applications of hyperbranched macromolecules are for drug delivery and biodegradable materials.[75,76] For drug delivery, the interior or terminal groups of hyperbranched macromolecules may covalently or no-covalently fix bio objectives. The hyperbranched macromolecules as drug delivery media have the functions of controlling drug concentration and delivery rate. A hyperbranched polyether polyols bearing protective poly(ethylene glycol) (PEG) chains with or without the folate targeting ligand at their ends is prepared.[77] Solubilisation of a fluorescent probe; pyrene, and an anticancer drug; tamoxifen, in these polymers is physicochemically investigated. It is found that PEG chains enhance the encapsulation efficiency. The release of pyrene and tamoxifen is only significant at concentrations exceeding the physiological extracellular concentration. Therefore, a high amount of the probe or drug remains solubilised inside the hyperbranched macromolecules.

10.2.3.3 *Polymer blending*

Hyperbranched macromolecules can be used as processing aids, and rheology modifiers of polymers. A slight amount of hyperbranched macromolecule and bromoterminated hyperbranched polyphenylene can significantly decrease the melt viscosity of polystyrene. Hyperbranched macromolecues has no affect on the thermal stability of the polymer. The functions of Boltorne-G4 (Perstorp Polyols, Inc) as a processing additive for polystyrene are studied. It is also reported that hyperbranched polyesters behave as lubricants during processing and have the function of toughening the polystyrene. The research on hyperbranched polyester/ low density polyethylene (LLPDE) shows that the hyperbranched macromolecules have a tendency to migrate to the surface.[78] Therefore, the surface properties of polymer films may be modified by using hyperbranched macromolecules with high density terminal groups.

The studies on the rheological properties of hyperbranched polyesters of different generations indicate that the hyperbranched macromolecules of low generation show shear-thinning behaviour, while those of higher generations display Newtonian characteristics.[79]

10.2.3.4 *Toughening for epoxy-based composites*

Epoxy resins are widely applied in coatings, electrical and electronic devices, adhesives and structural materials due to their good mechanical, thermal, electrical and impregnation properties. Unfortunately, epoxy resins cured with either diamines or anhydrides are brittle. As hyperbranched macromolecules exhibit low melt and solution viscosities, hyperbranched macromolecules with suitable terminal groups will experience high compatibility with epoxy.[80-82] Amine-terminated hyperbranched polyimide can be used to toughen the diglycidyl ether of bisphenol A (DGEBA) epoxy resin. The hyperbranched polyimide has no significant influence on thermal stabilities of the epoxy resins, whereas the T_gs (glass transition temperatures) of the epoxy increase in comparison with that of neat DGEBA as a result of restricted mobility of the chain segments by hyperbranched polyimide. The hyperbranched polyimide markedly increases the fracture toughness and impact strength of the epoxy resin.

10.2.3.5 *Nonlinear optics (NLO)*

To develop new three-dimensional structures for nonlinear optics (NLO), Zhang *et al.*[83] synthesised a hyperbranched macromolecule as a novel NLO material. During recent years, a number of papers have been presented where hyperbranched polymers are designed for NLO applications.[84,85] Two hyperbranched macromolecules with methyl ester (P1) and epoxy (P2) terminal groups with pendant azobenzene chromophores through an 'A_2 + B_3' strategy are synthesised.[86,87] These materials have good solubility and film-forming ability. The poled films show high second-harmonic generation coefficients (>50 pm/V) because of the three-dimensional spatial isolation effect due to the highly branched structures. Such three-dimensional macromolecule materials have good NLO properties.

10.2.4 *Applications in textiles*

10.2.4.1 *Dyeing*

Due to hydrophobic properties and high crystallinity, polypropylene fibres are difficult to be dyed. The dyeability of polypropylene can be improved with C.I. Disperse Blue 56 by incorporating hyperbranched polymer into polypropylene chips before fibre spinning.[88] The disperse Blue 56 is markedly enhanced by hyperbranched macromolecules Hybrance PS2550 (hyperbranched polyester-amide). This is attributed to the polar groups provided by the stearate-modified hyperbranched macromolecules. The fastness of dye to five repeated wash fastness tests at 60°C for modified fibres with 3% hyperbranched macromolecules is very good.

10.2.4.2 *Spinning*

Hyperbranched polyester can be used as a processing aid during the processing of LLDPE.[78] The hyperbranched polyester reduces melt fracture and shark skin. The hyperbranched polyester significantly decreases the LLDPE melt viscosity, even though the hyperbranched polyester and LLDPE are immiscible blends. Studies of the rheological behaviour of polyurethane (PU)/N,N-dimethylformamide (DMF) spinning solutions with hyperbranched polyesters as an additive indicate that the solution viscosity can be significantly reduced by polyesters even at low loadings of 0.5 wt%. The zero-shear viscosity of solutions and apparent activation energy of the viscous flow of the solution with hyperbranched polyesters is much lower in comparison with that of pure PU/DMF solution. Hyperbranched polyester has little influence on the mechanical properties of prepared spandex fibres.

10.2.5 *Summary*

Due to the simpler preparation process in comparison with perfect dendrimers, hyperbranched macromolecules have attracted increasing interest of scientists in recent years not only in chemistry, but also in many other areas. Hyperbranched macromolecules have high density terminal groups, which may be subject to a variety of modifications by reaction with different moieties by end capping, terminal grafting, surface growing, hypergrafting and hybrid blending, to achieve tailorable physical and chemical properties. Hyperbranched macromolecules and their derivatives can find many applications in textiles for functional finishing and smart textiles. Interdisciplinary work by chemists, physical scientists and textile scientists will mean intensive and extensive studies of hyperbranched macromolecule applications in textiles in the future.

10.3 Structure, Properties of Chitosan and its Applications

The principle derivative of chitin is chitosan, produced by alkaline deacetylation of chitin. Chitin is one of the most abundant organic materials, being second only to cellulose in the amount produced annually by biosynthesis. The terms chitin and chitosan mainly refer to a continuum of copolymers of N-acetyl-D-glucosamine and D-glucosamine resides. Chitosan is important in scientific and technical studies because of its multiple unique characteristics and potential implications in the industry. Chitosan can carry a large number of amine groups on its chain, and thus form multiple complexes. Chitosan can be fabricated into versatile forms, such as gels, solutions, membranes, beads, powders and so on. As a type of biodegradable material, chitosan has good bio-compatibility, non-toxicity, antibacterial activity, ability to accelerate wound healing and absorbency. Many functional groups can be grafted into the chitosan molecular chain, so chitosan and its derivatives have many applications in many fields, including medical, textile, skin care, environmental and many others.

10.3.1 *Definition and structure of chitosan*

Chitosan is derived from a natural product; chitin, which is quite abundant in the shells of shrimps. Chitin is a non toxic, biodegradable polymer of high molecular weight. Like cellulose, chitin is a fibre, and in addition, it presents exceptional chemical and biological qualities that can be used in many industrial and medical applications.

Chitin and chitosan have similar chemical structures. Chitin is made up of a linear chain of acethylglucosamine groups. Chitosan is obtained by removing enough acethyl groups (CH_3-CO) for the molecule to be soluble in most diluted acids. This process, called deacetylation, releases amine groups (NH) and gives the chitosan a cationic characteristic. This is especially interesting in an acid environment where the majority of polysaccharides are usually neutral or negatively charged. Chitin is one of the most widespread of the naturally occurring polysaccharide which is an important reusable bioresource similar to cellulose.[89] Chitin widely exists in nature in the outer covering of crabs, shrimps and insects, as well as in the cellular walls of fungi. The annual natural production of chitin is estimated to be 100 billion metric tons. It is the second most abundant naturally occurring biopolymer (after cellulose). The chemical structure of chitin is a homopolysaccharide with (1-4) bonded 2-acetoamido-2-deoxy-b-D-glucose (N-acetyl-β-D-glusosamine). Also, chitin may be regarded as cellulose with hydroxyl at position C-2 replaced by an acetamido group. The structures of cellulose, chitin and chitosan are shown in Fig. 10.11.

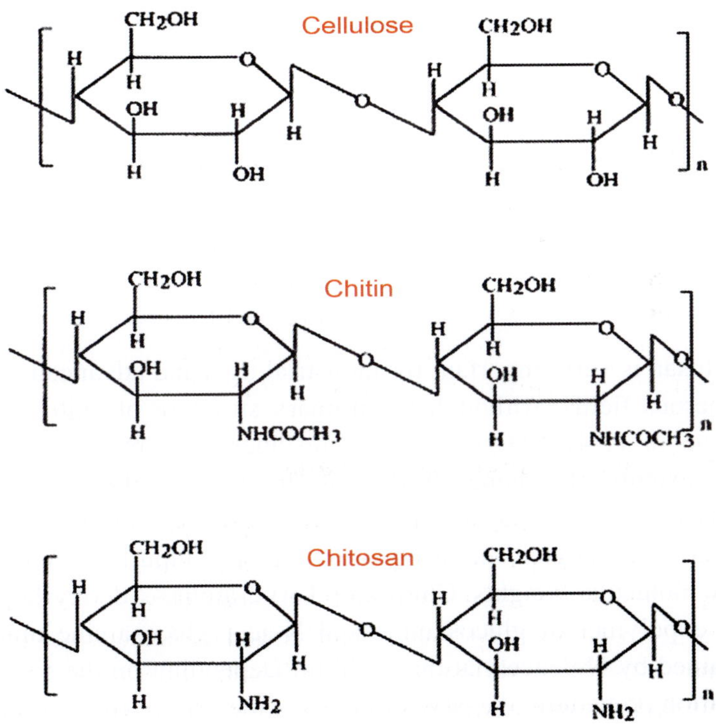

Fig. 10. 11 Chemical structure of cellulose, chitin and chitosan.[89]

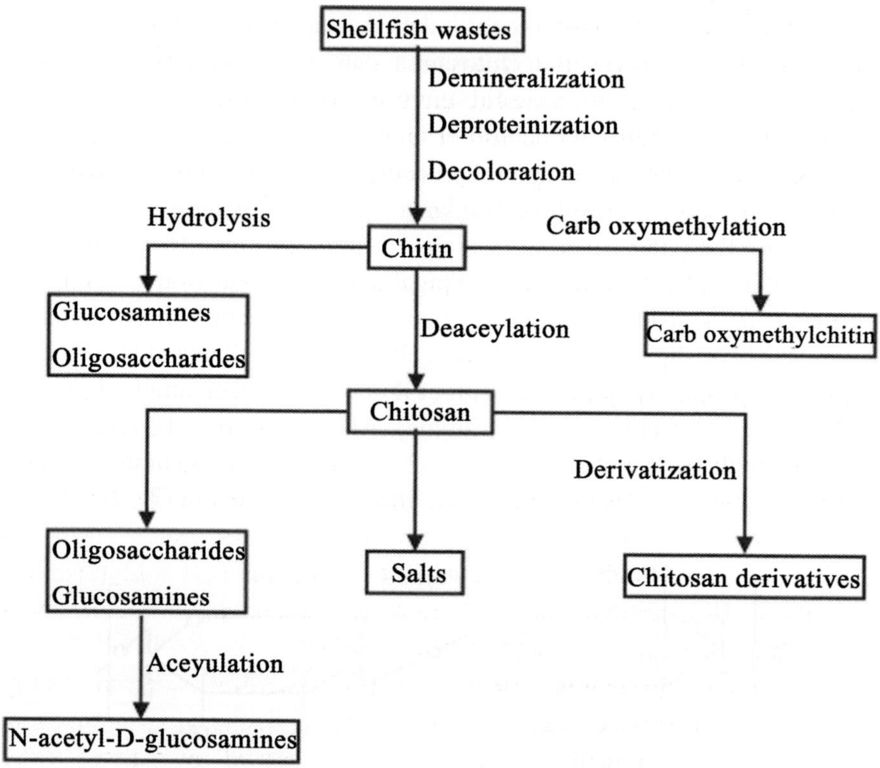

Fig. 10.12 Simplified representation of preparation of chitin, chitosan and their derivatives.[91]

Chitosan is very important for industrial uses and advanced research in numerous fields. Although the primary structure of chitosan is the backbone of (1-4)-β-D-glucosamine residues randomly acetylated to various extents, the name chitosan is, in fact, a collective term for deacetylated chitins differing in terms of crystallinity, optival characteristics, degree of acetylation (DA), impurity content and average molecular weights. Chitosan [(1-4) 2-amino 2-deoxy-D-glucan, CS], a copolymer of glucosamine and N-acetyglucosamine units, can be obtained by N-deacetylation of chitin. Depending on the source and preparation procedure, the MW of chitosan may range from 300 to over 1000 kD with the degree of deacetylation from 30% to 95%. In its crystalline form, chitosan is normally insoluble in aqueous solutions above pH7. However, in dilute acids (pH<6.0), the protonated free amino groups on glucosamine facilitate solubility of the molecule.[90]

Partial deacetylation of chitin results in the production of chitosan (Fig. 10.12), which is a polysaccharide composed by polymers of glucosamine and N-acetyl glucosamine.[91] The chitosan label generally corresponds to polymers with less than 25% acetyl content. Chitosan is soluble in acid solutions and chemically more versatile than chitin or cellulose.

10.3.2 *Properties of chitosan*

Due to its unique polycationic nature, chitosan has been used as an active material which is antifungal,[92-94] antibacterial[95-97] and antitumor[98,99] in nature. With many excellent properties, chitosan has numerous applications. For example, its forming feasibility[100] enables chitosan for use in various forms, such as zero dimension microsphere, two-dimension membrane and three-dimension pin or rod.

In solutions, chitosan is composed by a string of blocks of almost fully deacetylated polysaccharide stretched by electrostatic repulsion, intercalated with micelle-like agglomerates formed by almost fully acetylated polysaccharide. These agglomerates include $–NH_3{}^+$ groups which produce electrostatic swelling of the agglomerates, giving a radius proportional to the deacetylation degree. The length of the strings is also proportional to the deacetylation degree. These strings are extended because of the electrostatic repulsion between charged ammonium groups.[101]

The excellent intrinsic properties of chitosan which have been reviewed are biodegradability, biocompatibility, film-forming ability, bioadhesivity, polyfunctionality, hydrophilicity and adsorption properties (Table 10.1).[91] Most of the properties of chitosan can be related to its cationic nature, which is unique among abundant polysaccharides and natural polymers. These numerous properties lead to the recognition of this polyamine as a promising raw material for versatile purposes.

Table 10.1. Intrinsic properties of chitosan.

Physical and chemical properties	• Linear aminopolysaccharide with high nitrogen content • Rigid d-glucosamine structure; high crystallinity; hydrophilicity • Capacity to form hydrogen bonds intermolecularly; high viscosity • Weak base; the deprotonated amino group acts a powerful nucleophile (pK_a 6.3) • Insoluble in water and organic solvents; soluble in dilute aqueous acidic solutions • Numerous reactive groups for chemical activation and crosslinking • Forms salts with organic and inorganic acids • Chelating and complexing properties • Ionic conductivity
Polyelectrolytes (at acidic pH)	• Cationic biopolymer with high charge density (one positive charge per glucosamine residue) • Flocculating agent; interacts with negatively charged molecules • Entrapment and adsorption properties; filtration and separation • Film-forming ability; adhesivity • Materials for isolation of biomolecules
Biological properties	• Biocompatibility • Non-toxicity • Biodegradable • Adsorbable • Bioactivity • Antimicrobial activity (fungi, bacteria, viruses) • Antiacid, antiulcer, and antitumoral properties • Blood anticoagulants • Hypolipidemic activity • Bioadhesivity

10.3.2.1. *Degree of acetylation*

DA can be defined as the mole fraction of acetylated units in the polymer chain. DA of chitosan has an influence on all of the physicochemical properties (molecular weight, viscosity, solubility, etc.) Hence, DA is one of the most important parameters of chitosan. Many techniques have been tried in order to determine the DA more precisely, such as UV, Infrared (IR), Nuclear Magnetic Resonance (NMR), and Gel Permeation Chromatography (GPC).

10.3.2.2. *Solubility*

In the form of free amine, chitosan is insoluble in water to a pH near neutrality, concentrated acids with the exception of sulphuric acid, blases and organic solvents. In practice, the terms 'chitin' and 'chitosan' refer to a sample of polymers that vary in DA, with 'chitin' referring to high (ideally 100%) DA polymers and 'chitosan' referring to low (ideally 0%) DA. When the DA is 50% or lower, the polymer becomes water-soluble due to the protonation of the –NH2 groups of the glucosamine unit. Their acid base properties allow easy dissolution. Therefore, chitosan is soluble in dilute acids on accounts of the protonation of free amine groups, including HCL, HBr, HI, HNO_3, $HClO_4$ and also acetic acid. Chitosan is also slightly soluble in dilute H_3PO_4 but is insoluble in dilute H_2SO_4 at room temperature although chitosan sulphate dissolves in water upon heating and reforms upon cooling.

10.3.2.3. *Crystallinity*

A higher crystallinity of chitosan often has a greater degree of deacetylation. Two major polymorphs have been firmly established for chitin in nature, namely α- and β-chitin. Each can be readily distinguished on the basis of its X-ray diffraction patterns. α-chitin is the most stable and ubiquitous form of chitin that occurs in the exoskeleton of arthropods and fungi. β-chitin is often present in deep sea organisms. β-chitin has the ability to undergo conversion into the α-chitin polymorph by precipitation from a formic acid solution or treatment with cold 6 M HCl. Guibal *et al.*[102] showed that the maximum sorption capacity for the molybdate depends mainly on the crystallinity and degree of deacetylation of chitosan, and the worst sorbents are characterised by a higher crystallinity index.

10.3.2.4. *Swelling/deswelling (pH sensitivity)*

A pH sensitive hydrogel can be formed by chitosan.[103] Chitosan can be modified easily to form porous hydrogels for the existence of reactive amino groups.[104] The amino groups can be protonated in different pH solvents and display selling/deswelling properties.[104,105] Chitosan shows pH-sensitive behaviour due to the large quantities of amino groups on its chains. When it is in acid and alkali conditions, its molecular chains will be in extending and curling states respectively.

10.3.2.5. *Antibacterial activity*

Chitosan inhibits the growth of a wide variety of bacteria and fungi[106] (see Table 10. 2), and exhibits broad spectra of antibacterial activity, high killing rate, and low toxicity toward mammalian cells. There are two proposed modes of action for microbial growth inhibition. In one mechanism, the polycationic nature of chitosan interferes with the negatively charged residues of macromolecules at the cell membrane surface. Chitosan interacts with the membrane to alter cell permeability, and even cause the leakage of the intracellular components of bacteria. Another mechanism involves the binding of chitosan with DNA to inhibit RNA synthesis. The antibacterial capability of chitosan is influenced by the MW of the chitosan. When the MW increases from 5000 to 9.16×10^4, the antibacterial activity also increases. As the MW of the chitosan is further increased from 9.16×10^4 to 1.08×10^6, the antibacterial activity of the chitosan is decreased slowly. Moreover, the antibacterial activity of chitosan increases with an increase in the degree of deacetylation.[97] Carboxymethylated chitosan also has antibacterial activity. Fuji Spinning Co. Ltd. has developed a type of antibacterial fibre; Chitopoly, by blending chitosan with viscose fibres.[107]

Table 10.2 Antimicrobial activities of chitosan.[106]

Bacteria	MIC[a] (ppm)	Fungi	MIC[a] (ppm)
Agrobacterium tumefaciens	100	Botrytis cinerea	10
Bacillus cereus	1000	Drechstera sorokiana	100
Corinebacterium michiganence	10	Fusarium oxysporum	10
Erwinia ssp.	500	Micronectriella nivalis	10
Erwinia carotovora ssp.	200	Piricularia oryzae	5000
Escherichia coli	20	Rhizoctonia solani	1000
Klebsiella pneumoniae	700	Trichophyton equinum	2500
Micrococcus luteus	20		
Pseudomonas fluorescens	500		
Staphylococcus aureus	20		
Xanthomonas campestris	500		

[a] MIC: minimum growth inhibitory concentration

10.3.3 *Applications of Chitosan*

Chitosan is a biological product with cationic (positive electrical charge) properties. Table 10.3 shows the chitin derivatives and their propoased uses. By controlling the molecular weight, the degree of deacetylation and purity, it is possible to produce a broad range of chitosans and derivatives that can be used for industrial, dietary, cosmetic and biomedical purposes. Together, these properties have led to the development of hundreds of applications. Applications of chitosan can be classified mainly in three categories according to the requirements for purity:

- technical grade for agriculture and water treatment
- pure grade for the food and cosmetics industries
- ultra-pure grade for biopharmaceutical uses

In agriculture, chitosan can be used in seed-coating, frost protection, bloom and fruit-setting stimulation and timed release of product into the soil. In water treatment, it can be used as a greener method to purify water and avoid water pollution. In food, chitosan is already used as an ingredient in Japan, Europe and the United States as a lipid trap. Since chitosan is not digested by the human body, it acts as fibre, which is a crucial diet component. In the medical field, possible applications are ointments for wounds, surgical sutures and so on.

Table 10.3 Chitin derivatives and their proposed uses.[108]

Derivative	Examples	Potential
N-Acyl chitosans	Formyl, acetyl, propionyl, butyryl, hexanoyl, octanoyl, decanoyl, dodecanoyl, tetradecanoyl, lauroyl, myristoyl, palmitoyl, stearoyl, benzoyl,monochloroacetoyl, dichloroacetyl, trifluoroacetyl, carbamoyl, succinyl, acetoxybenzoyl	Textiles, membranes and medical aids
N-Carboxyalkyl (aryl) chitosans	*N*-Carboxybenzyl, glycine-glucan (*N*-carboxy-methyl chitosan), alanine glucan, phenylalanineglucan, tyrosine glucan, serine glucan, glutamicacid glucan, methionine glucan, leucine glucan	Chromatographic media and metal ion collection
N-Carboxyacyl chitosans	From anhydrides such as maleic, itaconic, acetyl-thiosuccinic, glutaric, cyclohexane 1,2-dicarboxylic,phthalic, *cis*-tetrahydrophthalic, 5-norbornene-2,3-dicarboxylic, diphenic, salicylic, trimellitic, pyromellitic anhydride	Drug encapsulation
o-Carboxyalkyl chitosans	*o*-Carboxymethyl, crosslinked *o*-carboxymethyl	Molecular sieves, viscosity builders, and metal ion collection
Metal ion chelates	Palladium, copper, silver, iodine	Catalyst, photography, health products, and insecticides
Semisynthetic resins of chitosan	Copolymer of chitosan with methyl methacrylate, polyurea-urethane, poly(amideester), acrylamidemaleic anhydride	Textiles
Natural polysaccharide complexes, miscellaneous	Chitosan glucans from various organisms	Flocculation and Flocculation

Chitosan beads can be used for chromatography packing and fertilisers. As various properties of chitosan have been discovered, the development efforts have increased. Various chitosan derivatives have been studied to learn about the changes in solubility and degree of crystallinity etc.[109-112] When a small mammal (mouse) was studied for toxicity,[113] chitosan was found to be as non-toxic as sugar and salt. Hence, chitosan has no toxicity and is a highly safe material.

As a natural renewable resource with a number of unique properties, chitosan is now attracting increasingly more scientific and industrial interest from multiple fields, such as chemistry, biochemistry, medicine, pharmacology, biotechnology, food science and textile science.[114] In addition, many unique products have been developed for various applications. These include for example, surgical sutures, artificial skin, cosmetics and dietary food. As a type of biocompatible, biodegradable, and antigenic cationic biopolymer, chitosan represents a promising candidate for employment in the design of antibacterial activity and controlled release systems.[108] The characteristics of chitosan as a medical material are moldability, bioabsorptivity, acceleration of wound healing,[115] inhibition of bleeding, and immune response recovery.[16] In addition to the above characteristics, chitosan has other advantages, including having both reactive amino and hydroxyl groups which can be used to modify its properties under mild conditions.[116] Bacteriolysis enzymes can break down chitosan at different speeds depending on its difference in acylation degree. We can say that chitosan is a biocompatible polymer, which can accelerate wound healing and the rebuilding of soft and hard tissues.

10.3.3.1. *Chitosan based wound-dressing*

Stevens *et al.* used a chitosan membrane as wound dressing material and found that it can accelerate the wound healing process.[117] Wang *et al.*[118] used chitosan-alginate polyelectrolyte complex (PEC) membranes as wound dressing. In their research, flexible, thin, transparent, novel chitosan-alginate PEC membranes, cast from aqueous suspensions of chitosan-alginate coacervates with $CaCl_2$, were evaluated as potential wound-dressing materials. Compared with conventional gauze dressing, the PEC membranes caused an accelerated healing of incision wounds in a rat model. On the basis of its biocompatibility and wound-healing efficacy, the chitosan-alginate PEC membrane can be considered for wound-dressing applications.

10.3.3.2. *Chitosan modified PNIPAAm hydrogels*

During the past years, stimuli responsive hydrogels that combine the properties of chitosan with those of PNIPAAm have received very little attention.[119] Verestiuc *et al.* prepared a series of semi-interpenetrating polymer networks by using the free radical polymerisation of NIPAAm under the existence of chitosan and using tetraethylene glycol diacrylate(TEGDA) as the crosslinking agent.[120] It was found that the structure of the network seems to be influenced by the quantity of crosslinker and chitosan used. The proportion of chitosan that could be entrapped into the matrix increases when the crosslinking density of the network increases. Poly[*N*-isopropylacrylamide (NIPAAm)-chitosan] crosslinking copolymer particles were synthesised by the soapless emulsion copolymerisation of NIPAAm and chitosan as reported by Lee *et al.*[121] The chitosan-NIPAAm copolymer particles were processed to form copolymer disks. The effects of various variables, such as the chitosan/NIPAAm weight ratio, the concentration of the crosslinking agent, and pH values on the swelling ratio of chitosan-NIPAAm copolymer disks were investigated. Furthermore, caffeine was used as the model drug to study the characteristics of drug loading of the chitosan-NIPAAm copolymer disks. Two factors (pore size and swelling ratio) affected the behaviour of caffeine release from the chitosan-NIPAAm copolymer disks. Our research group has completed some research on this issue.[122] In our research project, thermosensitive poly(N-isopropyl acrylamide)/polyurethane hydrogel and chitosan modified nonwoven fabrics were obtained. From the experimental results, poly(N-isopropy lacrylamide)/polyurethane hydrogels can endow the nonwoven fabrics with temperature sensitivity without heavy syneresis; and the hydrogel modified nonwoven fabrics have controlled release properties and also antibacterial activity after chitosan modification. The two most important characteristics of the controlled release and antibacterial activity make the modified nonwoven fabrics suitable for medical and skin care uses. The antibacterial mechanism is shown in Fig. 10.13.

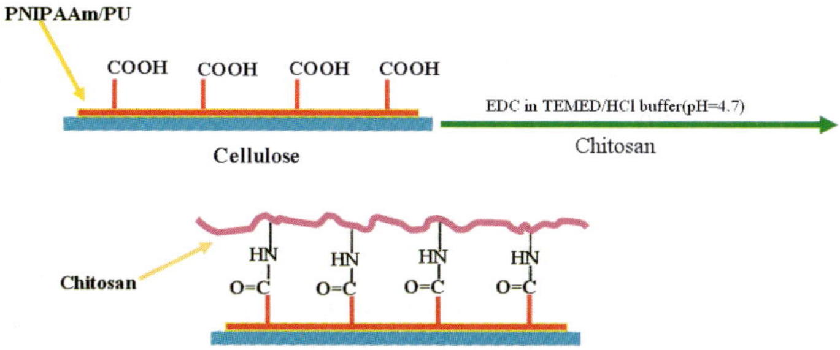

Fig. 10. 13 Schematic illustration of surface modification with chitosan.

In addition, as the chitosan content and crosslinking density increased, the phase transition temperature of the chitosan interpenetrated hydrogel became less well defined and shifted towards lower temperatures. The synthesised hydrogel with the incorporation of chitosan into the structure was highly sensitive to temperature variations. The preliminary drug release study involving pilocarpine demonstrated that the system under consideration shows little promise as a controlled delivery vehicle for the delivery of positively charged drugs due to the heavy syneresis of pure PNIPAAm hydrogels.[123]

10.3.3.3. *Chitosan antibacterial finishing on cotton*

Due to antibacterial activity and minimal skin reaction over a wide range of biomedical investigations, chitosan can be used in antibacterial next-to-skin fabrics. Cotton fibre and viscose rayon are the undergarment materials of choice because they have demonstrated safety to the body and demonstrate comfort resulting from their high moisture retentivity. Cellulose, the main ingredient of cotton and viscose rayon, has a molecular structure similar to that of chitosan. The difference between the two polysaccharides is that chitosan has the amino group instead of the hydroxyl group at the C2 position of every pyranose ring. This similarity is expected to give high compatibility for these two polymers. Moreover, the formation of intermolecular hydrogen bonding between them also contributes to biocompatibility. Blending cellulose and chitosan is expected to be a useful method for introducing antibacterial activity into cellulosic materials. Nasr *et al.*[124] researched cotton fabrics that were treated with acrylate monomers, chitosan and PEG to provide antibacterial properties, UV-protection as well as improvement of dyeing properties with direct, acid and reactive dyes. The obtained data showed that the tested fabrics have appropriate antibacterial activity with high UV-protection properties by increasing chitosan concentration up to 3%. The mechanical properties expressed as tensile strength and abrasion resistance, increase after the finishing treatment.

10.3.3.4. *Chitosan durable press finishing on cotton*

Huang *et al.* investigated the effects of using low molecular weight chitosan (LWCS) for anti-creasing treatment of cotton fabrics.[125] LWCS was mixed with dimethylol dihydroxyl ethylene urea (DMDHEU) to form the finishing agent. The addition of LWCS increased the tensile strength retention (TSR) and creasing resistance of the treated fabrics. The yellowing index and the softness of the treated fabrics became worse when the molecular weight of LWCS decreased, and the concentration as well as the curing temperature increased. The anti-wrinkle properties of all processed fabrics decreased markedly after washing 20 times. The strength also decreased slightly. However, the softness of the fabric improved.

10.3.3.5. *Chitosan on dye applications*

Dragan *et al.* studied interactions that could occur during the dyeing of chitosan treated wool fibres by measuring the absorbance values of solutions containing dye and chitosan.[126] It was shown that there is a 1:1 stoichiometry between protonated amino groups and sulfonate acid groups on the dye ions in low concentrated chitosan solutions. With an excess of chitosan in the solution, the dye can be distributed between different chitosan molecules and the soluble chitosan/dye products will remain in the solution. The interaction involves the possibility of adsorbed dye molecules to be desorbed and redistributed between other components present in the system, depending on system parameters (pH, temperature and electrolyte presence).

10.3.3.6. *Chitosan finishing on silk*

Davarpanah *et al.*[127] grafted chitosan onto acrylated silk fibres. Silk fibres were degummed and acylated with two anhydrides, succinic anhydride (SA) and phthalic anhydride (PA), in different solvents (dimethyl sulfoxide (DMSO) and N,N-dimethyl formamide (DMF)). The dyeing of the chitosan grafted-acylated silk fibres indicated higher dyeability in comparison to the acylated and degummed silk samples. The mechanism of chitosan grafting over degummed silk through anhydride linkage was proposed. The findings of this research support the potential production of new environmentally friendly textile fibres. It is worth mentioning that the grafted samples have antibacterial potential due to the antibacterial property of chitosan molecules.

10.3.3.7. *Chitosan fibres*

Chitosan can be spun into fibres because of its high solubility. Alonso *et al.*[128] prepared chitosan crosslinked cellulose fibres by using non-toxic procedures in order to confer antimicrobial properties to the cellulose fibres. Citric acid was used as the crosslinker and NaH_2PO_4 as the catalyst in previously UV-irradiated cellulose fibres. A thermo-gravimetric analysis of the material with the highest chitosan content showed an increased thermal stability compared to cellulose and chitosan. The biomass and spore germination percentage of Penicillium chrysogenum and colony forming units per millilitre for Escherichia coli decreased significantly on the composed materials as compared to raw cellulose fibre. They were similar to that obtained with a commercial antimicrobial cellulose fibre.

10.3.4 *Chitosan used in cosmetics*

Chitosan forms a protective, moisturising, elastic film on the surface of skin with the ability to bind other ingredients that act on the skin. In this way, chitosan can be used in formulating moisturising agents, such as sunscreens, organic acids, etc. to enhance their bioactivity and effectiveness. Chitosan interacts with negatively charged biological surfaces, such as skin and hair. Other relevant characteristics of chitosan for cosmetics applications are its high molecular weight, water retention and film formation capacity as well as its heavy metal ion complexing ability. High molecular weight chitosan decreases the loss of trans-epidermic water, increasing the humidity of the skin which preserves its softness and flexibility. Chitosan is an excellent ingredient to treat skin with allergies as formulations with high molecular weight samples are found to reduce skin irritation.[129] Therefore, it is very convenient to use chitosan in formulations that contain alcohol, such as aftershave lotions and deodorants, again, in order to reduce skin irritation. Additionally, chitosan also inhibits inflammatory processes and promotes the regeneration of damaged tissues.[130] Sun-protecting emulsions that incorporate chitosan have a positive effect on water resistance, conferring increased protection to the skin.[131] High molecular weight chitosan is a valuable component in deodorants because it absorbs humidity, thus reducing transpiration. As well, its antibacterial properties protect the skin.[132] When added to hair care cosmetics preparations, chitosan interacts with keratin to form a uniform and elastic film which is more stable in high humidity than those usually formed with synthetic polymers. Moreover, it reduces electrostatic charges, so that the hair does not stand on its ends, preserving the hair style. Hair treated with chitosan formulations have less tendency to adhere and more easily brushed than with traditional fixers.[133]

10.3.5 *Summary*

In conclusion, the necessary steps to bring chitosan from its present state of use to those in adaptive materials have been shown. They may be applied to many areas, including the environmental field, textiles field, skin care and biomedical engineering field. Simultaneous material application is conducted to research the properties and structure. This takes into account, the processes to obtain the chitosan with various molecular weights and degrees of acetylation during the validation process. Biomedical grade chitosan products can be readily utilised as materials for medical devices and biomedical research. In the textile field, chitosan will have many notable applications. Chitosan and its derivates have multiple applications in adaptive fields, including being used as finishing, increasing colourfastness, and coating of fabrics to control their shrinkage and enhance their dye fastness. The excellent biomedical properties together with a safe toxicity profile make chitosan an exciting and promising material in diversified areas.

10.4 Cyclodextrins

Cyclodextrins (CDs) are a class of cyclic oligomers which are composed of α-1, 4 linked glucopyranose units and prepared from liquefied starch by means of enzymatic conversion. These unique substances with an internal hydrophobic cavity can form complexes with and improve solubility of many poorly water soluble guest molecules.[134] CDs have important functions in many areas for potential applications, including drugs,[135] food and flavors,[136] cosmetics[137] textiles,[138] separation technologies,[139] and fermentation and catalysis.[140] In particular, CDs are widely utilisable materials in textiles, cosmetics and other commodities industrially. This section first briefly introduces the structure and properties of CDs, followed by the main details of CD applications in textiles and skin/healthcare products. Recent applications of CDs in the textile industry focus on dyeing, finishing, functioning and monitoring aspects. The role of these compounds in skin/healthcare products is also extensively discussed.

10.4.1 *Structure and properties of CDs*

More than 100 years ago, CDs were mentioned in the literature by Villiers.[140] In 1903, Schardinger defined the cyclic polysaccharide characterisations of these substances. Then, he identified that three naturally occurring cyclodextrins: α-cyclodextrin, β-cyclodextrin and γ-cyclodextrin. They consist of six, seven and eight α-1,4 linked glucose residues respectively.[141] From 1911 to 1935, Pringsheim, a chief researcher in Germany, reasoned that these macrocyclic carbohydrates can form stable inclusion complexes with other guest components in an aqueous solution.[134] By the mid 1970's, the structure and chemical characterisations of the majority of natural CDs had been determined and a large number of inclusion complex systems had been explored.[141]

10.4.1.1 *Structure of CDs*

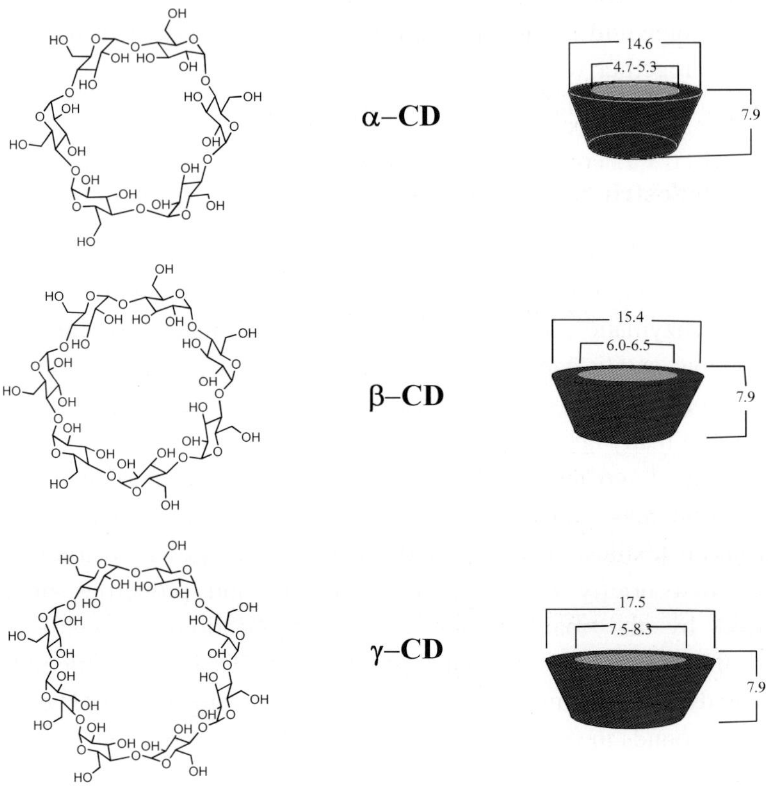

Fig. 10.14 Chemical structure of α-cyclodextrin, β-cyclodextrin and γ-cyclodextrin.

Commonly used CDs are composed of six to eight glucose units, which can be described as cyclic wherein the larger and smaller openings of the cyclic cavity are exposed to the solvent (Fig. 10.14). Each cyclodextrin is a rigid and doughnut shaped molecule structure. This structure with an hydro interior cavity is hydrophobic, whereas the external surface is hydrophilic. They act as a host for entrapping other chemicals either wholly or partially without the formation of covalent bonds. Many modifications of CDs can alter their chemical and physical properties. These methods can improve solubility of the guest molecule, stability against the effects of heat or light and help control the reaction activity of guest compounds. β-cyclodextrin (β-CD) has been known as the most readily accessible, the lowest in cost and the most widely useful in the CDs family.

CDs are composed of five or more α-D-glucopyranoside units, but the 5-membered macrocycle is not natural. Today, the largest well-characterised CD contains 32 1,4-anhydroglucopyranoside units. There is at least, a 150-membered cyclic oligosaccharides, even if they are a poorly characterised mixture.

10.4.1.2 *Physical and chemical properties of CDs*

Table 10.4 Certain physical properties of cyclodextrins.[140]

	α-cyclodextrin	β-cyclodextrin	γ-cyclodextrin
Number of glycopyranose units	6	7	8
Molecular weight(g/mol)	972	1135	1297
Solubility in water at 25°C(g/100ml)	14.5	1.85	23.2
Melting temperature range(°C)	255-260	255-265	240-255
Cavity diameter(Å)	4.7-5.3	6.0-6.5	7.5-8.3
Outer diameter(Å)	14.6	15.4	17.5
Height of torus(Å)	7.9	7.9	7.9
Volume of Cavity ($Å^3$)	174	262	427
Surface tension(MN/m)	71	71	71
Specific rotation$[\alpha]_D^{25}$	150.5+-0.5	162.5+0.5	177.4+0.5
Crystal water content(wt.%)	10.2	13-15	8-18
Water molecules in cavity	6	11	17

Physical properties of α-, β- and γ-cyclodextrins are summarised in Table 10.4. CDs have good heat stability from peaks of differential scanning calorimetry. Moreover, CDs can undergo acid hydrolysis in strong acids, such as hydrochloric or sulfuric acid. General speaking, bases do not hydrolyse CDs. CDs can be hydrolysed by cyclodextrin glucosyltransferase and alpha amylases. CDs are stable toward alkalis, even at higher temperatures.

CDs derivatives can be prepared by chemical or enzymatic reactions. In CDs, every glucopyranose unit has three active hydroxyl groups. Due to the availability of multiple reactive hydroxyl groups, the application of

CDs is greatly increased by chemical modification. In the structure of β-CD, 21 hydroxyl groups can be modified by substituting groups, such as amino- thio-, alkyl-, carboxyalkyl-, hydroxyalkyl-, tosyl-, glucosyl-, maltosyl-, etc. The aim of such derivatisations may be to improve the solubility of the cyclodextrin derivative and increase the fit between the CD and its guest. Then, the derivatives can also reduce the reactivity of the guest and form insoluble, immobilised CD-containing structures.

Among industrially produced and available β-CD derivatives, the most important ones are the highly water-soluble methylated β- cyclodextrins and 2-hydroxypropylated β- cyclodextrins. Due to their heterogeneity, amorphousness is an important advantage.

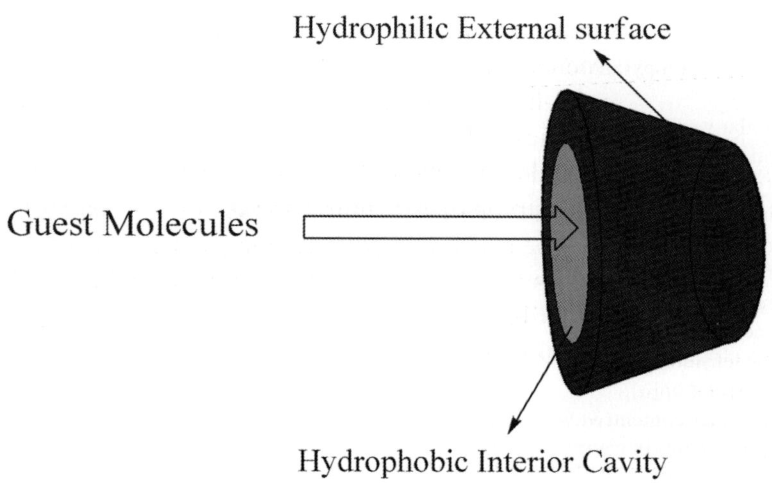

Fig. 10.15 Cyclodextrins inclusion complex formation.

CDs have a unique structure which enables them to form host-guest or inclusion complexes with a variety of materials.[140,142] In these complexes, guest molecules can be held within the cavity of the CD (Fig. 10.15). Complex formation is a dimensional and shaped fit between the host cavity and guest molecule. The formation of the inclusion complexes will greatly modify the chemical and physical properties of the guest molecule. They are used for masking taste and odours, and stabilising, delivering or solubilising active ingredients. This is the reason why CDs have attracted so much interest in many fields. The binding of guest molecules within the cavity of the CD is not fixed and the binding strength depends on specific local interactions of the surface atoms and how well the 'host–guest' complex fits together.

10.4.1.3 *Applications of CDs and CD complexes*

CDs are applied in the pharma industry, food, cosmetics, textile and other industrial fields as solubilising agents, stabilisers, emulsifiers, etc.[143] Nowadays, the most thoroughly studied field of application of CDs is the pharmaceutical industry. However, less than 10% of all CDs produced are consumed by the pharmaceutical industry. The largest amount of CDs produced are consumed by the food and cosmetics industries.[141] In Japan, about 78.8% and 11.5% of all CDs are consumed by the food and cosmetics industries, respectively. Nearly 3.2% of CDs produced are used for industrial chemicals. Around 4.6% are dedicated to the pharmaceutical industry and agricultural chemicals.[144] CDs also have various possibilities in the textile industry, but the blockbuster will be the chemical binding of CDs onto the surface of natural and synthetic fibres. The use of CDs in textile processes was reviewed some years ago.[145] During recent years, further textile applications of CDs have been developed.

10.4.2 ***Applications of CDs to textiles***

CDs and CD derivatives have found a number of applications in textile and clothing fields.[138] So far, CDs have played a significant role in the textile industry. The hydrophobic cavity of CDs can bind volatile molecules, such as unpleasant components, for instance, sweat, cigarette smoke, etc. Also, fragrances, insect repellents and even drugs can be possibly bonded by CD containing garments. Several important aspects might be used: (1) application of CDs to textile dyeing; (2) CDs used in finishing textiles; (3) CDs used in textile functioning fibres and (4) CDs in textile monitoring. The permanent fixation of CDs offers new textiles with interesting properties.

10.4.2.1 *Applications in textile dyeing*

CDs have been used for dyeing fabrics.[146] Using CDs, more dye goes onto the fabric, reducing the amount of dye remaining in the wastewater. The solubility of the dye in water is increased, and no other auxiliaries are needed to solubilise the dye.[147] Preformed complexes perform better than adding the dye and CD to the dyeing solution, and the complexes dissolve rapidly. Hydrophobic tosyl derivatives of β-CD have also been used.[148] Using these derivatives, about a three-fold increase in the binding of fluorescent dye to polyester fibre is observed.

β-cyclodextrin modifies the interaction between cotton and direct dyes.[149] Most of time, disperse dyes are used for colouration of polyester fibres. Unfortunately, disperse dyes are very poorly soluble in water, and uniform dyeing is impossible without using solubility-enhancing agents. CDs can replace the surfactant in colouration of polyester fibres.[138] Research has found that nylon 66 and microfibre nylon 6 fabrics in the presence of CDs have 4- to 10-fold improvement of colour uniformity and minor changes in colour yield.[146]

The sorption of dyes on β-CD-epichlorohydrin crosslinked gels has been tested, and it is found that cyclodextrin gels are more efficient than hydroxypropyl- β-CD gels. The influence of pH can be ignored while that of ionic strength is predominant on the sorption rate. It can be deduced that the mechanism of sorption of the dyes on the gels is physical absorption, and the adsorption sites lie in the polymer network and/or host-guest inclusion complex formation.[103]

There is a need for the development of a textile printing system in the textile industry. Digital printing has the potential to provide such a system when combined with the fabric formation process, and CDs can provide the key component for successful integration of the two processes. Water soluble CD oligomers could be used as warp sizes for natural and synthetic fibres. The dye will be left on the fibre surface after the water soluble CD oligomers are dissolved away with hot water. Also, the National Textile Center have carried out experiments with reactive- disperse dyes on nylon and polyester/cotton substrates.[150]

10.4.2.2 *Applications in textile finishing*

Textile finishing is another application in which CDs are increasingly attracting attention. CDs provide excellent textile finishing to cottons, woolens and blended materials. One possibility of modifying the surfaces of fibres to alter the fibre properties is to bind the CD molecules permanently onto the fibres.[151]

It is demonstrated that CD derivatives can be bonded onto polymer surfaces by using conventional textile processing technologies. As a result, the modification can process some special properties of the fibres. Conceivable fields of application are medical or functional textiles for clothing.[152] Thus, for example, synthetic fabrics and cellulosic or silk fabrics treated with deodorising agents that are formulated with CDs, will exhibit good wash resistance.[138]

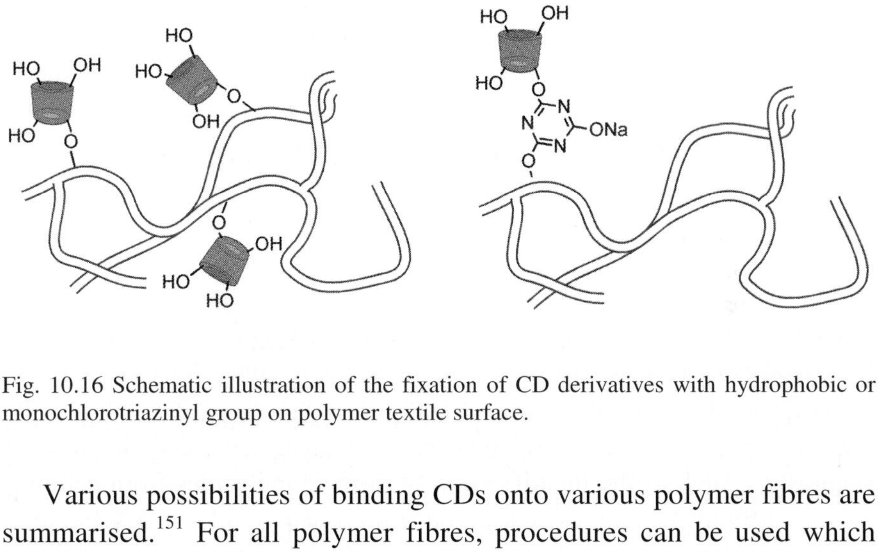

Fig. 10.16 Schematic illustration of the fixation of CD derivatives with hydrophobic or monochlorotriazinyl group on polymer textile surface.

Various possibilities of binding CDs onto various polymer fibres are summarised.[151] For all polymer fibres, procedures can be used which are similar to the dyeing processes of these polymer fibres in order to obtain fixation of the CDs onto the fibre surface. CD derivatives with a reactive group, e.g., the monochlorotriazinyl group, can react with the hydroxyl groups of cellulosic fibres.[145] Figure 10.16 is a schematic illustration of the fixation of CD derivatives with hydrophobic or monochlorotriazinyl group on polymer textile surface. Owing to the fact that the monochlorotriazinyi group derivative of β-CD does not possess any potential for allergies, the modified polymer fabrics may come into contact with skin without causing any side effect.

γ-cyclodextrin is a new auxiliary substance. γ-cyclodextrin can also be used to afford good textile finishing to woolens, cottons and other blended materials. When bound with fibres, it can afford enhanced hydrophilicity and form inclusion complexes to immobilise antimicrobial agents, insect repellents, perfumes, etc.[145] In addition, γ-cyclodextrin can increase the strength of polyester fibres which are used for reinforcement of rubbers.[138]

10.4.2.3 *Applications in textile functioning*

Considering the versatile properties when permanently transferring CDs into textiles, fabrics can be imbued with chemicals by means of CDs to give them novel properties.[145] CDs fixed onto textile materials are able to prepare complex substances into their cavities. Therefore, it is possible to prevent formation of malodour. If CDs are present on the textile, the odour from sportswear soaked with sweat is reduced or does not smell at all. At the same time, it is also valid for socks or other textiles used directly in contact with the body. Further applications include towels which leave pleasant smells after drying one's hands. Curtains treated with CDs can be used to improve the air in houses or offices with smokers. Also, bed linens may release a pleasant odour in the bedroom. It only takes mere imagination to find more applications for CDs.[138]

The complexation or release of substances by fixed CDs can also be used in medical applications. There are new related methods in medical diagnostics, such as the identification of organic molecules from patients. As it is difficult to take a sweat probe from a patient, blood or urine tests were normally used until now. Today, the complexation of organic compounds from sweat results in a preconcentration and identification becomes easier. Pharmaceutically active substances are complexed by fixed CDs. They are set free by wearing and can penetrate into the skin and the treatment of extensive skin diseases becomes simplified.[145] Wang *et al.* developed aromatherapic textiles by using fragrance with β-CD inclusion compounds. The fragrance inclusion compounds were fixed onto cotton with a low temperature binder by the conventional pad-thermofixed method. The results of sensorial evaluations showed that the perfume in the fabric was sensed for over 30 days. An aromatherapic textile could be achieved in this way.[153]

In our research group, we invented methods of manufacturing deodorants which are capable of releasing deodorant agents at certain temperatures and improving adherence to textiles.[154] In this invention, CDs and derivatives are preferred due to their capability of absorbing a broad spectrum of odours by forming complexes with them. The family of CDs includes: unsubstituted CDs containing from six to twelve glucose units, especially, alpha-, beta-, gamma-cyclodextrin and/or their derivatives and/or mixtures thereof. The fixation of CDs onto textile materials leads to new quality uses.[155] It is not possible to describe all of the applications that one may be able to conceive. The use of CDs together with textiles opens up new opportunities.

10.4.2.4 *Applications in textile monitoring*

The application of formaldehyde primarily concerns cotton, viscose, linen and their blends with synthetic fibres. Formaldehyde is strongly suspected of carcinogenic properties and can provoke allergy reactions, such as irritation in the eyes, skin and the respiratory system. In terms of human health or environmental protection considerations, the detection of formaldehyde in textiles is very necessary. Therefore, the development of a sensor system to measure the presence of formaldehyde in textiles with high sensitivity, selectivity and stability characteristics are highly desired. It is known that CDs enable encapsulation of diverse small organic compounds by forming inclusion complexes and serve as surfactants. Our research has indicated that CD based conductive polymer gas sensor materials are sensitive to formaldehyde by the drastic changes in conductivity. The variation of sensitivity of the CD based conductive polymer gas sensor material for different weights of fabric presented in the following figure (Fig. 10.17) proves that this material exhibits higher sensitivities to formaldehyde in the textile.

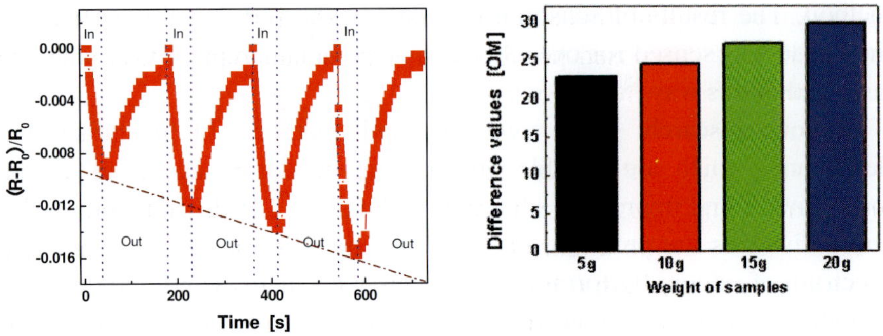

Fig. 10.17 The variation of sensitivity with CD based conductive polymer gas sensor material for the different weight of fabric.

10.4.3 *Applications of CDs to skin/healthcare products*

Improved stabilisation and other properties and targeted delivery of active materials are attractive propositions for the formulators of cosmetics and skin/healthcare products. Typical examples include the prolonged release of room fresheners, perfumes or detergents by controlled release of fragrances from inclusion molecules, increasing solubility of cold cream, enhancing stability of powdered hair bleach, carrying tocopherol of skin cleansers and increased stability or masking of odours of artificial tanning lotions. Nowadays, cosmetics preparation demands that a large amount of CD be used mainly in toothpastes, skin creams, liquid and solid fabric softeners, paper towels, tissues and underarm shields. The major benefits of CDs in this sector include: 1) improved stability and bioavailability; 2) changes solubility; 3) control release; 4) changes penetrability and 5) reduces inflammation and irritation. These are discussed below with different aspects of CDs in skin and healthcare fields.

10.4.3.1 *Applications in improving stability and bioavailability*

CDs can be used to stabilise compounds and improve bioavailability. Considering the finite cavity space of CDs, once a cavity has been occupied by a guest molecule, other molecules are excluded from occupying the space at the same time. Although the ends of the cavity are open, some steric hindrance is provided to prevent the approach of other molecules at the exposed portion of the molecule. Association with the cavity or hydroxyl groups surrounding the cavity can also stabilise the guest molecules.

Complexation with CDs can bring about stabilisation of active ingredients against oxidative, photolytic and thermal degradation. It can keep the molecules in a more rigid form, inhibit occurrences of reactive confirmation (e.g. vitamin E and vitamin C phosphates that are included in hydroxylated CDs show improved light stability compared with uncomplexed forms of the compound).[156] Stable ceramide powder can be made by using CDs. This ceramide powder is dispersable and suitable for cosmetics. Ceramide has a moisture-holding ability, which makes the skin moisture-rich and improves elasticity.[144] Evaporation and oxidation over a long period of time can be avoided when using CD-complexed fragrances in skin preparations, such as talcum powder, as the fragrance is stabilised. Fragrance is put into the CD cavity and complexed with calcium phosphate to stabilise the fragrance in manufacturing bathing preparations.[157] Skin cosmetics preparations are based on multi-unsaturated fats. Unsaturated fats, such as fish oils and vegetable oils, contain unsaturated fatty acids which are easily oxidised, resulting in unpleasant taste and odours. Complexation with CDs protects them from oxidation. Complexation also converts them from oils into solids containing 18% and greater oil in the complex.[147]

Hair dye preparations were made and evaluated immediately afterwards. After storing them for 6 months, no differences were found in the characteristics related to colour for the freshly prepared formulations made with or without CDs. After 6 months of storage, the preparations made with the complexed dye were judged by panelists where the verdict was that they have a deeper and richer colour, with increased resistance to being washed out of hair by shampoo or bleached by the sun.[147] CDs can increase the availability of triclosan (an antimicrobial) when used in silica-based toothpastes and induce an almost threefold enhancement of triclosan availability.[158]

10.4.3.2 *Applications in change solubility*

The solubility of a guest compound can be changed upon complexation with a CD. The guest molecule is generally a poorly water soluble active ingredient. The resulting inclusion complex benefits from the hydrophilic external part of the CD and has a higher solubility than the free active ingredient. When a guest compound is complexed by a cyclodextrin, the hydrophobic groups of the guest that would be in contact with the solvent in the free state, interact with the atoms of the cavity of the CD instead. The outer surface of the CD interacts with the solvent. Hence, the CDs can improve solubilisation of water insoluble active substances by the formation of water soluble complexes. For this purpose, they are used in cosmetics formulation as solubilising agents. The use of 2-hydroxypropyl-β-CD (2-HPβCD) for increasing the solubility of linalool and benzyl acetate fragrance materials was studied. The results showed that the inclusion complex with 2-HPβCD at 1:1 molar ratio can increase the solubility of linalool 5.9-fold and that of benzyl acetate 4.2-fold, whereas the inclusion complexes at a 1:2 molar ratio can increase the solubility of linalool and benzyl acetate 6.4- and 4.5-fold respectively.[159] As a result, it can be concluded that CDs (especially 2-HPβCD) are very appropriate for delivering fragrance materials to cosmetics. Parabens are widely used as typical preservatives for cosmetics, and if prolonged, the alkyl chain of parabens not only could increase their antiseptic action, but also their clinical safety. However, practical use of parabens with longer alkyl chains has been limited because of their low aqueous solubility. Therefore, Tanaka *et al.* studied the potential use of HP-b-CD to solubilise parabens and found that HP-β-CD can significantly increase the aqueous solubility of parabens.[160]

10.4.3.3 *Applications in control release*

Some volatile compounds can be complexed with CDs to reduce their volatility. Interaction between the guest and CDs produces a higher energy barrier to decrease volatilisers, and this behaviour can serve as producing long-lasting fragrances. As fragrance can surround the CD and form an inclusion compound, Holland *et al.* prepared cosmetics compositions compounded with CDs to keep fragrances long-lasting.

CD-based compositions were also used in various cosmetics products to reduce body odours.[140]

A dry CD powder can be used in products, such as menstrual products, diapers, tissues, paper towels, etc. As particle size becomes smaller, the surface area increases with respect to the volume, resulting in more rapid dissolution rates. Due to the rapid dissolution rate, only small amounts of water are needed for dissolution and dispersion of the molecules of CD, making them effective scavengers of unwanted and unpleasant odours.[147]

CDs can be used in hair care preparations to reduce volatility of odiferous mercaptans. Cationic polymers and CDs act synergistically to reduce odours from permanent wave solutions. α-, β-, or γ-cyclodextrins and polymers, such as cellulosics and quaternised vinylpyrolidine polymers, can be used. When polymer is used alone, only 40% of the panelists judged the results as good. More (40-60%) of the panelists judged results obtained with CD alone as good, and 80% of the panelists judged the preparation that used both CD and polymer as good. In some other formulations, CDs were used with a series of cyclic mercaptan compounds with electron attractive groups. In the presence of CDs, the formulations had no substantial odours and no odours were generated or less odour was generated when the formulation was used. The odour was selectively captured without reducing the effects or power.[147]

10.4.3.4 *Applications in change penetrability*

Moreover, CD complexation can increase the availability of active ingredients, either increasing or decreasing their permeability into the skin. HP-β-CD suppressed the *in vitro* cutaneous permeability of methyl paraben through the skin of hairless mice.[160] Both β-CD and DM-β-CD lowered the penetration of butyl paraben into the skin, and the decrease in butyl paraben penetration was explained by the good correlation between the partition coefficient and free butyl paraben fraction that was estimated based on the complex formation equilibrium.[161] Interestingly, by adding HP-β-CD into methyl paraben, this can promote the bioconversion of methyl paraben to the less toxic metabolite, *p*-hydroxybenzoic acid (*p*-HBA) in the epidermis. Since methyl paraben may be susceptible to enzymatic conversion in the epidermis, the promotion of HP-β-CD could be ascribed to the slower permeation rate of methyl paraben by an aqueous complex formation.

10.4.3.5 *Applications in reducing inflammation and irritation*

Generally, high safety standards and no side effects in cosmetics must be ensured, because cosmetics generally are placed at room temperature and used by people every day. Cosmetics often contain oleaginous raw materials, fragrance materials, colouring materials, sequestering agents, polymers, humectants, antioxidants, ultraviolet absorbents, etc. The penetration of these cosmetics materials with biologically active compounds into skin should be avoided. These substances include amino acids, hormones and vitamins. It has been reported that preservatives and fragrance materials may induce cutaneous allergies and dermatitis when they penetrate into skin.

When a guest is included in the cavity of CDs, the guest will be isolated and prevented from entering the surfaces of the body where it could cause unwanted side effects, such as irritation or dermatitis. Release of the guest from the complex is slow. As a result, the amount of free guest is lower than if the free guest is used alone so that the amount of guest to elicit a physiological response is reduced, resulting in masking or decreasing the intensity of the unwanted effects resulting from the guest.[147] There are many manufacturers who use CD inclusion technology to enable the inclusion of an object (active ingredient) to reduce vascular and skin irritations, such as gastrointestinal irritation and reduce side effects of drugs on the human body. Glycolic acid is used as an emollient and to induce superficial peeling of the skin. Glycolic acid is irritating to the skin. When complexed with CD, glycolic acid is found to be non-irritating due to its lower concentration in the free form.[147] CDs are used in the preparation of sunscreen lotions because the CD's cavity limits the interaction between the UV filter and the skin when they are in a 1:1 proportion (sunscreen/hydroxypropyl β-CD). Similarly, the performance and shelf life also can be improved when incorporating CDs into self-tanning emulsions or creams. More interestingly, the tan resulting from sunscreen lotions looks more natural than the yellow and reddish tinge when produced by traditional dihydroxyacetone products.[162]

10.4.4 *Summary*

The applications of CDs, CD derivatives and complexes are not limited to the above mentioned aspects. Due to the unique structure of CDs, they are now becoming an important part in drug development, chiral compound separations and as complexing agents in the textile, cosmetics and other industries. CDs deserve further exploration for potential applications in a variety of industry fields at both laboratory and industrial scales. We believe that these materials have many potential abilities, some that we already know and others that are still yet to be investigated.

References

1. Buhleier, E., Wehner, W., and Vogtle, F., "Cascade" and "Nonskid-Chain-like" Syntheses of Molecular Cavity Topologies. *Synthesis*, 1978. 1: pp. 155-158.
2. Tomalia, D. A., Baker, H., Dewald, J., Hall, M., Kallos, G., Martin, S., Roeck, J., J. Ryder, and Smith, P., A New Class of Polymers: Starburst-Dendritic Macromolecules. *Polymer Journal*, 1985. 17: pp. 117-132.
3. Newkome, G. R. Y., Z., Baker, G. R., Gupta, V. K., Cascade Molecules: A New Approach to Micelles. *Journal of Organic Chemistry*, 1985. 50: pp. 2003-2004.
4. Aulenta, F., Hayes, W., and Rannard, S., Dendrimers: a New Class of Nanoscopic Containers and Delivery Devices. *European Polymer Journal*, 2003. 39: pp. 1741-1771.
5. Matthews, O. A., Shipway, A. N., and Stoddart, J. F., Dendrimers—Branching out from curiosities into new technologies. *Progress in Polymer Science*, 1998. 23: pp. 1-56.
6. Klajnert, B. and Bryszewska, M., Dendrimers in Medicine. Vol. 1. 2008: Nova Science Publisher. pp. 131-132.
7. De, E. M. M., Berg, B. V. D., and Meijer, E. W., Poly-(propylenimin)-Dendrimere: Synthese in größerem Maßstab durch heterogen katalysierte Hydrierunge. Angewandte Chemie, 1993. 105: pp. 1370-1372.
8. Boas, U., Christensen, J. B., and Heegaard, P. M. H., Dendrimers in Medicine and Biotechnology New Molecular Tools. Vol. 1. 2006: *The Royal Society of Chemistry*. pp. 173-174.

9. Hawker, C. and Frechet, J. M. J., New Convergent Approach to Monodisperse Dendritic Macromolecules. *Chemical Communications*, 1990. 15: pp. 1010-1013.
10. Frechet, C. J., Preparation of Polymers with Controlled Molecular Architecture. A New Convergent Approach to Dendritic Macromolecules. *Journal of the American Chemical Society*, 1990. 112: pp. 7638-7647.
11. Hawker, C. J., Wooley, K. L., and Frechet, J. M. J., Unimolecular Micelles and Globular Amphiphiles: Dendritic Macromolecules as Novel Recyclable Solubilization Agents. *Journal of the Chemical Society. Perkin transactions*. I, 1993. 12: pp. 1287-1297.
12. Hawker, C. J., Dendritic and Hyperbranched Macromolecules-Precisely Controlled Macro-molecular Architectures. *Advances in Polymer Science*, 1999. 147: pp. 114-160.
13. Grayson, S. and Frechet, J., Divergent synthesis of dendronized poly(p-hydroxystyrene). *Macromolecules*, 2001. 34: pp. 6542-6544.
14. Grayson, S. and Frechet, J., Convergent Dendrons and Dendrimers: from Synthesis to Applications. *Chemical Reviews*, 2001. 101: pp. 3819-3867.
15. Ihre, H., Hult, A., Frechet, J., and Gitsov, I., Double-Stage Convergent Approach for the Synthesis of Functionalized Dendritic Aliphatic Polyesters Based on 2,2-Bis(hydroxymethyl)propionic Acid. *Macromolecules*, 1998. 31: pp. 4061-4068.
16. Pamela, J., Vande, V., Howard, W. T., Matthew, S. P., DeSilva, L. M., Wu, B., and Wooley, P. H., Evaluation of the biocompatibility of a chitosan scaffold in mice. *Journal of Biomedical Materials Research*, 2002. 59(3): pp. 585-590.
17. Inoue, T., Chen, G., Nakamae, K., and Hoffman, A. S., Temperature sensitivity of a hydrogel network containing different LCST oligomers grafted to the hydrogel backbone. *Polymer Gels and Networks*, 1997. 5(6): pp. 561-575.
18. Kawaguchi T, W. K., Wilkins, CL, Moore, JS, Double Exponential Eendrimer Growth. *Journal of the American Chemical Society*, 1995. 117: pp. 2159-2165.
19. Spindler, R. and Frechet, J. M. J., Two-step Approach Towards the Accelerated Synthesis of Dendritic Macromolecules. *Journal of the Chemical Society, Perkin Transaction* 1, 1993: pp. 913-918.
20. Zimmerman, S. C., Zeng, F., Reichert, D. E. C., and Kolotuchin, S. V., Self-Assembling Dendrimers. *Science*, 1996. 271: pp. 1095-1098.

21. Sampathkumar, S.-G. and Yarema, K. J. C. B., Targeting Cancer Cells with Dendrimers. *Chemistry and Biology*, 2005. 12: pp. 5-6.
22. Yang, W., Cheng, Y., Xu, T., Wang, X., and Wen, L.-p., Targeting Cancer Cells with Biotine-Dendrimer Conjugates doi:10.1-016/j.ejmech.2008.04.021. *European Journal of Medicinal Chemistry*, 2008. 44: pp. 862-868.
23. Supattapone, S., K. Nishina, and Rees, J. R. B. P., Pharmacological Approaches to Prion Research. *Biochemical Pharmacology*, 2002. 63: pp. 1383-1388.
24. Halkes, S. B., Vrasidas, I., Rooijer, G. R., van den Berg, A. J., Liskamp, R. M., and Pieters, R. J., Synthesis and Biological Activity of Polygalloyl-Dendrimers as Stable Tannic Acid Mimics. *Bioorganic and Medicinal Chemistry Letters*, 2002. 12: pp. 1567-1570.
25. Fischer, M. and Vögtle, F., Dendrimers: From Design to ApplicationÐA Progress Report. *Angewandte Chemie International Edition*, 1999. 38: pp. 884-905.
26. Reek, J. N. H., Groot, D. d., Oosterom, G. E., Kamer, P. C. J., and Leeuwen, P. W. N. M. v., Core and Periphery Functionalized Dendrimers for Transition Metal Catalysis: a Covalent and a Non-covalent Approach. *Reviews in Molecular Biotechnology*, 2002. 90: pp. 159-181.
27. Oosterom, G., Reek, J., Kamer, P., and Leeuwen, P. v., Transition metal catalysis using functionalized dendrimers. *Angewandte Chemie International Edition*, 2001. 40: pp. 1828-1849.
28. Caminade, A., Maraval, V., Laurent, R., and Majoral, J., Organometallic Derivatives of Phosphorus-containing Dendrimers. Synthesis, Properties and Applications in Catalysis. *Current Organic Chemistry*, 2002. 6: pp. 739-774.
29. Kassube, J. K. and Gade, L. H., Stereoselective Dendrimer Catalysis. *Topics in Organometallic Chemistry*, 2006. 20: pp. 61-96.
30. Hearshaw, M. A. and Moss, J. R., Organometallic and Related Metal-Containing Dendrimers. *Chemical Communications*, 1999. 11: pp. 1-8.
31. Xie, H., Howe, J. Y., Schwartz, V., Monnier, J. R., Williams, C. T., and Ploehna, H. J., Hydrodechlorination of 1,2-Dichloroethane Catalyzed by Dendrimer-Derived Pt-Cu/SiO_2 Catalysts. *Journal of Catalysis*, 2008. 259: pp. 111-122.

32. Balzani, V., Ccroni, P., Gcstcrmann, S., Kauffmann, C., Gorka, M., and Vogtlc, F., Dendrimers as Fluorescent Sensors with Signal Amplification. *Chemical Communications*, 2000: pp. 853-854.
33. Valerio, C., Fillaut, J., Ruiz, J., Guittard, J., Blais, J., and Astruc, D., The Dendritic Effect in Molecular Recognition: Ferrocene Dendrimers and Their Use as Supramolecular Redox Sensors for the Recognition of Small Inorganic Anions. *Journal of American Chemical Society*, 1997. 119: pp. 2588-2589.
34. James, T., Shinmori, H., Takeuchi, M., and Shinkai, S., A Saccharide 'Sponge'. Synthesis and Properties of a Dendritic Boronic Acid. *Chemical Communications*, 1996: pp. 705-706.
35. Beebe, D. J., Moore, J. S., Bauer, J. M., Yu, Q., Liu, R. H., Devadoss, C., and Jo, B. H., Functional hydrogel structures for autonomous flow control inside microfluidic channels. *Nature*, 2000. 404(6778): pp. 588-590.
36. Schenning, A., Martin, R., Ito, M., Dicdcrich, F., Boudon, C., Gissclbrccht, J., and Gross, M., Dendritic Rods with a Poly(triacetylene) Backbone: Insulated Molecular Wires. *Chemical Communications*, 1998: pp. 1013-1014.
37. Smith, D. and Diederich, F., Dendritic Hydrogen Bonding Receptors: Enantiomerically Pure Dendroclefts for the Selective Recognition of Monosaccharides. *Chemical Communications*, 1998: pp. 2501-2502.
38. Froehling, P. E., Dendrimers and Dyes--a Review. *Dyes and Pigments,* 2001. 48: pp. 187-195.
39. Baars, M. W. P. L., Froehling, P. E., and Meijer, E. W., Liquid-liquid Extractions Using Poly(propylene imine) Dendrimers with an Apolar Periphery. *Chemical Communications*, 1997: pp. 1959-1960.
40. de Brabander-van den Berg, E. M. M. and Meijer, E. W., Poly(propylene imine) Dendrimers: Large-Scale Synthesis by Heterogeneously Catalyzed Hydrogenations. *Angewandte Chemie International Edition*, 1993. 32: pp. 1308-1311.
41. Froehling, P. E., De Brabander-van, B. E. M. M., and Mostert, H. A. M., Process for Incorporating an Active Substance in a Moulded Plastic Part, PCT/NL1997/000527. 22.09.1997.
42. Froehling, P. E. and Burkinshaw, S. M., Dendritic Polymers: New Concept for Dyeable Poly(propylene) Fibers. *Chemical Fibers International*, 2000. 50: pp. 448-449.

43. Burkinshaw, S. M., Mignanelli, M., Froehling, P. E., and Bide, M. J., The use of Dendrimers to Modify the Dyeing Behaviour of Reactive Dyes on Cotton. *Dyes and Pigments*, 2000. 47: pp. 259-267.
44. Hu, J. L., Nie, K. M., Meng, Q. H., and Zheng, G. H., Multifunction Finishing Liquids Containing Dendrimers and the Application of the Liquids in Textile Finishing. 2007: US Patent 11748892.
45. Jikci, M. and Kakimoto, M.-a., Hyperbranchcd Polymers: A Promising New Class of Materials. *Progress in Polymer Science*, 2001. 26: pp. 1233-1285.
46. Yates, C. R. and Hayes, W., Synthesis and Applications of Hyperbranched Polymers. *European Polymer Journal*, 2004. 40: pp. 1257-1281.
47. Seiler, M., Hyperbranched Polymers: Phase Behavior and New Applications in the Field of Chemical Engineering. *Fluid Phase Equilibria*, 2006. 241: pp. 155-174.
48. Kim, Y. and Webster, O., Water Soluble Hyperbranched Polyphenylene: "A Unimolecular Micelle?" *Journal of the American Chemical Society*, 1990. 112: pp. 4592-4593.
49. Jikei, M. and Kakimoto, M.-a., Hyperbranched Polymers: A Promising New Class of Materials. *Progress in Polymer Science*, 2001. 26: pp. 1233-1285.
50. Kumar, A. and Ramakrishnan, S., Hyperbranched Polyurethanes with Varying Spacer Segments Between the Branching Points. *Journal of Polymer Science Part A: Polymer Chemistry*, 1996. 34: pp. 839-848.
51. Spindler, R. and Frechet, J., Synthesis And Characterization of Hyperbranched Polyurethanes Prepared from Blocked Isocyanate Monomers by Step-Growth Polymerization. *Macromolecules*, 1993. 26: pp. 4809-4813.
52. Versteegen, R., Sijbesma, R., and Meijer, E., Synthesis of [n]-Polyurethanes and Hyperbranched Polyurethanes. *Polymer Preprints*, 1999. 40: pp. 839-840.
53. Lach, C. and Frey, H., Enhancing the Degree of Branching of Hyperbranched Polymers by Postsynthetic Modification. *Macromolecules*, 1998. 31: pp. 2381-2383.
54. Yoon, K. and Son, D., Synthesis of Hyperbranched Poly-(carbosilarylenes). *Macromolecules*, 1999. 32: pp. 5210-5216.

55. Hobson, L. and Feast, W., Poly(amidoamine) Hyperbranched Systems: Synthesis, Structure and Characterization. *Polymer*, 1999. 40: pp. 1279-1297.
56. Londergan, T., You, Y., Thompson, M., and WP., R. W., synthesis of Cross-Conjugated Polymers and Hyperbranched Materials. Copoly(arylene/1,1-vinylene)s. *Macromolecules*, 1998. 31: pp. 2784-2788.
57. Lu, P., Paulasaari, J., and Weber, W., Hyperbranched Poly(4-acetylstyrene) by Ruthenium-Catalyzed Step-Growth Polymerization of 4-Acetylstyrene. *Macromolecules*, 1996. 29: pp. 8583-8586.
58. Kim, Y. and Webster, O., Hyperbranched Polyphenylenes. *Macromolecules*, 1992. 25: pp. 5561-5572.
59. Kim, Y. and Beckerbauer, R., Role of End Groups On the Glass Transition of Hyperbranched Polyphenylene And Triphenylbenzene Derivatives. *Macromolecules*, 1994. 27: pp. 1968-1971.
60. Coessens, V., Pintauer, T., and Matyjaszewski, K., Functional Polymers by Atom Transfer Radical Polymerization. *Progress in Polymer Science*, 2001. 26: pp. 337-377.
61. Liu, H. and Wilen, C.-E., Hyperbranched Poly[allyl Ether-alt-maleic Anhydride] Produced by the Self-Condensing Alternating Copolymerization Approach. *Macromolecules*, 2001. 34: pp. 5067-5070.
62. Fréchet, J. M. J., Henmi, M., Gitsov, I., Aoshima, S., Leduc, M. R., and Grubbs, R. B., Self-Condensing Vinyl Polymerization: An Approach to Dendritic Materials. *Science*, 1995. 269: pp. 1080-1083.
63. Matyjaszewski, K., Gaynor, S., Kulfan, A., and Podwika, M., Preparation of Hyperbranched Polyacrylates by Atom Transfer Radical Polymerization. 1. Acrylic AB* Monomers in `Living' Radical Polymerizations. *Macromolecules,* 1997. 30: pp. 5192-5194.
64. Matyjaszewski, K., Gaynor, S., and Miiller, A., Preparation of Hyperbranched Polyacrylates by Atom Transfer Radical Polymerization. 2. Kinetics and Mechanism of Chain Growth for the Self-condensing Vinyl Polymerization of 2-((2-bromopropionyl)oxy)ethyl Acrylate. *Macromolecules*, 1997. 30: pp. 7034-7041.
65. Matyjaszewski, K. and Gaynor, S., Preparation of Hyperbranched Polyacrylates by Atom Transfer Radical Polymerization. 3. Effect of Reaction Conditions on the Self-Condensing Vinyl Poly-

merization of 2-((2-bromopropionyl)oxy)ethyl Acrylate. *Macromolecules*, 1997. 30: pp. 7042-7049.

66. Slark, A. T., Sherrington, D. C., Titterton, A., and Martin, I. K., Branched Methacrylate Copolymers from Multifunctional Comonomers: The Effect of Multifunctional Monomer Functionality on Polymer Architecture and Properties. *Journal of Materials Chemistry*, 2003. 13: pp. 2711-2720.
67. O'Brien, N., McKee, A., Sherrington, D. C., Slark, A. T., and Titterton, A., Facile, Versatile and Cost Effective Route to Branched Vinyl Polymers. *Polymer*, 2000. 41: pp. 6027-6031.
68. Suzuki, M., Ii, A., and Saegusa, T., Multibranching Polymerization: Palladium-Catalyzed Ring-Opening Polymerization of Cyclic Carbamate. *Macromolecules*, 1992. 25: pp. 7071-7072.
69. Suzuki, M., Yoshida, S., Shiraga, K., and Saegusa, T., New Ring-Opening Polymerization via a π-Allylpalladium Complex. 5. Multibranching Polymerization of Cyclic Carbamate To Produce Hyperbranched Dendritic Polyamine. *Macromolecules*, 1998. 31: pp. 1716-1719.
70. Sunder, A., Hansclmann, R., Frcy, H., and Miilhaupl, R., Controlled Synthesis of Hyperbranched Polyglycerols by Ring-Opening Multibranching Polymerization. *Macromolecules*, 1999. 32: pp. 4240-4246.
71. Jikei, M., Chon, S.-H., Kakimoto, M.-A., Kawauchi, S., Imase, T., and Watenebe, J., Synthesis of Hyperbranched Aromatic Polyamides from Aromatic Diamines and Trimesic Acid. *Macromolecules*, 1999. 32: pp. 2061-2064.
72. Rannard, S. P., Davis, N. J., and Herbert, I., Synthesis of Water Soluble Hyperbranched Polyurethanes Using Selective Activation of AB_2 Monomers. *Macromolecules*, 2004. 37: pp. 9418-9430.
73. Grubbs, R., Hawker, C., Dao, J., and Frechet, J., A Tandem Approach to Graft and Dendritic Graft Copolymers Based on "Living" Free Radical Polymerizations. *Angewandte Chemie International Edition*, 1997. 36: pp. 270-272.
74. Hult, A., Johansson, M., and Malmström, E., Hyperbranched Polymers. *Advances in Polymer Science*, 1999. 143: pp. 1-34.
75. Frey, H. and Haag, R., Dendritic Polyglycerol: a New Versatile Biocompatible Material. *Reviews in Molecular Biotechnology*, 2002. 90: pp. 257-267.

76. Chen, S., Zhang, X.-Z., Cheng, S.-X., Zhuo, R.-X., and Gu, Z.-W., Functionalized Amphiphilic Hyperbranched Polymers for Targeted Drug Delivery. *Biomacromolecules*, 2008. 9: pp. 2578-2585.
77. Tziveleka, L.-A., Kontoyianni, C., Sideratou, Z., Tsiourvas, D., and Paleos, C. M., Novel Functional Hyperbranched Polyether Polyols as Prospective Drug Delivery Systems. *Macromolecular Bioscience*, 2006. 6: pp. 161-169.
78. Hong, Y., Coombs, S., Cooper-White, J., Mackay, M., Hawker, C., Malmstrom, E., and Rehnberg, N., Film Blowing of Linear Low density Polyethylene Blended with a Novel Hyperbranched Polymer Processing Aid. *Polymer*, 2000. 41: pp. 7705-7713.
79. Xie, J. B. and Hsieh, Y. L., Thermosensitive poly(N-isopropylacrylamide) hydrogels bonded on cellulose supports. *Journal of Applied Polymer Science*, 2003. 89(4): pp. 999-1006.
80. Mezzenga, R. and Manson, J.-A. E., Thermo-mechanical Properties of Hyperbranched Polymer Modified Epoxies. *Journal of Materials Science*, 2001. 36: pp. 4883-4891.
81. Plummer, C. J. G., Garamszegi, L., Leterrier, Y., Rodlert, M., and Manson, J.-A. E., Hyperbranched Polymer Layered Silicate Nanocomposites. *Chemistry of Materials*, 2002. 14: pp. 486-488.
82. Jin, F.-L. and Park, S.-J., Thermal Properties and Toughness Performance of Hyperbranched-Polyimide-Modified Epoxy Resins. *Journal of Polymer Science, Part B: Polymer Physics*, 2006. 44: pp. 3348-3356.
83. Zhang, Y., Wada, T., and Sasabe, H., A New Hyperbranched Polymer with Polar Chromophores for Nonlinear Optics. *Polymer*, 1997. 38: pp. 2893-2897.
84. Hussler, M., Liu, J., Zheng, R., Lam, J. W. Y., Qin, A., and Tang, B. Z., Synthesis, Thermal Stability, and Linear and Nonlinear Optical Properties of Hyperbranched Polyarylenes Containing Carbazole and/or Fluorene Moieties. *Macromolecules*, 2007. 40: pp. 1914-1925.
85. Sato, M., Nakashima, A., Sato, Y., and Yamaguchi, I., Linear and A_2+B_3-Type Hyperbranched Polyesters Comprising Phenylbenzothiazole Unit: Preparation, Liquid Crystalline, and Optical Properties. *Journal of Polymer Science: Part A: Polymer Chemistry*, 2008. 46: pp. 6688-6702.
86. Xie, J., Hu, L., Shi, W., Deng, X., Cao, Z., and Shen, Q., Synthesis and Nonlinear Optical Properties of Hyperbranched Polytriazole Containing Second-Order Nonlinear Optical Chromophore. *Journal*

of Polymer Science: Part B: Polymer Physics, 2005. 46: pp. 1140-1148.

87. Xie, J., Deng, X., Cao, Z., Shen, Q., Zhang, W., and Shi, W., Synthesis and second-order nonlinear optical properties of hyperbranched polymers containing pendant azobenzene chromophores. *Polymer*, 2007. 48: pp. 5988-5993.
88. Burkinshaw, S. M., Froehling, P. E., and Mignanelli, M., The Effect of Hyperbranched Polymers on the Dyeing of Polypropylene Fibres. *Dyes and Pigments*, 2002. 53: pp. 229-235.
89. Gooday, G. W., Physiology of microbial degradation of chitin and chitosan. *Biodegradation*, 1990. 1(2 - 3): pp. 177-190.
90. Di Martino, A., Sittinger, M., and Risbud, M. V., Chitosan: A versatile biopolymer for orthopaedic tissue-engineering. *Bio-materials,* 2005. 26(30): pp. 5983-5990.
91. Crini, G. and Badot, P. M., Application of chitosan, a natural aminopolysaccharide, for dye removal from aqueous solutions by adsorption processes using batch studies: A review of recent literature. *Progress in Polymer Science*, 2008. 33(4): pp. 399-447.
92. Kendra, D. F., Christian, D., and Hadwiger, L. A., Chitosan oligomers from Fusarium solani/pea interactions, chitinase/b-glucanase digestion of sporelings and from fungal wall chitin actively inhibit fungal growth and enhance disease resistance. *Physiological and Molecular Plant Pathology*, 1989. 35(3): pp. 215-230.
93. Roller, S. and Covill, N., The antifungal properties of chitosan in laboratory media and apple juice. *International Journal of Food Microbiology*, 1999. 47(1-2): pp. 67-77.
94. Ben-Shalom, N., Ardi, R., Pinto, R., Aki, C., and Fallik, E., Controlling gray mould caused by Botrytis cinerea in cucumber plants by means of chitosan. *Crop Protection*, 2003. 22(2): pp. 285-290.
95. Choi, B. K., Kim, K. Y., Yoo, Y. J., Oh, S. J., Choi, J. H., and Kim, C. Y., In vitro antimicrobial activity of a chito-oligosaccharide mixture against Actinobacillus actinomycetemcomitans and Streptococcus mutans. *International Journal of Antimicrobial Agents*, 2001. 18(6): pp. 553-557.
96. Lee, J.-Y., Nam, S.-H., Im, S.-Y., Park, Y.-J., Lee, Y.-M., Seol, Y.-J., Chung, C.-P., and Lee, S.-J., Enhanced bone formation by controlled growth factor delivery from chitosan-based biomaterials. *Journal of Controlled Release*, 2002. 78(1-3): pp. 187-197.

97. Liu, M., Bian, F., and Sheng, F., FTIR study on molecular structure of poly(N-isopropylacrylamide) in mixed solvent of methanol and water. *European Polymer Journal,* 2005. 41(2): pp. 283-291.
98. Suzuki, K., Mikami, T., Okawa, Y., Tokoro, A., Suzuki, S., and Suzuki, M., Anti-tumor effect of hexa-N-acetylchitohexaose and chitohexaose. *Carbohydrate Research*, 1986. 151: pp. 403-408.
99. Koide, S. S., Chitin-chitosan: Properties, benefits and risks. *Nutrition Research*, 1998. 18(6): pp. 1091-1101.
100. Hu, Q., Li, B., Wang, M., and Shen, J., Preparation and characterization of biodegradable chitosan/hydroxyapatite nanocomposite rods via in situ hybridization: a potential material as internal fixation of bone fracture. *Biomaterials*, 2004. 25(5): pp. 779-785.
101. Pedroni, V. I., Schulz, P. C., Gschaider, M. E., and Andreucetti, N., Chitosan structure in aqueous solution. *Colloid and Polymer Science*, 2003. 282(1): pp. 100-102.
102. Guibal, E., Dambies, L., Milot, C., and Roussy, J., Influence of polymer structural parameters and experimental conditions on metal anion sorption by chitosan. *Polymer International*, 1999. 48(8): pp. 671-680.
103. Guan, Y. L., Shao, L., and Yao, K. D., A Study on correlation between water state and swelling kinetics of chitosan-based hydrogels. *Journal of Applied Polymer Science*, 1996. 61(13): pp. 2325-2335.
104. Mao, J. S., Study of the near-neutral pH-sensitivity of chitosan/gelatin hydrogels by turbidimetry and microcantilever deflection. *Biotechnology and Bioengineering*, 2006. 95(3): pp. 333-341.
105. Khurma, J. R. and Nand, A. V., Temperature and pH sensitive hydrogels composed of chitosan and poly(ethylene glycol). *Polymer Bulletin*, 2008. 59(6): pp. 805-812.
106. Li, Z., Liu, X. F., Zhuang, X. P., Guan, Y. L., and Yao, K. D., Manufacture and properties of chitosan/N,O-carboxymethylated chitosan/viscose rayon antibacterial fibers. *Journal of Applied Polymer Science*, 2002. 84(11): pp. 2049-2059.
107. http://www.fujibo.co.jp/us/chitopoly/chito_01.html.
108. Ravi Kumar, M. N. V., A review of chitin and chitosan applications. *Reactive and Functional Polymers,* 2000. 46(1): pp. 1-27.

109. Kurita, K., Chemistry and application of chitin and chitosan. *Polymer Degradation and Stability*, 1998. 59(1-3): pp. 117-120.
110. Berth, G., Dautzenberg, H., and Peter, M. G., Physico-chemical characterization of chitosans varying in degree of acetylation. *Carbohydrate Polymers*, 1998. 36(2-3): pp. 205-216.
111. Majeti N. V. Ravi Kumar, A review of chitin and chitosan applications. *Reactive and Functional Polymers*, 2000. 46: pp. 1-27.
112. Peter, M. G., Albertsson, A. C., and Huang, S. J., Applications and environmental aspects of chitin and chitosan. *Journal of Macromolecular Science. Pure and Applied Chemistry*, 1995. 32(4): pp. 629-640.
113. Mita, K., Development and application of chitin and chitosan. *Kogyo Gijsutsu-kai*, 1987: pp. 248-269.
114. Lee, S., Cho, J. S., and Cho, G., Antimicrobial and blood repellent finishes for cotton and nonwoven fabrics based on chitosan and fluoropolymers. *Textile Research Journal,* 1999. 69(2): pp. 104-112.
115. Alemdaroglu, C., Degim, Z., Celebi, N., Zor, F., Ozturk, S., and Erdogan, D., An investigation on burn wound healing in rats with chitosan gel formulation containing epidermal growth factor. Burns, 2006. 32(3): pp. 319-327.
116. Lee, S. B., Ha, D. I., Cho, S. K., Kim, S. J., and Lee, Y. M., Temperature/pH-sensitive comb-type graft hydrogels composed of chitosan and poly(N-isopropylacrylamide). *Journal of Applied Polymer Science*, 2004. 92(4): pp. 2612-2620.
117. Azad, A. K., Sermsintham, N., Chandrkrachang, S., and Stevens, W. F., Chitosan membrane as a wound-healing dressing: characterization and clinical application. *Journal of Biomedical Materials Research - Part B Applied Biomaterials*, 2004. 69(2): pp. 216-222.
118. Wang, L. S., Khor, E., Wee, A., and Lim, L. Y., Chitosan-alginate PEC membrane as a wound dressing: Assessment of incisional wound healing. *Journal of Biomedical Materials Research*, 2002. 63(5): pp. 610-618.
119. Wang, M. Z., Qiang, J. C., Fang, Y., Hu, D., Cui, Y. L., and Fu, X. G., Preparation and properties of chitosan-poly(N-isopropylacrylamide) semi-IPN hydrogels. *Journal of Polymer Science, Part A: Polymer Chemistry*, 2000. 38(3): pp. 474-481.
120. Verestiuc, L., Ivanov, C., Barbu, E., and Tsibouklis, J., Dual-stimuli-responsive hydrogels based on poly(N-isopro-

pylacrylamide)/chitosan semi-interpenetrating networks. *International Journal of Pharmaceutics*, 2004. 269(1): pp. 185-194.

121. Lee, C. F., Wen, C. J., Lin, C. L., and Chiu, W. Y., Morphology and temperature responsiveness-swelling relationship of poly(N-isopropylamide-chitosan) copolymers and their application to drug release. *Journal of Polymer Science Part A: Polymer Chemistry*, 2004. 42(12): pp. 3029-3037.
122. Liu, B., Hu, J., and Meng, Q., Nonwoven supported temperature-sensitive poly(N-isopropylacrylamide)/polyurethane copolymer hydro-gel with antibacterial activity. *Journal of Biomedical Materials Research Part B: Applied Biomaterials*, 2009. 89B(1): pp. 1-8.
123. Vila, A., Sánchez, A., Janes, K., Behrens, I., Kissel, T., Jato, J. L. V., and Alonso, M. J., Low molecular weight chitosan nanoparticles as new carriers for nasal vaccine delivery in mice. *European journal of pharmaceutics and biopharmaceutics*, 2004. 57(1): pp. 123-131.
124. Nasr, H. E., Sayyah, S. M., Essa, D. M., Samaha, S. H., and Rabie, A. M., Utilization of acrylates emulsion terpolymer with chitosan as a finishing agent for cotton fabrics. *Carbohydrate Polymers*, 2009. 76(1): pp. 36-45.
125. Huang, K.-S., Wu, W.-J., Chen, J.-B., and Lian, H.-S., Application of low-molecular-weight chitosan in durable press finishing. *Carbohydrate Polymers*, 2008. 73(2): pp. 254-260.
126. Jocic, D., Vílchez, S., Topalovic, T., Navarro, A., Jovancic, P., Julià Maria, R., and Erra, P., Chitosan/acid dye interactions in wool dyeing system. *Carbohydrate Polymers*, 2005. 60(1): pp. 51-59.
127. Davarpanah, S., Mahmoodi, N. M., Arami, M., Bahrami, H., and Mazaheri, F., Environmentally friendly surface modification of silk fiber: Chitosan grafting and dyeing. *Applied Surface Science*, 2009. 255(7): pp. 4171-4176.
128. Alonso, D., Gimeno, M., Olayo, R., Vazquez-Torres, H., Sepuveda-Sachez, J. D., and Shirai, K., Cross-linking chitosan into UV-irradiated cellulose fibers for the preparation of antimicrobial-finished textiles. *Carbohydrate Polymers*. In Press, Corrected Proof.
129. Bonakdar E. A, Chitosan in cosmetics: Technical aspects when formulating. *Cosmet. Toiletries,* 2001. 116(8): pp. 73-80.
130. Wachter, R. and Stenberg, E., HYDAGEN CMF in Cosmetic applications. Efficacy in different in-vitro and in-vivo measurements. *Advances in Chitin Science,* ed. D. A., R. G.A.F., and V. K.M. Vol. I. 1997, Lyon: Jacques Andre Publisher. pp. 381-388.

131. Horner, V., Pittermann, W., and Wachter, R., Efficiency of high molecular weight chitosan in skin care application. Advances in Chitin Science, ed. D. A., R. G.A.F., and V. K.M. Vol. II. 1997, Lyon: Jacques Andre Publisher. 671-677.
132. Hohle, M. and V.Griesbach, Chitosan: A deodorizing component. *Cosmet. Toiletries,* 1998. 114(12): pp. 61-64.
133. Dee, G. J., Rhode, O., and Wachter, R., Chitosan multi-functional marine polymer. *Cosmet. Toiletries,* 2001. 116(2): pp. 39-44.
134. Szejtli, J., Introduction and General Overview of Cyclodextrin Chemistry. *Chemical Reviews,* 1998. 98(5): pp. 1743-1754.
135. Uekama, K., Hirayama, F., and Irie, T., Cyclodextrin Drug Carrier Systems. *Chemical Reviews,* 1998. 98(5): pp. 2045-2076.
136. Szente, L. and Szejtli, J., Cyclodextrins as food ingredients. *Trends in Food Science & Technology,* 2004. 15(3-4): pp. 137-142.
137. Buschmann, H. J. and Schollmeyer, E., Applications of cyclodextrins in cosmetic products: *A review. J Cosmet Sci,* 2002. 53(3): pp. 185-191.
138. Szejtli, J., Cyclodextrins in the textile industry. *Starch-Starke*, 2003. 55(5): pp. 191-196.
139. Juvancz, Z. and Szejtli, J., The role of cyclodextrins in chiral selective chromatography. *Trends in Analytical Chemistry,* 2002. 21(5): pp. 379-388.
140. Del Valle, E. M. M., Cyclodextrins and their uses: a review. *Process Biochemistry,* 2004. 39(9): pp. 1033-1046.
141. Szejtli, J., Past, present, and future of cyclodextrin research. *Pure and applied chemistry*, 2004. 76(10): pp. 1825-1846.
142. Saenger, W., Cyclodextrin inclusion compounds in research and industry. *Angewandte Chemie International Edition in English,* 1980. 19(5): pp. 344-362.
143. D'Souza, V. T. and Lipkowitz, K. B., Cyclodextrins: Introduction. *Chemical Reviews,* 1998. 98(5): pp. 1741-1742.
144. Nonaka, T., Hashimoto, K., and Kurihara, S., Preparation of thermosensitive cellophane-graft-N-isopropylacrylamide copolymer membranes and permeation of solutes through the membranes. *Journal of Applied Polymer Science,* 1997. 66(2): pp. 209-216.
145. Buschmann, H. J., Knittel, D., and Schollmeyer, E., New Textile Applications of Cyclodextrins. *Journal of Inclusion Phenomena and Macrocyclic Chemistry*, 2001. 40(3): pp. 169-172.

146. Savarino, P., Viscardi, G., Quagliotto, P., Montoneri, E., and Barni, E., Reactivity and effects of cyclodextrins in textile dyeing. *Dyes and Pigments,* 1999. 42(2): pp. 143-147.
147. Hedges, A. R., Industrial Applications of Cyclodextrins. *Chemical Reviews*, 1998. 98(5): pp. 2035-2044.
148. Ruppert, S., Knittel, D., Buschmann, H. J., Wenz, G., and Schollmeyer, E., Fixation of Cyclodextrinderivates on Polyester Fibers. *Starch-Stuttgart*, 1997. 49: pp. 160-164.
149. Denter, U., Buschmann, H. J., and Schollmeyer, E., Trichromatic dyeing of cotton: interactions between dyes and auxiliaries. *Textilveredlung,* 1991. 26(4): pp. 113-116.
150. George, B., Govindaraj, M., Ujiie, H., and Wood, D., National Tex tile Center Progress Report, November 2002.
151. Buschmann, H. J., Denter, U., Knittel, D., and Schollmeyer, E., The Use of Cyclodextrins in Textile Processes: An Overview. *Journal of the Textile Institute*, 1998. 89(3): pp. 554 - 561.
152. Denter, U. and Schollmeyer, E., Surface modification of synthetic and natural fibres by fixation of cyclodextrin derivatives. *Journal of Inclusion Phenomena and Macrocyclic Chemistry*, 1996. 25(1): pp. 197-202.
153. Wang, C. X. and Chen, S. L., Aromachology and its Application in the Textile Field. *Fibres & Textiles Eastern Europe*, 2005. 13(6): pp. 41-44.
154. Hu, J. L., Zeng, F., and Li, P., Methods of manufacturing deodorants, and deodorants resulting thereof, 2005. United States Patent Application 20050232880
155. Wang, J. and Cai, Z., Incorporation of the antibacterial agent, miconazole nitrate into a cellulosic fabric grafted with β-cyclodextrin. *Carbohydrate Polymers*. 2008. 72(4): pp. 695-700.
156. Patravale, V. B. and Mandawgade, S. D., Novel cosmetic delivery systems: an application update. *International Journal of Cosmetic Science*, 2008. 30(1): pp. 19-33.
157. Singh, M., Sharma, R., and Banerjee, U. C., Biotechnological applications of cyclodextrins. *Biotechnol Adv*, 2002. 20(5-6): pp. 341-59.
158. Loftsson, T., Leeves, N., Bjornsdottir, B., Duffy, L., and Masson, M., Effect of cyclodextrins and polymers on triclosan availability and substantivity in toothpastes in vivo. *Journal of Pharmaceutical Sciences*, 1999. 88(12): pp. 1254-1258.

159. Numanoğlu, U., Şen, T., Tarimci, N., Kartal, M., Koo, O., and Önyüksel, H., Use of cyclodextrins as a cosmetic delivery system for fragrance materials: Linalool and benzyl acetate. AAPS PharmSciTech, 2007. 8(4): pp. 34-42.
160. Tanaka, M., Iwata, Y., Kouzuki, Y., Taniguchi, K., Matsuda, H., Arima, H., and Tsuchiya, S., Effect of 2-hydroxypropyl-β-cyclodextrin on percutaneous absorption of methyl paraben. *Journal of pharmacy and pharmacology*, 1995. 47(11): pp. 897-900.
161. Iwata, H. and Matsuda, T., Preparation and properties of novel environment-sensitive membranes prepared by graft polymerization onto a porous membrane. *Journal of Membrane Science*, 1988. 38(2): pp. 185-199.
162. Scalia, S., Villani, S., and Casolari, A., Inclusion complexation of the sunscreen agent 2-ethylhexyl-p-dimethylaminobenzoate with hydroxypropyl-beta-cyclodextrin: effect on photostability. *The Journal of Pharmacy and Pharmacology*, 1999. 51(12): pp. 1367-1374.

Index

—2—

2,2-bis(hydroxymethyl)propionic acid, 323

—A—

active medical devices, 307
adaptive composites, 180, 195, 199, 204, 239
adaptive lenses, 305
adaptive polymer, 72, 142
adaptive polymer composites, 17, 181, 201, 203, 204, 205
adaptive polymeric gels, 10, 16, 82, 83, 90, 100
adaptive polymeric particles, 4, 6, 10, 12, 17, 107, 109, 111, 135
adaptive textiles, 2, 5, 12, 16, 17, 142, 159, 163, 164, 173, 205
aerosol filters, 241
all-organic dielectric-percolative composites, 194
alternating magnetic field, 32
ammonium persulfate, 89
amorphous regions, 27, 87
amphiphilic block copolymers, 298
amphoteric, 263, 274
aneurysm coils, 309
antibacterial activity, 18, 36, 340
antibody-mimics, 297
anti-inflammation, 265
antioxidation, 254, 262, 265
antisepsis, 254
artificial muscle, 96, 167, 180, 181, 203, 288, 291
ater immersion, 8
average pressure, 152
azobenzene, 7, 33, 198, 327, 338, 377
azobenzene moieties, 34, 198

—B—

bacterial cellulose nanofibers, 249
bagging recovery, 153, 155
bending test, 26
benzophenone, 266
bioadhesiveness, 99
biodegradability, 36, 75, 97, 262, 263, 271, 343
biodegradable hydrogels, 79
biodegradable intragastric implants, 308
biodegradable materials, 286, 337
biodegradable suture, 37
biological self-assembly, 224
biomedical application, 19, 37, 79, 84, 97, 105, 325
biomedical engineering, 79, 293, 354
bio-microelectromechanical system, 97

biomimetics, 180, 203, 209
bioresponses, 288
biosensor technology, 229, 301
bio-separation, 4
butyl methacrylate, 97

—C—

camphor, 254
carbon nanotube–elastomer nanocomposite, 189
catheters, 307
ceramide, 365
chameleon materials, 159, 161
chemical crosslink, 26, 32
chemiluminescence, 164
cholesterin types, 160
cinnamic groups, 198
cis and trans conformations, 198
coalescing filters, 241
coaxial electrospinning, 146
colour former, 160
complexation, 224, 263, 268, 362, 366, 367, 383
conducting electroactive polymers, 303
conducting polyaniline-polyisoprene, 302
convergent synthesis, 322
core yarn, 149, 153, 155
core-shell structure, 12, 121
couple-monomer methodology, 331
coupling agent, 12, 116, 323
crease retention, 38, 157
cross section, 228
crystal violet lactone, 160
crystalline regions, 27
crystallinity, 339, 342, 344, 345
cyclic tensile investigation, 26
cyclodextrin derivatives, 382

—D—

decomposition, 266
degradable implants, 37
degree of acetylation, 342, 379
dendritic subunits, 330
dendrons, 322, 323, 325
derivative, 340, 358, 362
diagnosis and medical treatment, 294
diagnostic assays, 299
diapers, 100, 367
dielectric elastomer, 193, 289, 290
dielectric electro adaptive polymers, 190
digoxin antibody-antigen system, 299
dimensional stability, 157
dimethylol propionic acid, 42
dip coating, 146
direct thermal sensitive, 26, 30
divergent synthesis, 321, 322
donors and acceptors, 188
drug targeting devices, 297
dry eyes, 304
dyeability, 339, 353
dynamic self-assembly, 224

—E—

elastic net-points, 26, 29
elastic network, 31, 47, 53
elastomer bearings, 204
electrical stimuli, 289
electrical voltage, 221
electro adaptive conductive polymers, 190
electro adaptive ionic polymer gels, 196
electro adaptive polymeric composites, 189

electroactive hydrogel composites, 303
electroactive polymers, 288
electrochemical devices, 180, 181, 203
electroluminescence, 163, 175
electrolyte, 177, 329, 352
electrolytic capacitors, 166
electromagnetic fields, 33
electromagnetic interference shielding effect, 187
electro-mechanical activity, 195
electrostrictive cellulose paper, 190, 191
electrostrictive graft elastomers, 190, 289
emollient, 275, 368
entropy elasticity, 27, 53
environmentally sensitive, 2, 74, 77, 97, 259
environmentally smart textiles, 205
enzyme sensing, 302
eye related materials, 294, 304

—F—

fabrication process, 81, 216
face masks, 100
feeding rate, 222
ferroelectric electro adaptive polymers, 190, 193
ferrofluid, 199, 200
ferrogels, 200
ferromagnetic fillers, 201
filtration efficiency, 242
fine particles, 109
flat appearance, 38, 157
flexible speakers, 180, 181, 203

—G—

garment finishing, 16, 38, 149, 157
gellants, 73
gelling agents, 73
gene therapy, 12, 22, 131, 141
geo-textiles, 5
glucopyranose, 265, 355, 357
glucosamine, 263, 340, 343, 344, 345
glucose oxidase, 295, 296, 302, 304
glucose sensor, 295, 301
graft copolymerisation, 85, 86, 89, 103, 161, 162
gravure, 146
grow tissues, 292

—H—

heat-shrinkable, 26
hematology related products and devices, 308
heterogeneous morphology, 323
high density terminal groups, 330, 337, 339
high-dielectric-constant organic particulates, 193
higher surface area/mass ratio, 109
hollow fibers, 174
huge surface area to volume ratio, 227
hybrid molecules, 299
hydrated state, 195
hydro-absorbency, 232
hydrogel-based photonic crystal, 301
hydrophobic interaction, 6, 224, 234, 326
hydrophobicity/hydrophilicity, 11, 112
hydroxypropyl methyl cellulose, 113
hyperbranched polymers, 4, 332, 338, 377
hypercore, 323

—I—

immunoassay, 299, 300
infrared irradiation, 31
inorganic/organic hybrid network, 329
in-situ microencapsulation method, 145
instant recoverability, 59
insulin release, 296
intelligent polymer systems, 301
intermolecular packing, 224, 225
interpenetrating network, 29, 63, 182
interpenetrating polymer network, 97, 100, 184, 267, 350
ionic electroactive polymers, 289
ionic migration effects, 191
ionisable comonomers, 196
ionizing radiation, 27, 30, 35
isotropic phase, 26, 30, 233

—K—

knife-over-air, 146
knife-over-roll, 146

—L—

lamellar structures, 77
laser-activated device, 308
light adaptive polymeric composites, 197
light sensitive shape memory effect, 33
lightfastness, 113
linalool, 267, 268, 366
linear macromolecule, 325, 330
liposomes, 133, 297
liquid crystal elastomers, 190, 194, 289
liquid crystalline polymer, 233
liquid cyrstalline elastomers, 233
local drug delivery, 202
lotus effect, 115
lower critical solution temperature, 75, 105
low-temperature oxygen plasma, 171
luminescent adaptive textiles, 142
luminescent materials, 163

—M—

magnetic adaptive gels, 200
magnetic particle, 199
magnetic resonance imaging, 326
magnetophoretic force, 199
measuring device, 309
melt spinning, 144, 145, 149, 161, 162, 168, 195
melt viscosity, 49, 52, 337, 339
melting point, 26, 27, 144, 148, 162, 231
mesophases, 77
methacrylate or acrylate, 333, 336
micellar drug carriers, 295
micro-actuator, 38, 66
micro-electro-mechanical systems, 12
microencapsulation, 12, 116, 122, 123, 147
microfluidics, 114, 298
microporous, 10
microporous structures, 94
miniature manipulators, 180, 181, 203
moisture retentivity, 121, 351
molecular imprinting polymers, 294
molecular imprinting technique, 298
molecular parameters, 84, 231
molecular template, 297

monoclinic, 192
monodisperse, 119, 138, 273
monodisperse particles, 114
monoprotected, 324
multifunctional cosmetics, 17, 253, 265
multiple noncovalent chemical interaction, 326

—N—

nanomachine, 225
nanospheres, 64, 134, 187, 189, 197, 240, 245, 270, 272, 282, 302
n-butyl methacrylate, 335
n-dopable polymers, 166
nematic director, 60, 189, 233
nematic phase, 26, 30, 233
nematic-isotropic transition, 60
noncentro-symmetric, 192
non-contact triggering, 32
non-covalent bond, 29,224
non-covalent bonding gels, 77
non-covalent interactions, 6, 49, 224, 226
nontoxicity, 262, 285, 310
nozzle, 219, 228

—O—

obesity, 308
one-way shape memory, 189
ophthalmic drug delivery, 305
optical microtweezers, 203
optical property, 36
organ printing, 294
organic gel, 77
organization, 223, 225, 226
orientation structure, 14, 216
orthogonal coupling strategy, 325
ovalbumin, 272, 273

—P—

pad-dry-cure, 127, 146, 161
paraben, 367, 383
paraelectric phases, 192
paraffin, 130, 145
payload, 273
p-dopable polymers, 166
penetrability, 364, 367
permanent shape, 8, 28, 31, 32, 35, 38, 269, 307, 308
phase change material, 22, 140, 143, 145, 148, 174
phase separation, 123, 217
photocatalyst, 170
photochemical phase transition, 33
photochromic, 161, 233
photo-contraction, 34
photoinduced ionic dissociation, 198
photoisomerisation, 198
photomechanical effects, 33
physical characteristics, 288
piezoelectric effect, 244
ply yarn, 153, 155
poly(cyclohexyl meth-acrylate), 31, 77
poly(ketone-co-alcohol)s, 30
poly(p-dioxanone), 32
poly(propyleneimine), 273
poly(styrene-co-butadiene), 26, 27
poly(tetramethyleneether glycol), 97
poly(vinylidene fluoride–trifluoroethylene), 192
poly(ε-caprolactone), 29, 31, 32, 130
polyamide, 21, 35, 70, 126, 140, 162, 220, 230, 232, 239, 248, 335
polyaniline, 162, 164, 167, 168, 194
polycyclooctene, 26, 30, 62
polydimethylsiloxane, 115

polyethylene glycol, 146, 148, 174
polyethylene-poly(vinyl acetate) copolymers, 27
polyhedral oligomeric silsesquioxane, 202
polymer blending, 183
polymer grafting percentage, 86
polymeric micelles, 296, 297, 298
polymerisation, 333
polynorbornene, 26, 62
polypyrrole, 168, 177
polypyrrole films, 302
polytetrafluoroethylene, 35
polyurethane anionomers, 42
polyvinylchloride, 35
pore structure, 88
PPy, 167, 168, 303, 304
principle, 175, 223, 290
printed tissues, 293
process parameters, 220
professional clothing, 147
protective clothing applications, 241
pseudocapacitance, 166
punctum plug, 304
pyridinium, 42, 48

—Q—

quadruple hydrogen bonding, 8, 49, 53, 56, 69
quaternary ammonium salts, 42

—R—

radiation-sensitive, 41
rapid prototyping, 293
recovery rate, 29, 33, 155, 270, 271
redox cycling, 194
regenerated cellulose, 191
rejuvenation, 263
remote sensing systems, 41
reversible particle system, 299
reversible trans-cis isomerisation, 198
reversible trans-cis isomerization, 33
rheological properties, 73, 338
rod-like trans-azobenzene moieties, 34
rotating cylindrical collectors, 229

—S—

sandalwood, 254
scaffold materials, 293
scaffolds, 23, 97, 177, 243, 292, 293
segmented polymer, 43
segmented polyurethane, 23, 24, 26, 47, 49, 67
segmented polyurethanes, 28, 69
self cleaning, 124, 140, 171
self cleaning modification, 126
self-complementary, 49, 50, 54, 69, 70
self-condensing ring-opening polymerisation, 331
self-condensing vinyl polymerisation, 331
self-deployable structures, 181
self-healing, 223, 239
self-regulating structures and performances, 202
semi-interpenetrating network, 29, 380
sensitive optical sensors, 244
shape memory behavior, 25, 40, 62, 195, 208
shape memory effect, 8, 14, 15, 20, 23, 25, 26, 28, 30, 31, 32, 47, 57, 58, 60, 61, 68, 70, 151, 153, 155, 178, 182, 183, 184, 185, 189, 201, 216, 233, 281

shape memory finishing, 12, 16, 149, 157, 158
shape memory function, 24, 25, 26, 28, 29, 30, 31, 34, 36, 37, 39, 41, 47, 50, 54, 55, 56, 57, 152
shape memory gels, 34
shape memory polymer stent, 309
shape recoverability, 8, 25, 181
shear-thinning behaviour, 338
shrinkage determination, 26
shrink-resistant apparel, 203
silane nanosol, 172
skin care particles, 134
skin/healthcare products, 355, 364
'smart' contact lenses, 305
smart hydrogels, 4, 22, 83
smart implants, 291
smart polymer, 18, 19, 246, 293, 298
smart polymer nanoparticles, 294, 299
smart surgical suture, 309
sol-gel method, 329
sol-gel process, 12, 120, 179
solid electrolyte, 197
solid lipid nanoparticles, 133, 141
solution spinning, 161
solution-driven shape memory polymer, 32
static self-assembly, 224
step-growth polycondensation, 331
stimuli-reponsive polymers, 19, 105
stimuli-responsive bioanalytical bead system, 299
stimuli-sensitive, 132
stimulus sensitive switches, 26
stimulus-adaptive polymeric composites, 205
storage modulus, 29
strain fixity rate, 29
styrene-butadiene-styrene tri-block copolymer, 29
subunits, 330
superparamagnetic nanoparticles, 33
superparamagnetism, 113
surface area/mass ratio, 108
swelling/deswelling equilibrium, 90
switch temperature, 15, 150, 152, 155, 156, 183, 186, 201
switch temperatures, 181
syneresis, 260, 350, 351
synthetic gels, 10
synthetic receptors, 297

—T—

tailorable, 37, 75, 181, 336, 340
tamoxifen, 266, 337
target moleculars, 297
telechelic crosslinking, 60
temperature, 139, 209, 259, 295
template synthesis, 217
temporary shape, 8, 25, 27, 34, 53, 232, 268, 307
tensile strength retention, 352
tere-phthalate segmented copolymer, 27
terminal groups, 34, 319, 322, 323, 325, 329, 336, 338
tetraethylene glycol diacrylate, 350
textile dyeing, 359, 360, 382
textile finishing, 107, 136, 141, 361, 362
textile functioning, 359, 362
textile monitoring, 359, 363
textile printing system, 360
thermal adaptive polymeric composites, 190
thermal sensitive, 31, 155

thermal stability, 144, 146, 203, 337, 353
thermoexpanded graphite, 184
thermomechanical deformation, 185
thermoplastic polymers, 36, 65, 281
thermoregulation, 142, 143, 201
thermo-sensitive polymers, 295
timuli-sensitive, 79
tissue or organ printing, 293
titanium dioxide, 132, 133, 170, 171
transfer coating, 146
trans-polyisoprene, 26
triblock copolymers, 15, 60
triblock liquid-crystal polymer, 189
triphenylmethane leuco derivatives, 197
triple-shape polymer, 30
tripro-pyleneglycol diacrylate, 114
two-phase heterogeneous structure, 28

—U—

ultrafine diameters, 218, 227
unintrusive system, 203
user-friendliness, 100
UV protection, 132, 169, 171, 187
UV-irradiation, 87, 94

—V—

vibration absorbers, 204
volume phase transition, 75, 83, 90, 91, 247

—W—

wash fastness tests, 339
water immersion, 31
water repellence, 169
water vapour permeability, 68
water-medium gels, 73
water-soluble polymers, 231
wet spinning, 5, 145, 149, 152, 168
wound closure, 309
wound dressing, 15, 87, 98, 100, 244, 349

—X—

zinc oxide whiskers, 172

—α—

α-cyclodextrin, 355, 356, 357

—β—

β-cyclodextrin, 281, 355, 356, 357, 360, 383

—γ—

γ-cyclodextrin, 355, 356, 357, 362, 367
γ-ray pre-irradiation, 86, 94